Genome Editing

Shabir Hussain Wani • Goetz Hensel
Editors

Genome Editing

Current Technology Advances
and Applications for Crop Improvement

Editors
Shabir Hussain Wani
Mountain Research Center for Field Crops
SKUAST-Kashmir
Anantnag, Jammu and Kashmir, India

Goetz Hensel
Institute of Plant Biochemistry
Heinrich Heine University Düsseldorf
Düsseldorf, Germany

ISBN 978-3-031-08074-6   ISBN 978-3-031-08072-2   (eBook)
https://doi.org/10.1007/978-3-031-08072-2

This Springer imprint is published by the registered company Springer Nature Switzerland AG
The registered company address is: Gewerbestrasse 11, 6330 Cham, Switzerland

Foreword

Plants have been enduring a series of biotic as well as abiotic stresses resulting in decreased yield and quality. Hence, it is the foremost objective of every established crop improvement program at global level to develop crop plants having resilience to these stresses and improved end use quality for better income generation. Crop improvement has received a huge impetus due to recent advancement in cutting-edge genome editing technologies. Noteworthy progress has been accomplished in the last three decades through the targeted transfer of recombinant DNA into crop plants. Nevertheless, the amalgamation of transgenes into the host genome has been arbitrary, which has seen lot of criticism. Developments in genome sequencing have delivered plentiful data on variances among susceptible and resistant varieties, which can now be directly targeted and altered using CRISPR/Cas technology. CRISPR/Cas9 is unadorned, handy, less exorbitant, and highly effective than its prior compeers. This is the reason CRISPR/Cas9 has become the most globally used technology for genome editing in many organisms such as bacteria, yeasts, animals, and plants. Since this technology showed fruitful results in trials conducted in recent years, hence it has been evidenced that CRISPR-Cas9 system can be metamorphosed into workable technology, which can edit the genetic code of almost any species. Due to its resourcefulness, it had been marked as "breakthrough of the year 2015" by *Science* magazine. In recent years, CRISPR-Cas9-based techniques have been pragmatic in various systems, including yeast, zebrafish, Drosophila, mice, and cultured human cell lines, and of late have been practiced in plants.

I am fortuitous to know that Dr. Shabir Hussain Wani and Dr. Goetz Hensel have edited this volume entitled *Genome Editing: Current Technology Advances and Applications for Crop Improvement* for the reputed publisher Springer Nature. The chapters included in this book are courteously written by potential scientists and researchers from globally reputed institutions. This book describes recent advances in genome editing technologies and their application in crop improvement.

This book is a timely reference material for academicians, researchers, and graduate students working in the area of plant breeding, genetics, and biotechnology. I applaud the editors for bringing up this volume and hope that it will be highly cited by scientists, students, and policy makers.

Nazir A. Ganai
Vice Chancellor, Sher-e-Kashmir University of Agricultural
Sciences and Technology,
Srinagar, Jammu and Kashmir, India

Preface

Plant and animal research has boomed due to the recent advancement in genome editing technology, particularly since the last two decades. In the last three decades, significant development has been achieved through the targeted transfer of recombinant DNA into crop plants. However, the integration of transgenes into the host genome has been random, which has been met with some unacceptance by the population. Advances in genome sequencing have provided enough information on differences between susceptible and resistant varieties, which can now be directly targeted and modified using CRISPR/Cas technology. CRISPR/Cas9 is more straightforward, much more versatile, less expensive, and highly efficient than its previous counterparts. That's why it has become the most widely used technology for genome editing in many organisms such as bacteria, yeasts, animals, and plants. Over several years of experimentation and successful results, it has been proved that the CRISPR-Cas9 system can be retooled into a viable technology, which can edit the genetic code of virtually any species. Due to its versatility, it had been marked as the "breakthrough of the year 2015" by *Science* magazine. For the past few years, CRISPR-Cas9-based techniques have been applied successfully in various systems, including yeast, zebrafish, Drosophila, mice, and cultured human cell lines. Compared to other organisms, comparatively, very few reports are present in plants. We have tried to compile this comprehensive book, which focuses initially on fundamental research in the area of the CRISPR-Cas system. The first few chapters in this book are based on the current status and challenges of plant genome editing using CRISPR/Cas technology. They first help readers get basic understanding and then explain how to design and implement the CRISPR-Cas9 system for different plant systems. The following section comprises chapters highlighting developments in the field of CRISPR/Cas technology. Further, the third section stresses on applications of CRISPR/Cas technology for biotic and abiotic stress tolerance. The last section of this book discusses the legal and bio-safety issues accompanied by the commercialization and patenting issues of this emerging technology. This book will serve as a complete guide for readers, helping them understand this rapidly expanding field.

We are highly thankful to the contributors of various chapters. The authors of this book were selected from diverse and reputed organizations based on their expertise in the subject. Also, the authors have profound insight on specific topics, which helped us make this volume a state-of-the-art reference material. Our sincere gratitude to the dynamic authors for their contribution and for sharing their current research. This book will serve as a ready reference for undergraduate and postgraduate students studying plant breeding, genetics, and plant biotechnology. Libraries of institutions dealing with crop science and biotechnology may need this book as reference material for students and teachers. The editors once again desire to show appreciation to all the contributors and the editorial workforce of Springer for their cooperation and speedy production of this book.

Srinagar, Jammu and Kashmir, India   Shabir Hussain Wani
Düsseldorf, Germany   Goetz Hensel

Contents

Part I
Current Status and Challenges of Plant Genome Editing Using CRISPR/Cas Technology

Genome Engineering as a Tool for Enhancing Crop Traits: Lessons from CRISPR/Cas9

Fozia Saeed, Tariq Shah, Sherien Bukhat, Fazal Munsif, Ijaz Ahmad, Hamad Khan, and Aziz Khan

Abstract Over the last few decades, several efforts were made to improve cereal and horticulture crops, mainly using conventional or molecular breeding methods. The current situation enabled researchers to successfully investigate molecular mechanisms by identifying target genes to produce desirable plants. High-yielding stress-resistant crops are needed to achieve food security for the world's increasing population. In this context, genetic engineering technologies like zinc fingers (ZFNs), transcription activator-like effector nucleases (TALENs), and clustered regularly interspaced short palindromic repeat (CRISPR) have shown enormous potential in the production of improved crop varieties with desired agronomic traits. Since 2013, genome editing using sequence-specific nucleases (SSNs), especially CRISPR/Cas9 ribonucleoprotein, has received extensive attention for plant research due to its specificity and ability to simultaneously silence multiple genes. CRISPR/ Cas9 has become a method of choice for researchers worldwide to improve crop yield, quality and reduce disease susceptibility. Many food crops, including maize, rice, wheat, tomato, potato and citrus have successfully undergone genome editing by CRISPR/Cas9 technology. The availability of complete genome sequences of various crops together with innovations in genome-editing techniques opens new opportunities for obtaining desirable traits. This

F. Saeed · S. Bukhat
Institute of Molecular Biology and Biotechnology, Bahauddin Zakariya University, Multan, Pakistan

T. Shah (✉) · F. Munsif · H. Khan
Department of Agronomy, Faculty of Crop Production Sciences, The University of Agriculture, Peshawar, Pakistan
e-mail: tariqshah@aup.edu.pk

I. Ahmad
Department of Plant Breeding, Faculty of Crop Production Sciences, The University of Agriculture, Peshawar, Pakistan

A. Khan (✉)
College of Agriculture, Guangxi University, Nanning, China

© The Author(s), under exclusive license to Springer Nature Switzerland AG 2022
S. H. Wani, G. Hensel (eds.), *Genome Editing*,
https://doi.org/10.1007/978-3-031-08072-2_1

3

chapter summarizes the crop improvement using CRISPR/Cas9-based genome-editing approaches with particular emphasis on cereals and horticulture crops.

Keywords Cereals · Disease resistance · Horticultural crops · TALENs · CRISPR/Cas

1 Introduction

Since crops were cultivated, researchers have developed high-yielding, disease-resistant and climate-resilient crops. Conventional crop improvement approaches such as hybridization and mutational breeding have played a crucial role in introducing the important agricultural and nutritional traits for crop improvement. However, the progressive decline in natural diversity in crop plants causes a significant effect on crop production (Govindaraj et al. 2015). In recent decades, biotechnology-based interventions such as transgenic and transformation technology have shown great potential to address the problems of traditional breeding methods and improved the specific traits of crops. Although transgenic technology has made significant progress around the world, the frequency of targeted insertion of the foreign gene was initially very low, which results in unspecific gene silencing (Puchta and Fauser 2013). Therefore, more specificity is needed to edit a gene for both function analysis and to find its role in crop improvement. Recently, genome modification has raised as an innovative method that uses sequence-specific nucleases (SSNs) to incorporate targeted alterations in the plant genome with high precision and efficiency. These artificially engineered SSNs include ZFNs, TALENs and CRISPR/Cas9-mediated endonuclease. These endonucleases have been widely utilized to edit various genes in animals and plants and have also been used to produce various commercial goods. The emergence of (CRISPR)/Cas9 systems provides a new dimension in plant science and contributes to remarkable progress in crop biotechnology in a relatively short time. Recently, CRISPR/Cas9-based genome alteration platforms have been quickly accepted by the research community to modify, regulate, and track genes in plants, animals and microbes as well (Zhang et al. 2020; Ma et al. 2016; Jiang et al. 2013a, b).

Considering the importance of CRISPR/Cas, the use of genome manipulation for the advancement of cereals and horticultural crops has dramatically progressed. In this chapter, we briefly introduced the different engineered nucleases and their working mechanism. Further, CRISPR/Cas9 technology was discussed because of its simplicity of design, acceptability and speed, relatively inexpensive and focused targeting. Moreover, the current implementations of CRISPR/Cas9 tool for the improvement of different cereals and horticulture crops with higher yield, better nutritional quality and disease tolerance have also been discussed.

2 Zinc-Finger Nucleases

ZFNs are the first generation endonucleases that serve as a vital technique for genome engineering. These are generated by joining the zinc fingers (ZFs) with the Fok1 DNA cleavage domain. The FokI endonuclease domain is typically used in nonspecific and independent DNA cleavage after dimerization (Carroll 2011). The DNA binding domain comprises of 3–6 ZF domains, and each domain recognizes a 3-bp targeted DNA sequence. A pair of ZFs is fused with a FokI endonuclease domain to produce a site-specific cleavage. Binding sites of two individual ZFs (each 18–24 bp in length) are separated by 5–8 bp spacer sequence. This spacing is a key component of the ZFN construct as it enables the Fok1 monomer to dimerize and produce a double-strand break in the targeted DNA sequence (Bibikova et al. 2003). ZFNs genome editing was initially reported in Arabidopsis and tobacco. Subsequently, ZFNs have successfully altered genomes in wheat, rice, maize and other plants. In the last couple of decades, various techniques have been proposed to increase the applicability, reliability and accuracy of ZFNs. However, ZFNs have specific limitations, such as their lower affinity for the AT-rich region; construction of ZFN units is complicated and costly. Furthermore, the limited number of available target sites leads ZFN application towards off-target editing due to non-specific DNA binding that brings off-target effects. These features of ZFNs restrict their application in genome editing (Khan 2019).

3 Transcription Activator-Like Effector Nucleases

TALENs belong to second-generation endonucleases used for precise genome editing. Like the ZFNs, TALENs are generated by combining a transcription activator-like effector (TALE) with the non-specific domain of FokI endonucleases (Miller et al. 2011). These effector proteins (TALEs) were discovered in bacteria *Xanthomonas* that infect their native plants and attach to the targeted DNA by DNA binding domain comprising 33–35 amino acid repeats (Doyle et al. 2012). The TALE proteins consist of a C-terminal conserved acidic transcription-activation domain, a nuclear localization signal (NLS) and a central DNA binding domain. TALEs' main binding domain is made up of 13–28 repetitive sequences. Each repeat, which contains a highly conserved 34-amino acid sequence, can bind directly to a single nucleotide through the variable di-residues at the 12th and 13th locations (Deng et al. 2012). The one-to-one pairing, and the lack of context-dependence on neighbouring repeats, facilitate TALENs to target particular sequences (Reyon et al. 2012). TALENs were applied to *Oryza sativa* for the first time in crop development. Progressively, TALENs were implied in tobacco, rapeseed, Arabidopsis, flax, potato, sugarcane, barley, soybean, rice, tomato, maize and wheat (Martínez-Fortún et al. 2017). TALENs outperform ZFNs in terms of success due to their uncomplicated design, higher binding affinity to their target sites and a higher level of cleavage activity. However, despite its potential benefits in crop improvement, TALENs have some limitations that restrict their application. A clear drawback of TALENs is their significantly greater size relative to ZFNs. A TALEN cDNA is around 3 kb in size which makes it difficult to transfer a set of TALENS in a cell. TALE repeats composition remains a challenge and the efficiency by which TALENS access a specific gene varies. Moreover, TALENS are extremely repetitive, which makes them incompatible with certain viral vectors. Researchers often choose to use certain more versatile genome editing methods to produce transgenics to avoid this complicated situation.

4   CRISPR/Cas9 System

CRISPR/Cas9 is considered the third generation and the most efficient genome editing technology. The success of CRISPR/Cas as a tool for genome editing in the present era can be dated back to the late 1980s at the time of its origin (Ishino et al. 1987). The progress of CRISPR/Cas9 adaptive-immune system of microbes results from the efforts of many researchers all over the world. The power of CRISPR/Cas method as a novel tool for genome-modification was deciphered by studying its function in archaea and bacteria. Various components of CRISPR/Cas and their role in ensuring adaptive immunity in microbial cells were revealed through several experiments using bioinformatics methods.

CRISPR loci are made up of CRISPR arrays and CRISPR-associated (Cas) clusters of genes on which all immunological-related memories are imprinted (Barrangou and Dudley 2016). CRISPR array consists of a genomic locus that includes a collection of 21–40 bp direct-repeat sequences separated by 25–40 bp spacers or variable sequences (Jansen et al. 2002). In 2005, three different research groups proposed that spacer elements serve as signs of past foreign DNA invasions, which provide resistance to the infection of phage (Bolotin et al. 2005; Mojica et al. 2005; Pourcel et al. 2005). They also found that spacers have the same end sequence, which is now referred to as PAM. PAM (Protospacer Adjacent Motif) is usually 2–6 bp short DNA sequence, which follows the targeted region of DNA for cleavage through CRISPR system like CRISPR-Cas9. It is necessary for Cas-nuclease for cutting and can be found 3–4 nucleotides at downstream of the cut site. Another experiment showed that CRISPR arrays, in combination with Cas genes, play a role in bacteriophage resistance (Barrangou et al. 2007).

At each infection, new DNA of phage is inserted in CRISPR array, enhancing the array's ability to combat the impending infection. Brouns et al. (2008) discovered that spacer sequences of phage are transcribed into the crRNAs (small RNAs), which direct Cas proteins towards the targeted DNA. Upon infection, crRNA guide the interference machinery known as protospacers, for cleaving complementary sequences in foreign genetic material. Another study also reported three nucleotides at PAM upstream as well as suggested Cas9 role in inserting DSBs in a specific location (Marraffini and Sontheimer 2008; Garneau et al. 2010). In addition, crRNA develops a duplex with tracrRNA (trans-activating CRISPR RNA) and directs Cas9 towards its target (Deltcheva et al. 2011). The method was further simplified by combining the tracrRNA and crRNA to produce a single and simple synthetic guide-RNA (gRNA) (Jinek et al. 2012). Lastly, Cong et al. (2013) discovered that Cas9 could promote homology-guided repair while causing minimal mutagenic activity.

5   Mechanism of CRISPR/Cas9 System

The CRISPR/Cas9 system adaptive immunity is grouped into three stages, including adaptation, interference and expression. Invading genetic material (DNA) from plasmids or viruses is cleaved in the small fragments and integrated into the locus of CRISPR during adaptation. The CRISPR loci transcribed and converted into crRNA use base complementarity to direct effector endonucleases for targeting viral material (Barrangou et al. 2007; Yosef et al. 2012). In Type-II CRISPR/Cas method, the single protein of Cas9 is needed for DNA interference (Hale et al. 2009; Zetsche et al. 2015). Cas9 being a large protein comprises many domains, such as HNH nuclease and RuvC domain, including tracrRNA and crRNA. Cas9 helps in adaptation, assists in the processing of pre-crRNA to crRNA, as well as incorporates targeted DSBs directed by tracrRNA and RNase III specific for double-stranded RNA (Mulepati et al. 2014). The RNA and DNA cleavage and its interaction with the Cas10 cleavage protein distinguish type III CRISPR from type II. The cleavage involves transcription-dependent modification of DNA sequence that also includes a promoter which is transcriptionally active (Samai et al. 2015). Under certain conditions, the Cas10 system allows bacteria to attain viral spacer components, allowing a form of resistance to foreign DNA. This resistance towards viral/foreign DNA inhibits the lytic pathway from being activated, which is harmful for the host organism. In this pathway, host cells are killed by viruses for producing a large number of their progeny. These sequences can also modify the physical attributes of the organism, giving the host cell a survival advantage (Samai et al. 2015).

In 2013, CRISPR/Cas was used for the first time in plants (Feng et al. 2013; Xie and Yang 2013). Feng et al. (2013) used agroinfiltration and protoplast transfection to target different endogenous transgenes and genes, resulting in healthy transgenic plants using both HR and NHEJ mechanisms. Another study tested the mutation effectiveness of 3–8% by introducing three guide-RNAs at different genomic loci of rice (Xie and Yang 2013). Off-target modifications were discovered as well, but with a minimum efficiency of genome editing compared to matched sites. Different studies on sorghum, wheat and maize demonstrated the best example of using CRISPR-based gene editing (Jiang et al. 2013a, b; Liang et al. 2014; Wang et al. 2014). These studies provided the first detailed information on parameters including cleavage specificity, mutation efficiency, large deletions of chromosomes, and locus structure resolution. The CRISPR/Cas9 technique has been improved continuously to increase the specificity and efficiency of gene editing and targeting. The requirement of reconfiguring CRISPR/Cas9 method to modify the genome of eukaryotes has demanded the introduction of signals of nuclear localization at protein ends. The advancement of orthogonal CRISPR/Cas9 mechanisms has significantly expanded the range of applications for this technology.

6 Targeted Improvement of Crop Traits Using CRISPR/Cas9-Based Genome Editing in Cereals

The CRISPR/Cas9 method has been efficiently used in cereal crops, because of its wider acceptability, easiness, speed, overall low-cost and focused targeting. Cereals are an essential part of our diet and provide significant health benefits. Cereals are the primary source of carbohydrates, proteins, fats, lipids, vitamins, minerals, and abundant fibers. The primary goal of cereal crops' genetic modification is to increase food production and give nutritional protection to the world's increasing population growth. Considering the significance, CRISPR/Cas9 tools have been extensively used to create novel crops with increased productivity, disease resilience and better nutritive values (Usman et al. 2021; Zhang et al. 2017). Table 1 enlists targeted genes and traits in different cereal crops that were modified using CRISPR/Cas9 technology.

Table 1 List of targeted genes and traits modified by CRISPR/Cas9-based system in cereal crops

Crop	Target gene	Target traits	References
Maize	*ALS2*	Herbicide resistance	Svitashev et al. (2015)
	ZmIPK1A, ZmIPK and ZmMRP4	Phytic acid synthesis	Liang et al. (2014)
	CLE	Enhanced grain yield	Liu et al. (2021)
	ZmSH2 and WX	Super sweet and waxy corn	Dong et al. (2019)
Wheat	*TaMLO*	Mildew resistance	Shan et al. (2014)
	TaEDR1	Powdery mildew resistance	Zhang et al. (2017)
	TaGASR7, TaDEP1	Increased thousand-kernel weight	Zhang et al. (2016)
	TaVIT2	Improved Fe content	Connorton et al. (2017)
	Alpha-gliadin	Low-gluten	Sánchez-León et al. (2018)
	TaGW2	Improved seed size with increased thousand-kernel weight	Wang et al. (2018)

(continued)

Table 1 (continued)

Crop	Target gene	Target traits	References
Rice	*Os8n3*	Bacterial resistance	Kim et al. (2019)
	SWEET11, SWEET13 and SWEET14	Bacterial blight resistance	Oliva et al. (2019)
	Xa13	Bacterial blight tolerance	Li et al. (2020)
	OsSWEET14	Bacterial blight resistance	Zafar et al. (2020)
	OsERF922	Blast resistance	Wang et al. (2016)
	eIF4G	Tungrospherical virus resistance	Macovei et al. (2018)
	ALS	Herbicide tolerance	Sun et al. (2017)
	OsALS1	Herbicide tolerance	Kuang et al. (2020)
	OsTubA2	Herbicide resistance	Liu et al. (2021)
	EPSPS	Improved aromatic amino acids	Li et al. (2016)
	OsGFR1	Increased grain filing	Dong et al. (2019)
	OsAH2	Improved grain size and hull development	Ren et al. (2019)
	Golder rice 1&2	Increased β-carotene	Dong et al. (2020)
	P450s and OsBADH2	Improved aroma in grain	Usman et al. (2020)
	OsSNB	Improved grain length and weight	Ma et al. (2019)
	OsC3'H	Increased biomass saccharification	Takeda et al. (2018)
	OsAAP5	Improved yield and tiller number	Wang et al. (2019a)
	OsNramp5	Low Cd accumulation	Tang et al. (2017)
	OsMYB108	Improved lignin accumulation	Miyamoto et al. (2019)
	Waxy	Altered amylose content	Zhang et al. (2018b)
	PYL1–PYL6 and PYL12(gp-1), PYL7–PYL11 and PYL13(gp-2)	Improved grain productivity and yield	Miao et al. (2018)
	OsLCT1 and OsNramp5	Low cadmium	Songmei et al. (2019)
	OsSPL16	Improved grain yield	Usman et al. (2021)
	OsBADH2	Introduction of aroma	Ashokkumar et al. (2020)
	OsTubA2	Herbicide resistance	Liu et al. (2021)
	Waxy	Altered amylose content	Liu et al. (2021)

(continued)

Table 1 (continued)

Crop	Target gene	Target traits	References
Barley	Hv d-Hordein	Increased β-glucan, starch and amylose content	Yang et al. (2020)
	HvCOMT1	Improved lignocellulose quality	Lee et al. (2021)
Cotton	Ghl14–3-3	Verticillium dahlia resistance	Zhang et al. (2018d)

6.1   Resistance Against Bacterial Disease

Bacteria cause many plant diseases and produce a range of metabolites, including toxins, pectic compounds, polysaccharides and hormones. The best example of CRISPR-based induction of bacterial disease protection is the use of *OsSWEET* genes to enhance resistance for bacterial blight resulting from *Xanthomonas oryzae pv. oryzae*. The rice *SWEET* clade III family comprises *OsSWEET11*, *OsSWEET13*, and *OsSWEET14* genes, that encode a glucose and sucrose transporter involved in the interaction between plant and pathogen. The CRISPR/Cas9-induced mutation of the *OsSWEET13* gene resulted in improved broad-spectrum resistance to bacterial blight in rice (Zhou et al. 2015). Another study recognized that CRISPR/Cas9-mediated editing of all three OsSWEET genes resulted in improved resistance against bacterial blight. (Oliva et al. 2019). In contrast to OsSWEET genes, the CRISPR/Cas9 mediated knockdown of the *Os8N3* gene also significantly increased resistance against the *Xanthomonas oryzae pv. oryzae* (Kim et al. 2019).

6.2   Resistance Against Fungal Disease

Fungal pathogens significantly impact crop yield and quality, producing a troubling situation for food and nutrition security. Several methods have been used to enhance plant-fungal resistance. Nowadays, potential candidate genes and gene products responsible for plant fungal resistance are ideal for CRISPR/Cas9 editing. Three genes *TaMLO-A1*, *TaMLO-B1*, and *TaMLO-D11* were mutated by using CRISPR/Cas9 tool to improve resistance to powdery mildew in wheat (Wang et al. 2014). A CRISPR/Cas9-based knockout of the ERF transcription factor gene *OsERF922* has been reported to develop a new source of resistance to rice blast (Wang et al. 2016). Blast disease-resistant rice plants were recently developed through CRISPR/Cas9-directed destruction of the rice genes *OsERF922* and *OsSEC3A* (Sánchez-León et al. 2018). *OsSEC3A* mutated plants also showed enhanced resistance to *Magnaporthe oryzae*, increased level of salicylic acid as well as up-regulation of pathogenesis and salicylic acid-responsive genes.

6.3   Resistance Against Viruses

Viruses infect nearly all crop plants, resulting in major losses in agricultural production around the world. Viruses are difficult to control chemically because of their stringent intracellular pathogenesis. Mainly, the prophylactic measures destroy diseased plants, and extreme pesticide usage is necessary to avoid the virus-carrying populations. Traditional breeding has developed many varieties with complete virus resistance. However, it is necessary to develop effective and long-lasting virus-resistant crops that can resist extreme genetic plasticity of viruses. Genome editing strategies are promising in this regard. Recently, mutagenesis of the rice *eIF4G* gene has generated resistance to the rice tungro spherical virus (RTSV) (Macovei et al. 2018). The transgene-free T2 plants were RTSV-resistant and did not display certain mutation in the off-target region. However, translation initiation factors are the prime candidates in the host genome. Infection transmitted by viruses can be efficiently controlled by altering the specific host genes encoding the responsive factors that the viruses need.

6.4   Resistance and Tolerance Against Herbicides

Weeds have consistently disrupted agricultural crops since their domestication, resulting in greater crop losses than diseases and insect pests that have necessitated the use of weed control strategies. Therefore, the production of herbicide-resistant crop plants is a practical method to control weeds and improve their capacity to interfere with competing weeds. Mutation in the rice *ALS* gene conferring herbicide-resistance was introduced effectively by CRISPR/Cas9 (Kuang et al. 2020). Zhang et al. (2019) used base-editing to modify acetolactate synthase (*ALS*) and acetyl-coenzyme A carboxylase genes to create herbicide-tolerant wheat variety. Carboxylase genes provide tolerance to herbicides, namely, aryloxyphenoxy propionate, sulfonylurea and imidazolinone. Li and colleagues created TIPS amino acid replacements into *OsEPSPS* and obtained endogenous gene substitution and targeted gene replacement at a rate of 2.0% and 2.20%, respectively. Targeted substitutions provided glyphosate tolerance in rice in *OsEPSPS* (Li et al. 2016). In rice, *OsTubA2* gene was effectively modified by using CRISPR-mediated adenine base editors. Edited rice plants showed high tolerance to pendimethalin and trifluralin herbicides compared to control plants (Liu et al. 2021). Herbicide tolerance has been considered as the most widely accepted and widely applied traits of cereal crops. In order to achieve long-term weed resistance, crops must be engineered with multiple herbicide resistance genes as a top priority.

6.5 Improved Quality and Yield

Many efforts have been made in cereal crops for genetic modification of quality and yield traits by employing the CRISPR/Cas9 technique. CRISPR/Cas9 technique was adopted in maize to target the *ARGOS8* locus, generating *ARGOS8* variants with enhanced grain yield. Wheat genes such as *TaGASR7*, which is linked to grain length and dense-erect panicle, were targeted with CRISPR/Cas9, and plants with a mutation in six alleles were developed in order to grow high yielding modified crops. Mutant plants showed improved thousand-kernel weight in wheat crop (Zhang et al. 2016). Increased iron contents of wheat was obtained by genome editing of the *TaVIT2* gene (Connorton et al. 2017). In rice, the *SBEIIb* gene was targeted to increase the amylase content (Sun et al. 2017). Likewise, Mao et al. (2018) developed potassium (K^+) deficiency tolerant rice by editing the *OsPRX2* gene. Transgenic rice plants showed stomatal closing and K^+-deficiency tolerance under the starvation of potassium (Mao et al. 2018). In another study, genome editing of rice *OsSPL16* gene improved the grain yield by inducing pyruvate enzymes and cell-cycle proteins (Usman et al. 2021). Heterosis is the main concern for wheat and rice. Recently, four yield-responsive rice genes, *IPA1*, *Gn1a*, *DEP1* and *GS3*, were edited using CRISPR/Cas9 (Huang et al. 2018). Similarly, the CRISPR/Cas9 approach was applied to target a number of genes that regulate yield component traits, including the number of panicles per plant, number of kernels per panicle/pod/bear and kernel weight (Upadhyay et al. 2013). Dong et al. (2019) produced a super waxy and sweet variety by modifying the *SH2* and *WX* genes associated with sugar and starch metabolism (Dong et al. 2019). These studies demonstrating the utilization of CRISPR/Cas9 tool to achieve the highest quality and yield by targeting multiple genes collectively in cereal crops are supposed to increase dramatically in the next few years.

7 CRISPR/Cas9 Mediated Genome Editing in Horticultural Crops

CRISPR/Cas9 is perhaps the most extensively used genome-editing method in the plant community. This technology has been used to modify genomes of various horticulture crops to achieve a variety of research goals, including nutritional and yield quality improvement, and disease resistance to bacterial, fungal and viral pathogens. This segment discusses how genome editing technology is being used to develop fruit trees, vegetable crops and other ornamental plants (Kwon et al. 2020; Veillet et al. 2019). Table 2 enlists targeted genes and traits in different horticulture crops modified using CRISPR/Cas9 technology.

7.1 Resistance Against Bacterial Disease

Citrus canker caused from *Xanthomonas citri* subsp. *citri* is an acute disease negatively impacting citrus fruit production and quality. Citrus *CsLOB1* genes were modified by using CRISPR/Cas9 method, and the mutated plants displayed enhanced citrus canker resistance (Peng et al. 2017). A high degree of bacterial resistance in homozygous mutants was obtained by deleting the entire sequence of effector binding elements (*EBEPtha4*) from the two alleles of *CsLOB1*. Wang et al. (2016) noticed that CRISPR/Cas9-mediated modification of *PthA4* effector cis-elements in the promoter of *CsLOB1* reduced the citrus canker infection on the mutant plants. Citrus *DMR6* orthologs were edited by the CRISPR/Cas9 platform, and the resultant silenced plants showed tolerance to *Huanglongbing* (HLB) disease produced by *Candidatus Liberibacter asiaticus* in citrus (Zhang et al. 2018c. In tomato, the CRISPR/Cas-based genome editing of *SlJAZ2* generates a resistant variety of tomato to counter bacterial speck infection caused by *Pseudomonas syringae* pv. *tomato* (Ortigosa et al. 2019). Similarly, fire blight resistance was achieved by targeting the *DIPM-1, DIPM-2*, and *DIPM-4* genes in the apple (Malnoy et al. 2016).

7.2 Resistance Against Fungal Disease

Fungal pathogens cause approximately 30% of plant diseases and damage a wide range of economically valuable food crops. Fungal disease resistance by CRISPR-mediated genome alteration is a new area of research in horticulture crops and vegetables that is gaining momentum now. A mutated tomato plant was generated using two sgRNAs that create a specific mutation in the *MLO1* locus, which increased the plant resistance against powdery mildew. Zhang et al. (2018d) employed CRISPR/Cas9 to disrupt a *VvWRKY52*, which encodes a transcription factor gene to improve resistance against *Botrytis cinerea* in grape (*Vitis vinifera*) variety. Similarly, Zhang et al. (2018b) reported that the CRISPR-based edition of the mitogen-activated protein kinase 3 (SlMAPK3) gene controls the production of reactive oxygen species (ROS), which resulted in tomato plants' resistance to *B. cinerea*. CRISPR/Cas9 mediated knockout and complementation of Solyc08g075770 gene revealed that it is important for *Fusarium* wilt resistance in tomato plants. CRISPR-Cas9-based edition of *powdery mildew resistance-4* (*Pmr4*) gene showed a substantial decrease in the symptoms of powdery mildew disease in tomatoes (Santillán Martínez et al. 2020).

Table 2 List of targeted genes and traits modified by CRISPR/Cas9-based technology in horticulture crops

Crop	Target gene	Target traits	References
Tomato	*SLMLO1*	Powdery mildew resistance	Nekrasov et al. (2017)
	TYLCV	Yellow Leaf Curl Virus resistance	Tashkandi et al. (2018)
	SlJAZ2	Bactcrial speck resistance	Ortigosa et al. (2019)
	ALS1 and ALS2	Herbicide resistance	Veillet et al. (2019)
	ALS1	Herbicide resistance	Danilo et al. (2019)
	MAX1	Root parasitic weed resistance	Bari et al. (2021)
	SlER, SP5, SP	early yielding plants	Kwon et al. (2020)
	ALC	Long shelf life	Yu et al. (2017)
	SGR1, LCY-E, Blc, LCY-B1, LCY-B2, LCY-B2	Increased lycopene content	Li et al. (2018b)
	SP5G	Rapid flowering and early yielding plants	Soyk et al. (2017)
	SP, SP5G, SlCLV3, SlWUS	Improved fruit production and Vitamin C content	Li et al. (2018a)
	SlCLV3	Improved fruit size and plant architecture	Rodríguez-Leal et al. (2017)
	SELF-PRUNING, OVATE, FRUITWEIGHT 2.2, LYCOPENE BETA CYCLASE	Fruit size, number and lycopene content	Zsogon et al. (2018)
	AP2a, NO, FUL1/ TDR4and FUL2/MBP7	Fruit development and ripening	Wang et al. (2019b)
Potato	*ALS1 and ALS2*	Herbicide resistance	Veillet et al. (2019)
	StALS1	Herbicide resistance	Butler et al. (2016)
	GBSS, EC 2.4.1.242	Develop amylopectin starch cultivars	Andersson et al. (2018)
Lettuce	*LsGGP2*	Increased vitamin C content	Zhang et al. (2018a)

(continued)

Table 2 (continued)

Crop	Target gene	Target traits	References
Soybean	*GmPRR3b*	Improved early flowering and growth	Li et al. (2020)
	GmPRR37	Improved growth	Yang et al. (2020)
	GmSPLs	Improved plant growth and architecture	Bao et al. (2019)
Oilseed rape	*BnWRKY11* and *BnWRKY70*	*Sclerotinia* resistance	Sun et al. (2018)
	BnITPK	Increased protein and low phytic acid in oil	Sashidhar et al. (2020)
	BnSFAR4 and BnSFAR5	Enhanced seed oil content	Karunarathna et al. (2020)
	FAD2	Increased fatty acid content	Okuzaki et al. (2018)
False flax	*CsFAD2*	Increased oleic acid content	Morineau et al. (2017)
	FAE1	Increased oleic acid and α-linolenic acid Content	Ozseyhan et al. (2018)
Wild strawberry	*TAA1 and ARF8*	Improved growth and auxin biosynthesis	Zhou et al. (2018)
	FvebZIPs1.1	Increased sugar content	Xing et al. (2020)
Strawberry	*FaTM6*	Improved quality	Martín-Pizarro et al. (2019)
Cucumber	*eIF4E*	Broad virus resistance	Chandrasekaran et al. (2016)
Sweet potato	*IbBBX24*	Fusarium wilt resistance	Zhang et al. (2020a)
Watermelon	*ClPSK1*	*Fusarium oxysporum* resistance	Zhang et al. (2020b)
	ClALS	Herbicide resistance	Tian et al. (2018)
Pear	*TFL1.1*	Early flowering	Charrier et al. (2019)
Cassava	*nCBP-1 and nCBP-2*	Brown streak resistance	Gomez et al. (2019)
Grapevine	*VvMLO3*	Powdery mildew resistance	Wan et al. (2020)
Sweet orange	*DMR6*	*Candidatus liberibacter* resistance	Zhang et al. (2018c)
	CsLOB1	Canker resistance	Peng et al. (2017)

(continued)

Table 2 (continued)

Crop	Target gene	Target traits	References
Grapefruit	*CsLOB1*	Canker resistance	Jia et al. (2017)
Pummelo	*CsLOB1*	Canker resistance	Jia and Wang (2020)
Apple	*DIPM*	Blight resistance	Malnoy et al. (2016)
	TFL1.1	Early flowering	Charrier et al. (2019)
Banana	*RGA2, Ced9*	Fusarium wilt resistance	Dale et al. (2017)
	MaeBSV	Streak virus resistance	Tripathi et al. (2019)
	MaLCYepsilon	Fortification of β-carotene	Kaur et al. (2020)
Kiwi fruit	*CEN4 and CEN*	Rapid terminal flowering and fruit production	Varkonyi-Gasic et al. (2019)
Garden petunia	*PhACO1*	Improved flower longevity	Xu et al. (2020)
Wishbone flower	*TfF3H*	Altered flower colure	Nishihara et al. (2018)
Japanese morning glory	*InEPH1*	Improved flower longevity	Shibuya et al. (2018)
	InCCD4	Altered flower color	Watanabe et al. (2018)
	InDFR	Anthocyanin biosynthesis	Watanabe et al. (2017)
Horse phalaenopsis	*MADS*	Floral initiation and development	Tong et al. (2020)
Orchid	*C3H, C4H, 4CL, CCR, IRX*	Lignocellulose biosynthesis	Kui et al. (2017)

7.3   Resistance Against Viruses

Plant viruses are mainly controlled by better agricultural practices and developing virus resistant crop cultivars. In cucumber, the CRISPR-Cas9-mediated *eIF4E* modification resulted in successful resistance towards different viruses, for example *TuMV, papaya ring spot mosaic virus-W, cucumber vein yellowing virus*, and *zucchini yellow mosaic virus* (Chandrasekaran et al. 2016). Tashkandi et al. (2018) applied the CRISPR/Cas tool to enhance immunity in defence to the *Tomato Yellow Leaf Cur virus* (*TYLCV*). Later studies mutated the cassava *eIF4E* isoforms, novel cap-binding protein-1 (nCBP-1) and nCBP-2, and demonstrated that this considerably decreased the severity of Cassava brown streak virus infectious disease (Gomez et al. 2019). Likewise, researchers could edit endogenous banana streak virus sequences in the banana B genome that prevent virus transmission throughout breeding events (Tripathi et al. 2019).

7.4   Resistance and Tolerance Against Herbicide

Weed infestation is a serious problem all around the world; however, selective herbicide control is an effective management tool. Tillage practices cause various complications with increased labour inputs. Herbicide-resistant crops are the cost-effective way to combat weeds in modern agriculture. For example, herbicide-resistant watermelon plants were developed using CRISPR/Cas9 to create a single point mutation in the acetolactate synthase (*ALS*) gene, which is a crucial enzyme in the biosynthesis of essential amino acids, including leucine, valine and isoleucine (Tian et al. 2018). Similarly, the *ALS1* and *ALS2* genes were mutated to increase herbicide-resistance (Veillet et al. 2019). Another innovative method is the use of geminivirus replicons, which were used in potato plants and modify CRISPR/Cas9 sections of the acetolactate synthase 1 (*ALS1*) gene-targeted site, while the transgenic free mutated plants showed reduced herbicide susceptibility (Butler et al. 2016).

7.5   Improved Quality and Yield

The quality characteristics of horticulture crops involve the production of flavour and nutrient components, structure and morphology, influence palatability and nutritive values. Efforts have been made for several years to enhance the productivity of horticultural crops. The latest development of genome editing technology with potential uses in horticultural crops offers an approach to achieve this aim quickly and efficiently. An interesting study was recently published by Zsogon et al. (2018), in which six main loci were engineered for production and productivity by exploiting the CRISPR/Cas9 technology to produce a novel tomato variety. This research found that modified lines have higher lycopene content (500%), three-times greater fruit size, and ten-time more fruit number than wild-type plants (Zsogon et al. 2018). Various genes involved in the carotenoid metabolic-pathway such as *SlLCY-B1*, *SlLCY-B2*, *SlLCY-E*, *SlBlc*, and *SlSGR1* have been altered to enhance the lycopene content of tomato fruits. The CRISPR/Cas9 technique was utilized to construct double mutants missing the *SlGAD2* and *SlGAD3* genes, raising the level of GABA (gamma-Aminobutyric acid) in tomato fruits (Nonaka et al. 2017). GABA is a non-proteinogenic amino acids, which is effective in lowering blood pressure in hypertensive-patients. Similarly, Miao et al. (2018) found significantly increased accumulation of GABA in genome-edited lines using a multiplex CRISPR/Cas9-method designed to target *GABA-TP1*, *GABA-TP2*, *GABA-TP3*, *SSADH* and *CAT9* genes simultaneously.

Genetic modification of *ACO* (*1-aminocyclopropane-1-carboxylate oxidase*) using CRISPR/Cas9 increased the flower longevity in contrast with wild-type petunia plants (Yang et al. 2020). The vitamin C content of lettuce was increased by editing the ORF (open reading frame) of *LsGGP2* gene, which is important in vitamin C biosynthesis (Zhang et al. 2018a). SGAs (steroidal glycoalkaloids) are highly toxic substances that are abundant in potato tuber sprouts and flowers. SGA-free potato variety have been created by modifying *St16DOX* gene with CRISPR technology (Nakayasu et al. 2018). CRISPR-mediated editing of floral repressor *TFL1.1* gene resulted an early-flowering phenotype in 9% of pear plants and 93% of resulted apple plants (Charrier et al. 2019). Tartaric acid synthesis genes related to other flower characteristics were also edited. A *carotenoid cleavage dioxygenase* (*CCD*) gene was modified which resulted in 20-fold improved carotenoid contents in mutated Japanese morning glory (Watanabe et al. 2018). Ethylene is an important post-harvest quality regulator for cut flowers. EPHEMERAL1 (*EPH1*), an ethylene-mediated NAC transcription factor gene, was edited to delay senescence and extend the flowering time in Japanese morning glory (Shibuya et al. 2018).

8   Conclusion

In recent decades, CRISPR/Cas-driven gene editing has become one of the most potent genetic engineering tools in crop biotechnology. CRISPR/Cas-based gene editing allows for effective, accurate and targeted genetic manipulation in cereals, fruits, vegetables and other ornamental crops to increase yield and nutritional value with also increasing disease resistance. Such applications offer an advantage to the latest genetic engineering strategies for food safety and combating undernourishment of vitamins, minerals, and nutraceuticals to the world's growing population. Although CRISPR/Cas9 tool has truly evolved as the most powerful technique for crop improvement, the genome editing of cereals and other important horticulture crops confronts various obstacles in its application. The unintended targets, genome stability, transformation efficiency and lack of specific PAM is a significant concern of wide use of the CRISPR-based genome editing for crop development. These hurdles must be eliminated to facilitate the successful implementation of these techniques with promising future prospects. Refinement of editing capacity, the establishment of transformation methods and advancement in delivering the CRISPR/Cas method will be the key prospects for progressing genome editing measures in important commercial crops. With these advancements, CRISPR/Cas genome editing would be a promising tool for creating novel stress-tolerant varieties, higher-yielding crops, improved pathogen resistance, and increased nutritional value in the upcoming decades.

References

Andersson M, Turesson H, Olsson N, Fält AS, Ohlsson P, Gonzalez MN, Samuelsson M, Hofvander P (2018) Genome editing in potato via CRISPR-Cas9 ribonucleoprotein delivery. Physiol Plant 164:378–384. https://doi.org/10.1111/ppl.12731

Ashokkumar S, Jaganathan D, Ramanathan V, Rahman H, Palaniswamy R, Kambale R, Muthurajan RJ (2020) Creation of novel alleles of fragrance gene *OsBADH2* in rice through CRISPR/Cas9 mediated gene editing. PLoS One 15:e0237018–e0237035. https://doi.org/10.1371/journal.pone.0237018

Bao A, Chen H, Chen L, Chen S, Hao Q, Guo W, Qiu D, Shan Z, Yang Z, Yuan S (2019) CRISPR/Cas9-mediated targeted mutagenesis of *GmSPL9* genes alters plant architecture in soybean. BMC Plant Biol 19:1–12. https://doi.org/10.1186/s12870-019-1746-6

Bari VK, Nassar JA, Aly R (2021) CRISPR/Cas9 mediated mutagenesis of *MORE AXILLARY GROWTH 1* in tomato confers resistance to root parasitic weed Phelipanche aegyptiaca. Sci Rep 11:1–11. https://doi.org/10.1038/s41598-021-82897-8

Barrangou R, Dudley EG (2016) CRISPR-based typing and next-generation tracking technologies. Ann Rev food Sci Tech 7:395–411. https://doi.org/10.1146/annurev-food-022814-015729

Barrangou R, Fremaux C, Deveau H, Richards M, Boyaval P, Moineau S, Romero DA, Horvath P (2007) CRISPR provides acquired resistance against viruses in prokaryotes. Science 315:1709–1712. https://doi.org/10.1126/science.1138140

Bibikova M, Beumer K, Trautman JK, Carroll D (2003) Enhancing gene targeting with designed zinc finger nucleases. Science 300:764–764

Bolotin A, Quinquis B, Sorokin A, Ehrlich SD (2005) Clustered regularly interspaced short palindrome repeats (CRISPRs) have spacers of extrachromosomal origin. Microbiology 151:2551–2561. https://doi.org/10.1099/mic.0.28048-0

Brouns SJ, Jore MM, Lundgren M, Westra ER, Slijkhuis RJ, Snijders AP, Dickman MJ, Makarova KS, Koonin EV, Van Der Oost J (2008) Small CRISPR RNAs guide antiviral defense in prokaryotes. Science 321:960–964. https://doi.org/10.1126/science.1159689

Butler NM, Baltes NJ, Voytas DF, Douches DS (2016) Geminivirus-mediated genome editing in potato (*Solanum tuberosum* L.) using sequence-specific nucleases. Front Plant Sci 7:1045–1057. https://doi.org/10.3389/fpls.2016.01045

Carroll D (2011) Genome engineering with zinc-finger nucleases. Genetics 188:773–782. https://doi.org/10.1534/genetics.111.131433

Chandrasekaran J, Brumin M, Wolf D, Leibman D, Klap C, Pearlsman M, Sherman A, Arazi T, Gal-On A (2016) Development of broad virus resistance in non-transgenic cucumber using CRISPR/Cas9 technology. Mol Plant Pathol 17:1140–1153. https://doi.org/10.1111/mpp.12375

Charrier A, Vergne E, Dousset N, Richer A, Petiteau A, Chevreau E (2019) Efficient targeted mutagenesis in apple and first time edition of pear using the CRISPR-Cas9 system. Front Plant Sci 10:40–65. https://doi.org/10.3389/fpls.2019.00040

Cong L, Ran FA, Cox D, Lin S, Barretto R, Habib N, Hsu PD, Wu X, Jiang W, Marraffini LAJS (2013) Multiplex genome engineering using CRISPR/Cas systems. Science 339:819–823. https://doi.org/10.1126/science.1231143

Connorton JM, Jones ER, Rodríguez-Ramiro I, Fairweather-Tait S, Uauy C, Balk J (2017) Wheat vacuolar iron transporter *TaVIT2* transports Fe and Mn and is effective for biofortification. Plant Physiol 174:2434–2444. https://doi.org/10.1104/pp.17.00672

Dale J, James A, Paul J-Y, Khanna H, Smith M, Peraza-Echeverria S, Garcia-Bastidas F, Kema G, Waterhouse P, Mengersen K (2017) Transgenic Cavendish bananas with resistance to Fusarium wilt tropical race 4. Nat Commun 8:1–8. https://doi.org/10.1038/s41467-017-01670-6

Danilo B, Perrot L, Mara K, Botton E, Nogué F, Mazier M (2019) Efficient and transgene-free gene targeting using Agrobacterium-mediated delivery of the CRISPR/Cas9 system in tomato. Plant Cell Rep 38:459–462. https://doi.org/10.1007/s00299-019-02373-6

Deltcheva E, Chylinski K, Sharma CM, Gonzales K, Chao Y, Pirzada ZA, Eckert MR, Vogel J, Charpentier E (2011) CRISPR RNA maturation by trans-encoded small RNA and host factor RNase III. Nature 471:602–607. https://doi.org/10.1038/nature09886

Deng D, Yan C, Pan X, Mahfouz M, Wang J, Zhu J-K, Shi Y, Yan N (2012) Structural basis for sequence-specific recognition of DNA by TAL effectors. Science 335:720–723. https://doi.org/10.1126/science.1215670

Dong L, Qi X, Zhu J, Liu C, Zhang X, Cheng B, Mao L, Xie C (2019) Supersweet and waxy: meeting the diverse demands for specialty maize by genome editing. Plant Biotechnol J 17:1853–1855. https://doi.org/10.1111/pbi.13144

Dong OX, Yu S, Jain R, Zhang N, Duong PQ, Butler C, Li Y, Lipzen A, Martin JA, Barry KWJNC (2020) Marker-free carotenoid-enriched rice generated through targeted gene insertion using CRISPR-Cas9. Nat Commun 11:1–10. https://doi.org/10.1038/s41467-020-14981-y

Doyle EL, Booher NJ, Standage DS, Voytas DF, Brendel VP, VanDyk JK, Bogdanove AJ (2012) TAL Effector-Nucleotide Targeter (TALE-NT) 2.0: tools for TAL effector design and target prediction. Nucleic Acids Res 40:W117–W122. https://doi.org/10.1093/nar/gks608

Feng Z, Zhang B, Ding W, Liu X, Yang D-L, Wei P, Cao F, Zhu S, Zhang F, Mao Y (2013) Efficient genome editing in plants using a CRISPR/Cas system. Cell Res 23:1229–1232. https://doi.org/10.1038/cr.2013.114

Garneau JE, Dupuis M-È, Villion M, Romero DA, Barrangou R, Boyaval P, Fremaux C, Horvath P, Magadán AH, Moineau S (2010) The CRISPR/Cas bacterial immune system cleaves bacteriophage and plasmid DNA. Nature 468:67–71. https://doi.org/10.1038/nature09523

Gomez MA, Lin ZD, Moll T, Chauhan RD, Hayden L, Renninger K, Beyene G, Taylor NJ, Carrington JC, Staskawicz BJ (2019) Simultaneous CRISPR/Cas9-mediated editing of cassava eIF 4E isoforms nCBP-1 and nCBP-2 reduces cassava brown streak disease symptom severity and incidence. Plant Biotechnol J 17:421–434. https://doi.org/10.1111/pbi.12987

Govindaraj M, Vetriventhan M, Srinivasan M (2015) Importance of genetic diversity assessment in crop plants and its recent advances: an overview of its analytical perspectives. Genet Res Int 2015:1–14. https://doi.org/10.1155/2015/431487

Hale CR, Zhao P, Olson S, Duff MO, Graveley BR, Wells L, Terns RM, Terns MP (2009) RNA-guided RNA cleavage by a CRISPR RNA-Cas protein complex. Cell 139:945–956. https://doi.org/10.1016/j.cell.2009.07.040

Huang L, Zhang R, Huang G, Li Y, Melaku G, Zhang S, Chen H, Zhao Y, Zhang J, Zhang Y (2018) Developing superior alleles of yield genes in rice by artificial mutagenesis using the CRISPR/Cas9 system. Crop J 6:475–481. https://doi.org/10.1016/j.cj.2018.05.005

Ishino Y, Shinagawa H, Makino K, Amemura M, Nakata AJ (1987) Nucleotide sequence of the iap gene, responsible for alkaline phosphatase isozyme conversion in Escherichia coli, and identification of the gene product. J Bacteriol 169:5429–5433. https://doi.org/10.1128/jb.169.12.5429-5433.1987

Jansen R, Embden JDV, Gaastra W, Schouls LM (2002) Identification of genes that are associated with DNA repeats in prokaryotes. Mol Microbiol 43:1565–1575. https://doi.org/10.1046/j.1365-2958.2002.02839.x

Jia H, Wang N (2020) Generation of homozygous canker-resistant citrus in the T0 generation using CRISPR-SpCas9p. Plant Biotechnol J 18:1990–1992. https://doi.org/10.1111/pbi.13375

Jia H, Xu J, Orbović V, Zhang Y, Wang N (2017) Editing citrus genome via SaCas9/sgRNA system. Front Plant Sci 8:2135–2144. https://doi.org/10.3389/fpls.2017.02135

Jiang W, Bikard D, Cox D, Zhang F, Marraffini LA (2013a) CRISPR-assisted editing of bacterial genomes. NAT Biotechnol 31:233–239. https://doi.org/10.1038/nbt.2508

Jiang W, Zhou H, Bi H, Fromm M, Yang B, Weeks DP (2013b) Demonstration of CRISPR/Cas9/sgRNA-mediated targeted gene modification in Arabidopsis, tobacco, sorghum and rice. Nucleic Acids Res 41:e188–e188. https://doi.org/10.1093/nar/gkt780

Jinek M, Chylinski K, Fonfara I, Hauer M, Doudna JA, Charpentier E (2012) A programmable dual-RNA–guided DNA endonuclease in adaptive bacterial immunity. Science 337:816–821. https://doi.org/10.1126/science.1225829

Karunarathna NL, Wang H, Harloff HJ, Jiang L, Jung C (2020) Elevating seed oil content in a polyploid crop by induced mutations in seed fatty acid reducer genes. Plant Biotechnol J 18:2251–2266. https://doi.org/10.1111/pbi.13381

Kaur N, Alok A, Kumar P, Kaur N, Awasthi P, Chaturvedi S, Pandey P, Pandey A, Pandey AK, Tiwari S (2020) CRISPR/Cas9 directed editing of lycopene epsilon-cyclase modulates metabolic flux for β-carotene biosynthesis in banana fruit. Metab Eng 59:76–86. https://doi.org/10.1016/j.ymben.2020.01.008

Khan SH (2019) Genome-editing technologies: concept, pros, and cons of various genome-editing techniques and bioethical concerns for clinical application. Mol Ther Nucleic Acids 16:326–334. https://doi.org/10.1016/j.omtn.2019.02.027

Kim Y-A, Moon H, Park C-J (2019) CRISPR/Cas9-targeted mutagenesis of Os8N3 in rice to confer resistance to Xanthomonas oryzae pv. oryzae. Rice 12:1–13. https://doi.org/10.1186/s12284-019-0325-7

Kuang Y, Li S, Ren B, Yan F, Spetz C, Li X, Zhou X, Zhou H (2020) Base-editing-mediated artificial evolution of *OsALS1* in planta to develop novel herbicide-tolerant rice germplasms. Mol Plant 13:565–572. https://doi.org/10.1016/j.molp.2020.01.010

Kui L, Chen H, Zhang W, He S, Xiong Z, Zhang Y, Yan L, Zhong C, He F, Chen J (2017) Building a genetic manipulation tool box for orchid biology: identification of constitutive promoters and application of CRISPR/Cas9 in the orchid, Dendrobium officinale. Front Plant Sci 7:2036–2048. https://doi.org/10.3389/fpls.2016.02036

Kwon J, Kasai A, Maoka T, Masuta C, Sano T, Nakahara KS (2020) RNA silencing-related genes contribute to tolerance of infection with potato virus X and Y in a susceptible tomato plant. Virol J 17:1–13. https://doi.org/10.1186/s12985-020-01414-x

Lee JH, Won HJ, Hoang Nguyen Tran P, Sm L, Kim HY, Jung JH (2021) Improving lignocellulosic biofuel production by CRISPR/Cas9-mediated lignin modification in barley. Glob Change Biol Bioenergy 13:742–752. https://doi.org/10.1111/gcbb.12808

Li J, Meng X, Zong Y, Chen K, Zhang H, Liu J, Li J, Gao C (2016) Gene replacements and insertions in rice by intron targeting using CRISPR–Cas9. Nat Plants 2:1–6. https://doi.org/10.1038/nplants.2016.139

Li T, Yang X, Yu Y, Si X, Zhai X, Zhang H, Dong W, Gao C, Xu C (2018a) Domestication of wild tomato is accelerated by genome editing. Nat Biotechnol 36:1160–1163. https://doi.org/10.1038/nbt.4273

Li X, Wang Y, Chen S, Tian H, Fu D, Zhu B, Luo Y, Zhu H (2018b) Lycopene is enriched in tomato fruit by CRISPR/Cas9-mediated multiplex genome editing. Front Plant Sci 9:559–570. https://doi.org/10.3389/fpls.2018.00559

Li C, Li Y-h, Li Y, Lu H, Hong H, Tian Y, Li H, Zhao T, Zhou X, Liu J (2020) A domestication-associated gene *GmPRR3b* regulates the circadian clock and flowering time in soybean. Mol Plant 13:745–759. https://doi.org/10.1016/j.molp.2020.01.014

Liang Z, Zhang K, Chen K, Gao C (2014) Targeted mutagenesis in Zea mays using TALENs and the CRISPR/Cas system. J Genet Genomic 41:63–68. https://doi.org/10.1016/j.jgg.2013.12.001

Liu L, Kuang Y, Yan F, Li S, Ren B, Gosavi G, Spetz C, Li X, Wang X, Zhou X (2021) Developing a novel artificial rice germplasm for dinitroaniline herbicide resistance by base editing of *OsTubA2*. Plant Biotechnol J 19:5–7. https://doi.org/10.1111/pbi.13430

Ma X, Zhu Q, Chen Y, Liu Y-G (2016) CRISPR/Cas9 platforms for genome editing in plants: developments and applications. Mol Plant 9:961–974. https://doi.org/10.1016/j.molp.2016.04.009

Ma X, Feng F, Zhang Y, Elesawi IE, Xu K, Li T, Mei H, Liu H, Gao N, Chen C (2019) A novel rice grain size gene *OsSNB* was identified by genome-wide association study in natural population. Plos genet 15:e1008191–e1008210. https://doi.org/10.1371/journal.pgen.1008191

Macovei A, Sevilla NR, Cantos C, Jonson GB, Slamet-Loedin I, Čermák T, Voytas DF, Choi IR, Chadha-Mohanty P (2018) Novel alleles of rice *eIF4G* generated by CRISPR/Cas9-targeted mutagenesis confer resistance to Rice tungro spherical virus. Plant Biotechnol J 16:1918–1927. https://doi.org/10.1111/pbi.12927

Malnoy M, Viola R, Jung M-H, Koo O-J, Kim S, Kim J-S, Velasco R, Nagamangala Kanchiswamy C (2016) DNA-free genetically edited grapevine and apple protoplast using CRISPR/Cas9 ribonucleoproteins. Front Plant Sci 7:1904–1913. https://doi.org/10.3389/fpls.2016.01904

Mao X, Zheng Y, Xiao K, Wei Y, Zhu Y, Cai Q, Chen L, Xie H, Zhang J (2018) OsPRX2 contributes to stomatal closure and improves potassium deficiency tolerance in rice. Biochem Biophys Res Commun 495:461–467. https://doi.org/10.1016/j.bbrc.2017.11.045

Marraffini LA, Sontheimer EJ (2008) CRISPR interference limits horizontal gene transfer in staphylococci by targeting DNA. Science 322:1843–1845. https://doi.org/10.1126/science.1165771

Martínez-Fortún J, Phillips DW, Jones HD (2017) Potential impact of genome editing in world agriculture. Emerging Top Life Sci 1:117–133. https://doi.org/10.1042/ETLS20170010

Martín-Pizarro C, Triviño JC, Posé D (2019) Functional analysis of the *TM6 MADS-box* gene in the octoploid strawberry by CRISPR/Cas9-directed mutagenesis. J Exp Bot 70:885–895. https://doi.org/10.1093/jxb/ery400

Miao C, Xiao L, Hua K, Zou C, Zhao Y, Bressan RA, Zhu J-K (2018) Mutations in a subfamily of abscisic acid receptor genes promote rice growth and productivity. Proceed Nat Acad Sci 115:6058–6063. https://doi.org/10.1073/pnas.1804774115

Miller JC, Tan S, Qiao G, Barlow KA, Wang J, Xia DF, Meng X, Paschon DE, Leung E, Hinkley SJ (2011) A TALE nuclease architecture for efficient genome editing. Nat Biotechnol 29:143–148. https://doi.org/10.1038/nbt.1755

Miyamoto T, Takada R, Tobimatsu Y, Takeda Y, Suzuki S, Yamamura M, Osakabe K, Osakabe Y, Sakamoto M, Umezawa T (2019) *Os MYB 108* loss-of-function enriches p-coumaroylated and tricin lignin units in rice cell walls. Plant J 98:975–987. https://doi.org/10.1111/tpj.14290

Mojica FJ, García-Martínez J, Soria E (2005) Intervening sequences of regularly spaced prokaryotic repeats derive from foreign genetic elements. J Mol Evol 60:174–182. https://doi.org/10.1007/s00239-004-0046-3

Morineau C, Bellec Y, Tellier F, Gissot L, Kelemen Z, Nogué F, Faure JDJ (2017) Selective gene dosage by CRISPR-Cas9 genome editing in hexaploid Camelina sativa. Plant Biotechnol J 15:729–739. https://doi.org/10.1111/pbi.12671

Mulepati S, Héroux A, Bailey S (2014) Crystal structure of a CRISPR RNA–guided surveillance complex bound to a ssDNA target. Science 345:1479–1484. https://doi.org/10.1126/science.1256996

Nakayasu M, Akiyama R, Lee HJ, Osakabe K, Osakabe Y, Watanabe B, Sugimoto Y, Umemoto N, Saito K, Muranaka T (2018) Generation of α-solanine-free hairy roots of potato by CRISPR/Cas9 mediated genome editing of the *St16DOX* gene. Plant Physiol Biochem 131:70–77. https://doi.org/10.1016/j.plaphy.2018.04.026

Nekrasov V, Wang C, Win J, Lanz C, Weigel D, Kamoun S (2017) Rapid generation of a transgene-free powdery mildew resistant tomato by genome deletion. Sci Rep 7:1–6. https://doi.org/10.1038/s41598-017-00578-x

Nishihara M, Higuchi A, Watanabe A, Tasaki K (2018) Application of the CRISPR/Cas9 system for modification of flower color in Torenia fournieri. BMC Plant Biol 18:1–9. https://doi.org/10.1186/s12870-018-1539-3

Nonaka S, Arai C, Takayama M, Matsukura C, Ezura HJSR (2017) Efficient increase of γ-aminobutyric acid (GABA) content in tomato fruits by targeted mutagenesis. Sci Rep 7:1–14. https://doi.org/10.1038/s41598-017-06400-y

Okuzaki A, Ogawa T, Koizuka C, Kaneko K, Inaba M, Imamura J, Koizuka N (2018) CRISPR/Cas9-mediated genome editing of the fatty acid desaturase 2 gene in *Brassica napus*. Plant Physiol Biochem 131:63–69. https://doi.org/10.1016/j.plaphy.2018.04.025

Oliva R, Ji C, Atienza-Grande G, Huguet-Tapia JC, Perez-Quintero A, Li T, Eom J-S, Li C, Nguyen H, Liu B (2019) Broad-spectrum resistance to bacterial blight in rice using genome editing. Nat Physiol 37:1344–1350. https://doi.org/10.1038/s41587-019-0267-z

Ortigosa A, Gimenez-Ibanez S, Leonhardt N, Solano R (2019) Design of a bacterial speck resistant tomato by CRISPR/Cas9-mediated editing of *Sl JAZ 2*. Plant Biotechnol J 17:665–673. https://doi.org/10.1111/pbi.13006. Epub 2018 Oct 5

Ozseyhan ME, Kang J, Mu X, Lu C (2018) Mutagenesis of the *FAE1* genes significantly changes fatty acid composition in seeds of Camelina sativa. Plant Physiol Biochem 123:1–7. https://doi.org/10.1016/j.plaphy.2017.11.021

Peng A, Chen S, Lei T, Xu L, He Y, Wu L, Yao L, Zou X (2017) Engineering canker-resistant plants through CRISPR/Cas9-targeted editing of the susceptibility gene *Cs LOB 1* promoter in citrus. Plant Biotechnol J 15:1509–1519. https://doi.org/10.1111/pbi.12733

Pourcel C, Salvignol G, Vergnaud G (2005) CRISPR elements in Yersinia pestis acquire new repeats by preferential uptake of bacteriophage DNA, and provide additional tools for evolutionary studies. Microbiology 151:653–663. https://doi.org/10.1099/mic.0.27437-0

Puchta H, Fauser F (2013) Gene targeting in plants: 25 years later. Int J Dev Biol 57:629–637. https://doi.org/10.1387/ijdb.130194hp

Ren D, Cui Y, Hu H, Xu Q, Rao Y, Yu X, Zhang Y, Wang Y, Peng Y, Zeng D (2019) *AH 2* encodes a MYB domain protein that determines hull fate and affects grain yield and quality in rice. Plant J 100:813–824. https://doi.org/10.1111/tpj.14481

Reyon D, Tsai SQ, Khayter C, Foden JA, Sander JD, Joung JK (2012) FLASH assembly of TALENs for high-throughput genome editing. Nat Biotechnol 30:460. https://doi.org/10.1038/nbt.2170

Rodríguez-Leal D, Lemmon ZH, Man J, Bartlett ME, Lippman ZB (2017) Engineering quantitative trait variation for crop improvement by genome editing. Cell 171:470–480. https://doi.org/10.1016/j.cell.2017.08.030

Samai P, Pyenson N, Jiang W, Goldberg GW, Hatoum-Aslan A, Marraffini LA (2015) Co-transcriptional DNA and RNA cleavage during type III CRISPR-Cas immunity. Cell 161:1164–1174. https://doi.org/10.1016/j.cell.2015.04.027

Sánchez-León S, Gil-Humanes J, Ozuna CV, Giménez MJ, Sousa C, Voytas DF, Barro F (2018) Low-gluten, nontransgenic wheat engineered with CRISPR/Cas9. Plant Biotechnol J 16:902–910. https://doi.org/10.1111/pbi.12837

Santillán Martínez MI, Bracuto V, Koseoglou E, Appiano M, Jacobsen E, Visser RGF, Wolters AA, Bai Y (2020) CRISPR/Cas9-targeted mutagenesis of the tomato susceptibility gene *PMR4* for resistance against powdery mildew. BMC plant biology 20:284. https://doi.org/10.1186/s12870-020-02497-y

Sashidhar N, Harloff HJ, Potgieter L, Jung C (2020) Gene editing of three *BnITPK* genes in tetraploid oilseed rape leads to significant reduction of phytic acid in seeds. Plant Biotechnol J 18:2241–2250. https://doi.org/10.1111/pbi.13380

Shibuya K, Watanabe K, Ono M (2018) CRISPR/Cas9-mediated mutagenesis of the *EPHEMERAL1* locus that regulates petal senescence in Japanese morning glory. Plant Physiol Biochem 131:53–57. https://doi.org/10.1016/j.plaphy.2018.04.036

Shan Q, Wang Y, Li J, Gao C (2014) Genome editing in rice and wheat using the CRISPR/Cas system. Nat protoc. 9(10):2395-410. https://doi.org/10.1038/nprot.2014.157.

Songmei L, Jie J, Yang L, Jun M, Shouling X, Yuanyuan T, Youfa L, Qingyao S, Jianzhong H (2019) Characterization and evaluation of *OsLCT1* and *OsNramp5* mutants generated through CRISPR/Cas9-mediated mutagenesis for breeding low Cd rice. Rice Sci 26:88–97. https://doi.org/10.1016/j.rsci.2019.01.002

Soyk S, Müller NA, Park SJ, Schmalenbach I, Jiang K, Hayama R, Zhang L, Van Eck J, Jiménez-Gómez JM, Lippman ZB (2017) Variation in the flowering gene *SELF PRUNING 5G* promotes day-neutrality and early yield in tomato. Nat Genet 49:162–168. https://doi.org/10.1038/ng.3733

Sun Y, Jiao G, Liu Z, Zhang X, Li J, Guo X, Du W, Du J, Francis F, Zhao Y (2017) Generation of high-amylose rice through CRISPR/Cas9-mediated targeted mutagenesis of starch branching enzymes. Front Plant Sci 8:298. https://doi.org/10.3389/fpls.2017.00298

Sun Q, Lin L, Liu D, Wu D, Fang Y, Wu J, Wang YJ (2018) CRISPR/Cas9-mediated multiplex genome editing of the *BnWRKY11* and *BnWRKY70* Genes in Brassica napus L. Int J Mol Sci 19:2716–2734. https://doi.org/10.3390/ijms19092716

Svitashev S, Young JK, Schwartz C, Gao H, Falco SC, Cigan AM (2015) Targeted mutagenesis, precise gene editing, and site-specific gene insertion in maize using Cas9 and guide RNA. Plant Physiol 169:931–945. https://doi.org/10.1104/pp.15.00793

Takeda Y, Tobimatsu Y, Karlen SD, Koshiba T, Suzuki S, Yamamura M, Murakami S, Mukai M, Hattori T, Osakabe KJTPJ (2018) Downregulation of *p-COUMAROYL ESTER 3-HYDROXYLASE* in rice leads to altered cell wall structures and improves biomass saccharification. Plant J 95:796–811. https://doi.org/10.1111/tpj.13988

Tang L, Mao B, Li Y, Lv Q, Zhang L, Chen C, He H, Wang W, Zeng X, Shao Y (2017) Knockout of *OsNramp5* using the CRISPR/Cas9 system produces low Cd-accumulating indica rice without compromising yield. Sci Rep 7:1–12. https://doi.org/10.1038/s41598-017-14832-9

Tashkandi M, Ali Z, Aljedaani F, Shami A, Mahfouz MM (2018) Engineering resistance against Tomato yellow leaf curl virus via the CRISPR/Cas9 system in tomato. Plant Signal Behavior 13:e1525996–e1526004. https://doi.org/10.1080/15592324.2018.1525996

Tian S, Jiang L, Cui X, Zhang J, Guo S, Li M, Zhang H, Ren Y, Gong G, Zong M (2018) Engineering herbicide-resistant watermelon variety through CRISPR/Cas9-mediated base-editing. Plant Cell Rep 37:1353–1356. https://doi.org/10.1007/s00299-018-2299-0

Tong CG, Wu FH, Yuan YH, Chen YR, Lin CS (2020) High-efficiency CRISPR/Cas-based editing of Phalaenopsis orchid *MADS* genes. Plant Biotechnol J 18:889–892. https://doi.org/10.1111/pbi.13264

Tripathi JN, Ntui VO, Ron M, Muiruri SK, Britt A, Tripathi L (2019) CRISPR/Cas9 editing of endogenous banana streak virus in the B genome of *Musa spp.* overcomes a major challenge in banana breeding. Commun Biol 2:1–11. https://doi.org/10.1038/s42003-019-0288-7

Upadhyay SK, Kumar J, Alok A, Tuli RJGG (2013) RNA-guided genome editing for target gene mutations in wheat. G3 Genes Genomes Genetics 3:2233–2238. https://doi.org/10.1534/g3.113.008847

Usman B, Nawaz G, Zhao N, Liu Y, Li R (2020) Generation of high yielding and fragrant rice (Oryza sativa L.) Lines by CRISPR/Cas9 targeted mutagenesis of three homoeologs of cytochrome P450 gene family and OsBADH2 and transcriptome and proteome profiling of revealed changes triggered by mutations. Plants 9:788–814. https://doi.org/10.3390/plants9060788

Usman B, Nawaz G, Zhao N, Liao S, Qin B, Liu F, Liu Y, Li R (2021) Programmed Editing of Rice (*Oryza sativa* L.) *OsSPL16* gene using CRISPR/Cas9 improves grain yield by modulating the expression of pyruvate enzymes and cell cycle proteins. Int J Mol Sci 22:249. https://doi.org/10.3390/ijms22010249

Varkonyi-Gasic E, Wang T, Voogd C, Jeon S, Drummond RS, Gleave AP, Allan AC (2019) Mutagenesis of kiwifruit *CENTRORADIALIS*-like genes transforms a climbing woody perennial with long juvenility and axillary flowering into a compact plant with rapid terminal flowering. Plant Biotechnol J 17:869–880. https://doi.org/10.1111/pbi.13021

Veillet F, Perrot L, Chauvin L, Kermarrec M-P, Guyon-Debast A, Chauvin J-E, Nogué F, Mazier MJ (2019) Transgene-free genome editing in tomato and potato plants using agrobacterium-mediated delivery of a CRISPR/Cas9 cytidine base editor. Int J Mol Sci 20:402–411. https://doi.org/10.3390/ijms20020402

Wan D-Y, Guo Y, Cheng Y, Hu Y, Xiao S, Wang Y, Wen Y-Q (2020) CRISPR/Cas9-mediated mutagenesis of *VvMLO3* results in enhanced resistance to powdery mildew in grapevine (*Vitis vinifera*). Hortic Res 7:1–14. https://doi.org/10.1038/s41438-020-0339-8

Wang Y, Cheng X, Shan Q, Zhang Y, Liu J, Gao C, Qiu J-L (2014) Simultaneous editing of three homoeoalleles in hexaploid bread wheat confers heritable resistance to powdery mildew. Nat Biotechnol 32:947–951. https://doi.org/10.1038/nbt.2969

Wang F, Wang C, Liu P, Lei C, Hao W, Gao Y, Liu Y-G, Zhao K (2016) Enhanced rice blast resistance by CRISPR/Cas9-targeted mutagenesis of the ERF transcription factor gene *OsERF922*. Plos One 11:e0154027–e0154044. https://doi.org/10.1371/journal.pone.0154027

Wang W, Pan Q, He F, Akhunova A, Chao S, Trick H, Akhunov E (2018) Transgenerational CRISPR-Cas9 activity facilitates multiplex gene editing in allopolyploid wheat. CRISPR J 1:65–74. https://doi.org/10.1089/crispr.2017.0010

Wang J, Wu B, Lu K, Wei Q, Qian J, Chen Y, Fang ZJPP (2019a) The amino acid permease 5 (*OsAAP5*) regulates tiller number and grain yield in rice. Plant Physiol 180:1031–1045. https://doi.org/10.1104/pp.19.00034

Wang R, da Rocha Tavano EC, Lammers M, Martinelli AP, Angenent GC, de Maagd RA (2019b) Re-evaluation of transcription factor function in tomato fruit development and ripening with CRISPR/Cas9-mutagenesis. Sci Rep 9:1–10. https://doi.org/10.1038/s41598-018-38170-6

Watanabe K, Kobayashi A, Endo M, Sage-Ono K, Toki S, Ono M (2017) CRISPR/Cas9-mediated mutagenesis of the dihydroflavonol-4-reductase-B (*DFR-B*) locus in the Japanese morning glory Ipomoea (Pharbitis) nil. Sci Rep 7:1–9. https://doi.org/10.1038/s41598-017-10715-1

Watanabe K, Oda-Yamamizo C, Sage-Ono K, Ohmiya A, Ono M (2018) Alteration of flower colour in Ipomoea nil through CRISPR/Cas9-mediated mutagenesis of carotenoid cleavage dioxygenase 4. Trangenic Res 27:25–38. https://doi.org/10.1007/s11248-017-0051-0

Xie K, Yang Y (2013) RNA-guided genome editing in plants using a CRISPR–Cas system. Mol Plant 6:1975–1983. https://doi.org/10.1093/mp/sst119

Xing S, Chen K, Zhu H, Zhang R, Zhang H, Li B, Gao CJ (2020) Fine-tuning sugar content in strawberry. Genome Biol 21:1–14. https://doi.org/10.1186/s13059-020-02146-5

Xu J, Kang BC, Naing AH, Bae SJ, Kim JS, Kim H, Kim CK (2020) CRISPR/Cas9-mediated editing of *1-aminocyclopropane-1-carboxylate oxidase1* enhances Petunia flower longevity. Plant Biotechnol J 18:287–297. https://doi.org/10.1111/pbi.13197

Yang Q, Zhong X, Li Q, Lan J, Tang H, Qi P, Ma J, Wang J, Chen G, Pu Z (2020) Mutation of the D-hordein gene by RNA-guided Cas9 targeted editing reducing the grain size and changing grain compositions in barley. Food Chem 311:125892. https://doi.org/10.1016/j.foodchem.2019.125892

Yosef I, Goren MG, Qimron U (2012) Proteins and DNA elements essential for the CRISPR adaptation process in *Escherichia coli*. Nucleic Acids Res 40:5569–5576. https://doi.org/10.1093/nar/gks216

Yu Q-h, Wang B, Li N, Tang Y, Yang S, Yang T, Xu J, Guo C, Yan P, Wang Q (2017) CRISPR/Cas9-induced targeted mutagenesis and gene replacement to generate long-shelf life tomato lines. Sci Rep 7:1–9. https://doi.org/10.1038/s41598-017-12262-1

Zafar K, Khan MZ, Amin I, Mukhtar Z, Yasmin S, Arif M, Ejaz K, Mansoor S (2020) Precise CRISPR-Cas9 mediated genome editing in super basmati rice for resistance against bacterial blight by targeting the major susceptibility gene. Front Plant Sci 11:575. https://doi.org/10.3389/fpls.2020.00575

Zetsche B, Gootenberg JS, Abudayyeh OO, Slaymaker IM, Makarova KS, Essletzbichler P, Volz SE, Joung J, Van Der Oost J, Regev A (2015) Cpf1 is a single RNA-guided endonuclease of a class 2 CRISPR-Cas system. Cell 163:759–771. https://doi.org/10.1016/j.cell.2015.09.038

Zhang Y, Liang Z, Zong Y, Wang Y, Liu J, Chen K, Qiu J-L, Gao C (2016) Efficient and transgene-free genome editing in wheat through transient expression of CRISPR/Cas9 DNA or RNA. Nat Commun 7:1–8. https://doi.org/10.1038/ncomms12617

Zhang Y, Bai Y, Wu G, Zou S, Chen Y, Gao C, Tang D (2017) Simultaneous modification of three homoeologs of *Ta EDR 1* by genome editing enhances powdery mildew resistance in wheat. Plant J 91:714–724. https://doi.org/10.1111/tpj.13599

Zhang S, Shi Q, Duan Y, Hall D, Gupta G, Stover E (2018a) Regulation of citrus DMR6 via RNA interference and CRISPR/Cas9-mediated gene editing to improve Huanglongbing tolerance. In: International Congress of Plant Pathology (ICPP) 2018: Plant Health in A Global Economy, 2018c. APSNET

Zhang H, Si X, Ji X, Fan R, Liu J, Chen K, Wang D, Gao C (2018b) Genome editing of upstream open reading frames enables translational control in plants. Nat Biotechnol 36:894–898. https://doi.org/10.1038/nbt.4202

Zhang Z, Ge X, Luo X, Wang P, Fan Q, Hu G, Xiao J, Li F, Wu J (2018d) Simultaneous editing of two copies of *Gh14-3-3d* confers enhanced transgene-clean plant defense against Verticillium dahliae in allotetraploid upland cotton. Front Plant Sci 9:842. https://doi.org/10.3389/fpls.2018.00842

Zhang R, Liu J, Chai Z, Chen S, Bai Y, Zong Y, Chen K, Li J, Jiang L, Gao C (2019) Generation of herbicide tolerance traits and a new selectable marker in wheat using base editing. Nat Plants 5:480–485. https://doi.org/10.1038/s41477-019-0405-0

Zhang D, Zhang Z, Unver T, Zhang B (2020a) CRISPR/Cas: a powerful tool for gene function study and crop improvement. J Adv Res. https://doi.org/10.1016/j.jare.2020.10.003

Zhang H, Zhang Q, Zhai H, Gao S, Yang L, Wang Z, Xu Y, Huo J, Ren Z, Zhao N (2020b) *IbBBX24* promotes the jasmonic acid pathway and enhances fusarium wilt resistance in sweet potato. Plant Cell 32:1102–1123. https://doi.org/10.1105/tpc.19.0064

Zhou J, Peng Z, Long J, Sosso D, Liu B, Eom JS, Huang S, Liu S, Vera Cruz C, Frommer WB (2015) Gene targeting by the TAL effector *PthXo2* reveals cryptic resistance gene for bacterial blight of rice. Plant J 82:632–643. https://doi.org/10.1111/tpj.12838

Zhou J, Wang G, Liu ZJPBJ (2018) Efficient genome editing of wild strawberry genes, vector development and validation. Plant Biotechnol J 16:1868–1877. https://doi.org/10.1111/pbi.12922

Zsogon A, Čermák T, Naves ER, Notini MM, Edel KH, Weinl S, Freschi L, Voytas DF, Kudla J, Peres LEP (2018) De novo domestication of wild tomato using genome editing. Nat Biotechnol 36:1211–1216. https://doi.org/10.1038/nbt.4272

Vegetable Crop Improvement Through CRISPR Technology for Food Security

Joorie Bhattacharya, Alamuru Krishna Chaitanya, Niranjan Hegde, Satnam Singh, Amardeep Kour, and Rahul Nitnavare

Abstract In the current agricultural scenario, high nutritional value and yield are the major goals to deal with increasing population and malnutrition. Vegetable crops include a wide variety of species having high genetic diversity and form a crucial part of the human diet worldwide. Recent biotechnological interventions have allowed for incorporation of desired traits into target varieties. Among these interventions, genome editing has gained widespread application at a rapid pace. Tools like zinc finger nucleases (ZFNs), transcription activator-like effector nucleases (TALENs) and CRISPR/Cas have been utilized extensively and curated accordingly for the improvement of vegetable crops. This chapter focuses on the progress and challenges pertaining to CRISPR/Cas and will also shed light on the biosafety aspect of genome editing in vegetable germplasms globally.

J. Bhattacharya
International Crops Research Institute for the Semi-Arid Tropics (ICRISAT), Patancheru, India, Telangana

Department of Genetics, Osmania University, Hyderabad, Telangana, India

A. K. Chaitanya
Department of Genetics and Plant Breeding, Lovely Professional University, Phagwara, Punjab, India

World Vegetable Center, South Asia, Hyderabad, India

N. Hegde
McGill University, Montreal, Quebec, Canada

S. Singh
Punjab Agricultural University (PAU), Regional Research Station, Faridkot, Punjab, India

A. Kour
Punjab Agricultural University (PAU), Regional Research Station, Bathinda, India

R. Nitnavare (✉)
Division of Plant and Crop Sciences, School of Biosciences, University of Nottingham, Leicestershire, UK

Department of Plant Sciences, Rothamsted Research, Harpenden, UK
e-mail: rahul.nitnavare@rothamsted.ac.uk; rahul.nitnavare1@nottingham.ac.uk

S. H. Wani, G. Hensel (eds.), *Genome Editing*,
https://doi.org/10.1007/978-3-031-08072-2_2

Keywords Genome editing · Vegetable crops · CRISPR/Cas9 · Abiotic stress · Disease resistance · Nutritional improvement · Pest and disease resistance · Biosafety

1   Introduction

Vegetable crops provide an extensive plethora of nutritional value in the form of vitamins, minerals, antioxidant compounds, fibre, and carbohydrates. However, vegetables are sensitive to pest and diseases, as well as to abiotic stresses. Along with this, owing to the high demand of food as well as deficiencies, it has become a necessity to explore and further enhance the nutritional components of these crops. In this framework, continuous efforts have been made in breeding and molecular techniques to improve the ever-growing global food demand in an efficient and cost-intensive way. Even though over the years, trait improvement in vegetable crops has taken major strides, time constraints have created severe obstacles for vegetable breeders. Hybridization-based breeding approaches are quite useful for creating variability and combining the desirable traits into adoptable genetic backgrounds. However, the trait introgression becomes challenging in case of linkage drag with the undesirable traits, such as plant type, low yield, nutritional quality, and market preferred traits such as seed size and color. Additionally, hybridization breeding techniques can take several years for the generation of final outputs (Xiong et al. 2015). As a result of these shortcomings, biotechnological techniques such as mutation breeding and transgenic technology saw a rise. While mutation breeding technology ensures specific variations for new characteristics, it is highly inheritable and low in frequency (Parry et al. 2009). The advent of transgenic technology paved the way for modern biotechnological techniques and provided a rather precise technology for trait improvement without altering the genetic components of superior varieties. While the field of biotechnology provides a rather wholesome approach for crop improvement, the biosafety aspect and legal regulation limits the exploration on the field scale. Also, recalcitrancy in several crops also is a hindrance to successful genetic transformation (Xiong et al. 2015). Therefore, in order to reduce the time-intensive process of traditional breeding and regulatory constraints of modern biotechnological techniques, genome editing using zinc finger, TALENs and CRISPR/Cas9 is being utilized. Genome editing utilizes engineered endonucleases, which bind to specific DNA or RNA sequences and cleave in a sequence-specific manner. These sequence breaks then are repaired by cellular damage repair mechanisms such as homology-directed repair (HDR) and non-homologous end-joining (NHEJ) breaks consequently causing gene modifications at desired sites. Among the genome editing techniques, ZFNs and TALENs are the early gene editing techniques, which were comparatively more time-consuming and less efficient (Chen et al. 2020). All these tools provide a viable approach to targeting specific genes in the plant genome. However, among these, CRISPR/ Cas9 is the most widely utilized genome editing tool (Tian et al. 2021).

CRISPR/Cas9 is based on the *Streptococcus pyogenes* system which contains the Cas9 endonuclease and binds to two naturally occurring molecules, viz, crRNA and tracrRNA. These two molecules are known as single-guide RNA (sgRNA) and contain a binding site for Cas9 along with a sequence complementary to the target gene sequence (protospacer adjacent motif, PAM) (Corte et al. 2019). CRISPR/Cas9 has been utilized for various vegetable crops such as tomato, potato, cucumber and cabbage among many (Chen et al. 2020). The incorporation of the sgRNA is done mainly via *Agrobacterium*-mediated transformation, which is the most effective method now. In the T_0 generation, four major genotypes are obtained, namely, homozygotes, heterozygotes, chimeras and biallelic. The sgRNA is removed in the subsequent generations through segregation which gives rise to a transgene-free plant with desired mutations (Corte et al. 2019).

The trait improvement for vegetables focuses on goals such as extended shelf life, improving yield and quality, enhancing pest and disease resistance, enhancement of secondary metabolites, abiotic stress tolerance and improvement of nutritional value along with phytohormones. These can be achieved via targeted mutagenesis, gene knockout, microRNA knockdown screening which will also provide a cumulative perspective about the gene function (Kamburova et al. 2017). One of the key necessities of genome editing is the availability of whole genome sequence of the concerned crop. Due to this, genome editing has been explored comparatively more in major cereals. Genomes of model plants have been used as reference genome for editing in vegetable crops. However, over the years, the genome sequence of vegetable crops such as cabbage, Chinese cabbage, radish, spinach, Spanish pepper, tomato, eggplant and potato have been assembled and made available in the public domain (Chen et al. 2019). The availability of genomic sequences would further allow to elucidate the underlying mechanisms of various traits and in turn facilitate the identification of genes for specific targeting using CRISPR/Cas9 (Xiong et al. 2015).

In vegetable crops, the most trait improvement has been explored in tomato and potato which can be owed to availability of genomic resources and calcitrancy of crops to *Agrobacterium*-mediated transformation. Development of improved delivery reagents, other variants of nucleases, plant regeneration methods and transcriptional regulation would enhance genome editing in vegetable crops in the near future (Cardi et al. 2017). The current chapter elucidates the status of genome editing for traits such as pest and disease resistance, nutritional trait improvement and abiotic stress tolerance along with the biosafety regulations in major vegetable crops.

2   Applications of Genome Editing in Improvement of Vegetable Crops

Due to various consequences in climate change, crops are attacked by new pathogens (bacteria, viruses, fungi, etc.) and pests that cause drastic reduction in the yield and renders them susceptible to different biotic and abiotic stress. Plants, which are biologically resilient to numerous pests and diseases, can impact pesticide drop and crop protection. Although, they contribute the resistance, yield and other yield contributing characters like plant height might face consequences (Gao et al. 2020; Van Butselaar and Van den Ackerveken 2020). Thus, alternate resistance mechanism sources which coordinate yield, growth and defence are highly essential. With all these challenges, breeding programs aim to develop resistant varieties and genome editing is one of the powerful technologies to meet this. Qualitative traits include traits such as improvement of shelf life, enhancement of active compounds and minerals, starch quality regulation, delayed ripening, and pigment biosynthesis modifications. In vegetable crops, the major species which has been explored for nutritional trait enhancement using genome-editing tools are tomato, potato, carrot, yam, and sweet potato. While CRISPR/Cas9 has been applied to vegetables such as cauliflower and cabbage, there are no reported studies of trait enhancement in these vegetables. Abiotic stresses, including drought, salinity, extreme temperature, heavy metal toxicity, are the major yield-limiting factors and greatly affect vegetable crop production worldwide (Canter 2018). All major stresses including extreme temperature, drought, and salinity have been discussed in the past to describe how they lead to significant yield losses in major food crops (Mantri et al. 2012). Physical stress sensing may start anywhere in the cell, in different cell organelles or any cellular compartments. In plants, initial stress signals are perceived on the cell surface or at the transmembrane. Signalling events are controlled by secondary massengers and regulatory proteins (Ca^{+2} and protein kinases) and will be transmitted downstream to activate transcriptional regulators such as transcription factor binding proteins (TFBPs), hormones, and transcription factors (TFs) (Zhu 2016). Major stress responses are marked as, increase in the primary metabolism to meet the high energy demand, decrease in the secondary metabolism for efficient energy conservation, increase in the accumulation of callose, lignin, and polyphenols for strengthening secondary cell wall, rapid anti-oxidative mechanism, changes in cell cycle activity for cell size maintenance, and increase in the cytoskeleton, chaperon and signalling proteins (Zhang et al. 2021). Most importantly, all the abiotic stresses are often interconnected and induce the excessive production of reactive oxygen species (ROS) which causes cellular damage, alters major physiological processes, and inhibits overall plant growth (Choudhury et al. 2017). Abiotic stress is a complex trait consisting of different molecular components and metabolic pathways. Stress responses resulted in induced expression of the genes involved in a series of molecular and metabolic pathways. Stress-induced genes can be categorized into structural and regulatory genes. The class structural genes mainly consist of genes, which encode transmembrane proteins, enzymes involved in the osmolyte biosynthesis and detoxification. Genes encoding Late Embryogenesis Abundant (LEA) protein, anti-freezing proteins, chaperons, and miRNA binding protein can also be considered as the structural genes. Genes-encoding transcription factors (TFs) including dehydration-responsive element-binding proteins (DREB'S), transduction proteases (e.g phospholipase C), protein kinases such as ribosomal protein kinases and receptor kinases fall into the regulatory genes category (Dos Reis et al. 2012). A wide range of genes have been identified across major vegetable crops in order to obtain desired traits using CRISPR technology (Table 1).

2.1 Qualitative Traits

2.1.1 Starch Content

Potato (*Solanum tuberosum*) is one of the most widely consumed vegetables from the tuber family. Some of the main focuses of breeding pipeline pertaining to potatoes are in reduction of toxic steroidal glycoalkaoids (SGAs), low starch content and reduced polyphenol oxidases to minimize browning. CRISPR/Cas9 technology has been significantly studied in this crop for the above-mentioned traits. Regulation of starch properties is crucial in potato consumption to reduce the health risk associated with it. Potato starch is readily digested in the body and therefore effects blood sugar level as well as contributes to obesity. Therefore, efforts are being made to make the potato starch less digestible thereby reducing the health risks. CRISPR/Cas9 technology has been employed for generating mutations in starch branching. The granule bound starch synthase I (*GBSSI*) gene has been targeted in potato along with a translational enhancer, dMac3 owing to tetraploid nature of potato. The translational enhancer was seen to enhance the digestion efficiency of Cas9 and consequently improve the mutation frequency. Targeting the *GBSSI* gave rise to progenies which exhibited reduced amylose content in potato starch. Amylose-free potato is a valuable component of the paper industry (Kusano et al. 2018). Potato starch might be deleterious effects on human and animal health as it is readily digestible and subsequently causes increased blood sugar levels along with obesity. Therefore, less digestible starch in the form of amylose-rich potato is being preferred and is now crucial trait studied in the breeding pipeline. While amylose is a large α-1, 4 linked glucose polymer, amylopectin consists of α-1, 4 linked glucose and α-1, 6 linkages. Upon gelatinization and cooling, the long linear chains which become packed together are resistant to α-amylase which is predominantly seen in amylose-rich starch and long branched amylopectin chains. Starch-branching enzymes (SBEs) aid in the formation of α-1, 6 linkages. In order to increase the amylose content, the SBEs concentration need to be reduced which was attained by targeted mutagenesis of *SBE1* and *SBE2*. The study demonstrated that *SBE* mutagenesis through CRISPR/Cas9 technology can aid development of a wide range of genotypes with varying starch properties (Tuncel et al. 2019). Another isoform which has been examined is *SBE3* gene attached to translational enhancer dMAC3. The resultant mutants

Table 1 Summary of genome editing studies targeting gene/s governing qualitative and abiotic stress tolerance traits for vegetable crop improvement

Stress/trait	Plant	Targeted gene (s)	Effect	Reference
Drought, heat, high light	Tomato	*SlLBD40*	Enhanced drought tolerance	Liu et al. (2020)
	Tomato	*SlMAPK3*	Reduced drought tolerance	Wang et al. (2017)
	Tomato	*SlNPR1*	Reduced drought tolerance	Li et al. (2019)
	Tomato	*SlAGL6*	The mutants could produce fruit under heat stress	Klap et al. (2017)
	Potato	*StCDF1* and *StFLORE*	Increased drought tolerance	Ramírez Gonzales et al. (2021)
	Potato	*StAOX*	Resistant to high-light stress	Hua et al. (2020)
	Soybean	*GmLCLs*	Reduced drought tolerance	Yuan et al. (2021)
	Soybean	*GmNAC8*	Reduced heat tolerance	Yang et al. (2020)
	Soybean	*GmHsp90A2*	Decreased tolerance to heat stress	Huang et al. (2019a)
	Lettuce	*LsNCED4*	Increase in the temperature for seed germination	Bertier et al. (2018)
Salinity	Tomato	*SlARF4*	Increased salinity tolerance	Bouzroud et al. (2020)
	Tomato	*SlHyPRP1*	Enhanced salinity tolerance	Tran et al. (2021)
	Tomato	*SlHAK20*	Enhanced salt and osmotic stress tolerance	Wang et al. (2020)
	Pumpkin and Cucumber	*RBOHD*	Enhanced salinity tolerance	Huang et al. (2019a)
Cold and chilling	Tomato	*SlCBF1*	Decreased chilling tolerance	Li et al. (2018b)
Mineral deficiency	Potato	*StMYB44*	Increased phosphate transport	Zhou et al. (2017)
	Tomato	*SlPHO1;1*	Decreased phosphate transport	Zhao et al. (2019)
Multiple abiotic stress	Tomato	Class II glutaredoxin (GRX) family (S14, S15, S16, and S17)	Increased heavy metal accumulation and increased heat sensitivity	Kakeshpour et al. (2021)

(continued)

Table 1 (continued)

Stress/trait	Plant	Targeted gene (s)	Effect	Reference
Qualitative traits	Oilseed rape	*FAD2*	Desaturation of oleic acid	Okuzaki et al. (2018)
	Chinese cabbage	*Bra003491, Bra007665, Bra014410*	Generation of genic male sterility	Xiong et al. (2019)
	Chinese Kale	*BoaCRTISO*	Alteration of carotenoids and chlorophyll content	Sun et al. (2020)
	Carrot	*F3H*	Regulation of anthocyanin biosynthesis pathway	Klimek-Chodacka et al. (2018)
	Potato	*GBSSI*	Regulation of amylose content in starch	Kusano et al. (2018)
		SBE1, SBE2	Regulation of amylose content in starch	Tuncel et al. (2019)
		SBE3	Regulation of amylose content in starch	Takeuchi et al. (2021)
		St16DOX	Removal of toxic SGAs	Nakayasu et al. (2018)
		StPPO1-9	Reduction of enzymatic browning	González et al. (2020)
	Sweet potato	*IbGBSSI* and *IbSBEII*	Starch content alteration	Wang et al. (2019a)
	Tomato	*Ap2a, NOR, FUL1 and FUL2*	Regulation of fruit ripening	Wang et al. (2019b)
		SlNAC4	Regulation of fruit ripening	Gao et al. (2021)
		Alc	Enhanced shelf-life	Yu et al. (2017)
		PL, PG2a, TBG4	Enhanced shelf-life	Wang et al. (2019c)
		GABA-TP1,2,3, CAT9 and SSADH	GABA regulation	Li et al. (2018a)

demonstrated loss-of-function of the SBE3 gene causing reduced amount of amylose indicating towards the role of SBE3 as a major starch branching enzyme (Takeuchi et al. 2021). Sweet potato (*Ipomoea batatas*) is another crucial high-starch content tuber consumed globally containing high nutritional value. Apart from the above-mentioned SBEs and GBSSI genes, the other major enzymes associated with starch biosynthetic pathway are ADP-glucose pyrophosphorylase (ADPG), starch synthases (SS) and starch debranching enzymes (DBEs). Genetic studies pertaining to sweet potato are limited due to inefficiency in genetic transformation and limited genomic information. Targeted mutagenesis has been performed in sweet potato against two genes, *IbGBSSI* and *IbSBEII*. Knockout of both the genes by CRISPR/Cas9 system did not significantly reduce the starch content in sweet potato; however, the amylose percentage was found to be altered in both cases. Further selective knockout studies would help in regulation of amylose levels in other cultivars of sweet potato. Additionally, multiple targeting of genes involved in the starch biosynthetic pathway through CRISPR/Cas9 would help in improvement of sweet potato varieties (Wang et al. 2019a).

2.1.2 Pigmentation

Mutagenesis studies performed in Chinese kale (*Brassica oleracea* var. *alboglabra*) to alter the pigment synthesis was successfully carried out using CRISPR/Cas9 technology. Chinese kale has high antioxidant and anticancer properties along with possessing high levels of vitamin C and carotenoids. Carotenoids facilitate the antioxidant process and therefore is crucial to the vegetable crop. Carotenoid isomerase (CRTISO) enzyme plays a role in conversion of lycopene precursors to lycopene. The loss of function of CRTISO leads to synthesis of yellow or orange colour in various crop species such as tomato, rice, and Chinese cabbage (Isaacson et al. 2002; Chai et al. 2011; Su et al. 2015). Chinese kale is usually devoid of colour except for light green leaves. Targeted editing via knockdown of the *BoaCRTISO* gene, of Chinese kale, was observed to cause alteration in pigmentation of the vegetable. The mutation of *BoaCRTISO* gene leads to down regulation of genes associated with carotenoid as well as chlorophyll biosynthesis pathway leading to yellowing of the *crtiso* mutants. However, it was also observed that two genes, *Psy3* and *PDS2* were upregulated which might suggest a compensatory mechanism of the plant which needs to be further explored to determine the functional variations among the genes. The technique, however, shows promise for commercial purposes for generation of Chinese kale with new colours as it also effects the sensory perceptions (Sun et al. 2020).

Carrots (*Daucus carota*) are another important vegetable crop due to its high content of Vitamin-C carotenoids and have been extensively explored for plant tissue culture and related genetic transformations. However, owing to its biennial cycle and high-inbreeding depression effects, it has not been studied as extensively using modern biotechnological techniques. Nevertheless, carrot callus grows fast and therefore serves as an excellent material for genetic research using in vitro culture studies. Additionally, the recent genome sequencing data of carrot (Iorizzo et al. 2016) provides further foundation for genomic research and subsequently in crop improvement. Anthocyanins are a group of pigment proteins whose colours vary based on side group arrangement of the compound. A colourless flavanone is the precursor in the pathway which is subsequently converted to unstable dihydroflavonol, which is then hydroxylated to anthocyanin. Thus the hydroxylation process is crucial for anthocyanin biosynthesis which is regulated by the Flavanone-3-hydroxylase (F3H) enzyme. Inhibition of expression of *F3H* restricts the expression of the pigments. Loss of function has been induced through multiplex CRISPR/Cas9 vectors resulting in white calli of carrot. This proved the functionality of F3H gene in the anthocyanin biosynthesis pathway as well as a visual marker to determine the mutants. Further, it also validated the callus regeneration system in carrot for application of genome editing in other traits as well (Klimek-Chodacka et al. 2018).

2.1.3 Saturated Fatty Acid Content

Oilseed rape (*Brassica napus*) is one of the most important oil crops containing an abundance of nutritional fatty acids. One of the fundamental traits being explored through various breeding and biotechnological methods in this crop is high oleic content as it imparts increased stability to the oil. Oleic acid, linoleic acid and linolenic acid unsaturated fatty acid compounds present in oilseed whose desaturation is regulated by fatty acid desaturase genes. CRISPR/Cas9 system has been utilized to modify the fatty acid desaturase gene *FAD2* which catalyzes the desaturation of oleic acid. Targeting the *FAD2* gene resulted in progenies which contained higher levels of oleic acid as compared to the wild type implying towards the effectiveness of the technology in providing viable trait enhancement results (Okuzaki et al. 2018).

2.1.4 Improvement of Shelf-Life and Quality

Tomato (*Solanum lycopersicum*) has been mainly studied in the context of increasing the shelf life of the ripening vegetable. However, apart from this, other traits which have been explored in tomato are nutritional in nature. Several factors have been determined which are responsible for tomato ripening and the subsequent change in taste and texture. The RIPENING INHIBITOR (RIN), NON RIPENING (NOR), APETELA2a (AP2a), COLORLESS NON RIPENING (CNR), alcobaca (ALC) and FRUITFUL (FUL) are some of the important transcription factors which regulate the early or late ripening in tomatoes. Spontaneous mutations have been observed across genotypes which have been further explored through forward genetics. CRISPR/ Cas9 has proven to be an efficient tool for the exploration of mutant generation in tomato. Knockout mutants of these genes were carried out and the phenotype of these mutants was observed. The *ap2a* mutation exhibited distinct phenotype while *nor* mutants showed mild phenotype. Additionally, *FUL1* and *FUL2* of *FUL* mutants showed partially redundant phenotypes collectively with FUL2 demonstrating role in early fruit development (Wang et al. 2019b). Apart from this, the NAC transcription factors are also known to play a vital role in tomato ripening as well as carotenoid accumulation. CRISPR/Cas9 technology-based knockout mutants have been developed against *SlNAC4* transcription factor. *SlNAC4* mutants were found to have softer fruit as compared to wild type which indicates that the transcription factor is a positive regulator of tomato ripening. The transcription factor was also found to be directly associated with SlEXP1 (expansin) and SlCEL2 (endo-β-1,4 glucanase) genes which function in cell wall metabolism (Gao et al. 2021). The ALC mutation has also been explored through CRISPR/Cas9 for enhancing the shelf life of tomato lines. Targeted gene replacement of the *alc* gene was performed and the phenotypic changes were observed. Heterozygote plants in

T_0 generation and homozygotes in T_1 was obtained. While the T_1 mutants demonstrated comparable plant height, stem diameter and soluble solid content as that of the wild type, it exhibited enhanced shelf life at room temperature. This would be the first successful study of allelic substitution in tomatoes using CRISPR/Cas9 technology suggesting the potential of the technology in tomato breeding (Yu et al. 2017). Shelf-life in tomatoes has also been studied taking into consideration the genes encoding pectin-degrading enzymes pectate lyase (PL), polygalacturonase 2a (PG2a), and b-galactanase (TBG4). Mutants of these genes generated using CRISPR/Cas9 were analyzed for pectin localization in pericarp of ripe fruit. While softening of fleshy fruits and vegetables is important for palatability, it significantly affects the shelf-life of the same. Pectin is a cell wall polysaccharide and is classified into three major classes of polymers, *viz*, homogalacturonan (HG), rhamnogalacturonan-I (RG-I), and rhamnogalacturonan-II (RG-II). CRISPR/Cas9-based mutant lines showed that only PL mutants showed firmer fruits while PG2a and TBG4 mutants affected the colour and weight of tomatoes. Additionally, PL silencing caused loss of de-esterified HG and degradation of RG-I and HG by the other two enzymes (PG2a and TBG4) (Wang et al. 2019c).

Enzymatic browning occurring in vegetables such as potato is another important trait which needs to be investigated thoroughly as it compromizes the quality of the crop. Polyphenol oxidases (PPOs) are compounds present in higher plants, which converts phenolic compounds to quinones which in turn self-polymerize to impart the dark brown colour often exhibited in potato. This undesirable trait is controlled usually via physical or chemical agents which are unfavourable for long-term solutions. Therefore, the exploitation of modern technologies is vital for the reduction of enzymatic browning. PPOs are encoded by multigene families and in potato itself 9 PPOs have been identified, viz, *StPPO1-9*. Mutations via genome editing in the *StPPO2* gene of tetraploid potato cultivar, Desiree, was seen to cause reduction of PPO activity to about 69%. Additionally, the enzymatic browning was found to reduce 73% in mutants as compared to wild type. The study provided valuable insight into the functionality of a single gene of the PPO family as well as the potential of CRISPR/Cas9 system in development of superior potato varieties (González et al. 2020). The accumulation of SGAs in potato, namely, α-solanine and α-chaconine, are known to confer bitter taste to the tuber and also cause toxicity to other organisms. Regulation of genes associated with SGAs have led to generation of potato with reduced SGAs. Knockout of *St16DOX* gene encoding steroid 16α-hydroxylase in SGA present in hairy root of potato, using CRISPR/Cas9, led to mutation of the gene and subsequently no accumulation of SGAs. However, glycosides of 22, 26-dihydroxycholesterol were found, which is the substrate of *St16DOX*. Site-directed mutagenesis successfully carried out in tetraploid root demonstrated the efficacy of multiplex gRNA expression for mutagenesis studies in other crops as well (Nakayasu et al. 2018).

2.1.5 Other Qualitative Traits

The development of genic male sterile lines using multiplex genome editing has been performed targeting *Bra003491, Bra007665*, and *Bra014410* genes of Chinese cabbage (*Brassica campestris*). The aforementioned genes are orthologs of the *VGDH2* gene of *Arabidopsis* which encodes pectin methylesterases (PME) has functional role in pollen and pollen tube development. Stable mutagenesis frequency was observed in T_0 generation. While the study did not demonstrate generation of genic male sterile transformants, the high mutation frequency implied towards the potential of multiplex genome editing in *B. campestris* for the development of heritable mutations (Xiong et al. 2019).

In a rather novel CRISPR/Cas9 system, pYLCRISPR/Cas9 has been constructed which consists of six sgRNA cassettes. Tomatoes consist of vital nutrients such as vitamin C and E, alkaloids, lycopenes and ɣ-aminobutyric acid (GABA). GABA is essential for plant growth and fruit development. GABA is regulated by GABA shunt metabolic pathway. In this, GABA is synthesized from glutamate and then converted to succinic semialdehyde (SSA) by pyruvate- dependent GABA transaminase (GABA-TP) which is then oxidized to succinate via Succinate semialdehyde dehydrogenase (SSADH). The cat9 protein helps in transportation of GABA to mitochondria for catabolism. pYLCRISPR/Cas9 was utilized to target 5 genes (*GABA-TP1, GABA-TP2, GABA-TP3, CAT9 and SSADH*) in GABA shunt leading to 6 GABA mutants. Following this, the GABA content in leaves was found to be 19-folds greater than the wild type. Also, this technique showed the potential of pYLCRISPR/Cas9 system in multisite knockout mutation studies (Li et al. 2018a).

2.2 Abiotic Stress Tolerance

Progress in genetic engineering techniques has been offering a wide range of applications to improve the vegetable crops for stress tolerance and quality vegetable production. CRISPR-Cas9-based genome editing is a simple, efficient, and potential tool for crop improvement. CRISPR-Cas9 is considered the new breeding technique and has been extensively utilized to improve different traits in vegetable crops. Most of the genome editing for abiotic stress tolerance have been performed in tomato (Wang et al. 2017).

Structural genes can be targeted to achieve the high-level specific features of stress tolerance. New terminologies have been introduced to define negative and positive regulators of abiotic stresses. Structural genes were called positive regulators (sensitivity genes or S genes) and negative regulators (tolerance gene or T genes) (Zafar et al. 2020). Genes encoding detoxification enzymes such as catalases (CATs), superoxide dismutase (SOD), glutathione reductases (GRs), glutathione S-transferases (GSTs), and many peroxidases (PODs) are referred to as T genes. T genes are the choice to be targeted, as they have a direct impact on all the major abiotic stresses. The majority of the T genes encode antioxidant enzymes which are involved in the detoxification of ROS, which is a common response to multiple stressors (Zhang et al. 2021). Several other positive regulators that can be targeted, to achieve high-level stress tolerance, are papain-like cysteine proteases (PLCPs) and melatonin biosynthetic genes. Biosynthetic T genes can increase the antioxidant melatonin production during the different stresses to scavenge ROS and reactive nitrogen species (RNS). Strategies to target these T genes have been reviewed before (Zafar et al. 2020). On the other hand, knocking down S genes encoding negative regulators of abiotic stresses can be explored to enhance stress tolerance in vegetable crops. A good example for the S genes is those that enhance the ROS species (H_2O_2) production in response to various stresses. Genes like stress-related RING finger protein 1 (*SRFP1*), DIS1 (induced SINA protein 1) and DST (drought and salt tolerance) are belonging to the E3 ubiquitin ligase class and know to reduce the activity of antioxidant enzymes. Several S genes belonging to this class were targeted in rice. However, these S genes can also be mutated in vegetable crops using CRISPR-Cas9 to enhance abiotic stress tolerance (Fang et al. 2015). Regulatory genes including transcription factors have been extensively studied and are always the preferred target for improving stress tolerance. Stress-induced regulatory factors can increase the expression of stress-related genes by physically binding in their prompter regions. Generally, positive regulators increase the expression of the antioxidant enzymes to facilitate ROS detoxification. Negative regulators including stress-responsive transcription factors, on the other hand, decrease the expression of the stress-responsive genes (Javed et al. 2020). *Cis*-regulatory sequences and small non-coding RNAs like miRNAs can also be an important target to achieve high-level stress tolerance. Expression of stress-responsive genes may alter by targeting their promoter regions where mostly cis-regulatory sequences are present. Position and sequence variation of cis-regulatory sequence within the promoter region resulted in increased, reduced and no expression. Besides, miRNAs could positively and negatively regulate stress tolerance by regulating the expression of a single of multiple stress-responsive genes. CRISPR-Cas9 system provides an opportunity to knock-down negative regulators (miRNAs, cis-regulatory sequences) to improve tolerance to different stresses in vegetable crops.

2.2.1 Drought and Extreme Temperature

Drought is one of the serious abiotic stresses responsible for heavy yield losses in vegetable crops across rainfed ecologies. Adequate water supply to crops is challenging across many vegetable-growing nations due to severe climate change. There are multiple reports available in tomato where CRISPR-Cas9 is used to edit structural and regulatory genes to generate drought-tolerant tomato. The LATERAL ORGAN BOUNDARIES DOMAIN (LBD)-containing gene *SlLBD40* was knocked-out using the CRISPR-Cas9 and edited tomato plants showed the increased tolerance to drought compared to wild type and overexpressing transgenic tomato plants (Liu et al. 2020). Mitogen-activated protein kinases (MAPKs) are important signalling molecules induced upon various stresses including heat and drought. In tomato, *SlMAPK3* gene mutation using CRISPR-Cas9 lead to varied expression of other stress-responsive genes such as *SlLOX*, *SlGST*, and *SlDREB* (Wang et al. 2017). CRISPR-Cas9-induced mutation within the *SlNPR1* gene reduced the drought tolerance in tomato. The *SlNPR1* gene is a master regulator which controls the expression of stress-responsive genes and mutant plants showed the lower activity level of stress-responsive genes including *SlGST, SlDHN, and SlDREB* (Li et al. 2019). Single, double, and triple combinations of *Slgrxs14, 16, and 17* mutated tomato plants generated by using multiplex CRISPR-Cas9 system showed the increased tolerance to multiple stresses including heat and drought (Kakeshpour et al. 2021). Role of mutation within SlAGAMOUS-LIKE 6 (*SlAGL6*) gene was confirmed, fruit production under heat stress, using CRISPR-Cas9 in tomato (Klap et al. 2017). In potato, transcription factor StCDF1 association with *StFLORE* is known to enhance drought tolerance. This gene association was studied by generating mutations within the *StFLORE* promoter. This is the classical example of using CRISPR-Cas9 for gene association studies apart from routine loss-of-function and gain-of-function analysis (Ramírez Gonzales et al. 2021). In soybean, the circadian rhythm gene (*GmLCLs*) was characterized as a negative regulator of ABA signalling during the drought stress (Yuan et al. 2021). The CRISPR-Cas9 mutagenesis system assisted in characterizing positive regulators (TFs) of abiotic stresses. The role of BRASSINAZOLE RESISTANT 1 (BZR1) TF in response to heat was studied by generating a bzr1 mutant using the CRISPR-Cas9 system. The same study in tomato also showed the binding of BZR1 to the promoter region of FERONIA2 (*FER2*) and *FER3* genes. These genes encode FERONIA Receptor-Like Kinase involved in ROS signalling in response to heat stress (Yin et al. 2018). CRISPR-Cas9-based soybean mutants of *GmLCLs* showed improved response to drought conditions. The study confirmed that *GmLCLs* negatively regulates ABA perception and signalling (Yuan et al. 2021). In another study, CRISPR-Cas9 was used to knock-down transcription factor *GmNAC8* and characterized as the positive regulator of drought stress (Yang et al. 2020). Similarly, the role of *GmHsp90A2*, a positive regulator of heat stress, was characterized using CRISPR-Cas9 mutagenesis (Huang et al. 2019a). In lettuce, CRISPR-Cas9 induced homozygous mutation within the *LsNCED4* (9-cis-EPOXYCAROTENOID DIOXYGENASE4) gene, increased the overall temperature required for seed germination (Bertier et al. 2018). The role of cyanide-resistant respiration against the high-light stress was studied by knocking out the alternative oxidase (*AOX*) gene using CRISPR-Cas9 in potato (Hua et al. 2020). Chilling is another important abiotic stress factor in vegetable crops. C-Repeat Binding Factors (CBFs) are transcription factors and are familiar for their role in low temperature or chilling stress in plants. The freezing tolerance gene *SlCBF1* was characterized in tomato by generating *SlCBF1* gene mutants. Mutants generated by CRISPR-Cas9 reduced the expression of downstream cold stress relate genes and showed significant decrease in the chilling tolerance (Li et al. 2018b). CRISPR-Cas9 knock-out of Class II glutaredoxin (SlGRX) – previously discussed for drought stress – showed increased sensitivity to cold stress in tomato (Kakeshpour et al. 2021).

2.2.2 Salinity and Mineral Deficiency

In plants, there are similar physiological consequences to salt stress in comparison with drought stress. Only a few studies, wherein CRISPR-Cas9 genome-editing tool has been used to generate heritable mutations. *SlHyPRP1* gene encoding proline-rich protein 1 (HyPRP1), a previously characterized negative regulator of salt stress tolerance, was manipulated to alter the protein domain region. CRISPR-Cas9-mediated manipulation of the *SlHyPRP1* gene exhibited high salt stress tolerance in tomato (Tran et al. 2021). In another study, it was shown that the Auxin Response Factor 4 (ARF4) is involved in the salt and osmotic stress response. The gene-encoding ARF4 was mutated using CRISPR-Cas9, and the mutant tomato plants showed increased tolerance to salinity and osmotic stress (Bouzroud et al. 2020). CRISPR-Cas9 induced mutation with the coding region of *SlHAK20* (clade IV HAK/KUP/KT transporters) revealed its function as it is associated with the change in NA+/K+ ratio. Also, mutant plants confer salt tolerance in tomato (Wang et al. 2020). Cucumber (*Cucumis sativus*) is known for having low tolerance to salt as compared to the other vegetables in the same family, like pumpkin. K^+ uptake by roots is regulated by H_2O_2 signalling under salt stress. Under salt stress, few of the differentially expressed genes (DEGs) in both the species were found as NADPH oxidase (respiratory burst oxidase homolog D; RBOHD), 14-3-3 protein (GRF12), plasma membrane H^+-ATPase (AHA1), and potassium transporter (HAK5) which were exhibiting higher expression in pumpkin when under salt stress. Mutation of RBOHD using CRISPR/Cas9 resulted in lowered H_2O_2 and K^+ content as well as GRF12, AHA1 and HAK5 expression subsequently leading to salt-sensitive phenotype. This suggested the role of RBOHD in salt tolerant phenotype of pumpkin. HAK5 expression is known to enhance K^+ uptake and accumulation in root apex, thus preventing programmed cell death and consequently demonstrating salinity tolerance. The study sheds light on the differential tolerance of salt within the Cucurbitaceae family which can be further analyzed to comprehend the mechanism (Huang et al. 2019b).

There are only limited number of genes which are studied to describe mineral homeostasis in vegetable crops. Recently, two studies used CRSIPR-Cas9 system to characterize *StMYB44* in potato and *SlPHO1;1* in *tomato*. In the first study, it was shown that the StMYB44 negatively regulate phosphate transport, since *StMYB44* mutation failed to increase the expression of PHOSPHATE1. The second study in tomato, validated the phosphate transport role of gene encoding PHOSPHATE 1 (PHO1) (Zhao et al. 2019; Zhou et al. 2017).

As it is having been described earlier, different abiotic stresses have similar physiological responses. There are few studies where CRISPR-Cas9 gene editing resulted in multiple stress tolerance in vegetable crops. In potato, gene-encoding coilin, the main structural protein functionally associated with subnuclear Cajal bodies (CBs) was mutated using CRISPR-Cas9. The coilin mutants showed the tolerance to salt and osmatic stresses (Makhotenko et al. 2019). Similarly, targeting class II glutaredoxin (GRX) gene family generated multiple mutant tomato plants.

Few of these gene mutants exhibited multiple stress responses (Kakeshpour et al. 2021).

2.3   Pest and Disease Resistance

There are various tools in genome-editing tools now-a-days which are widely used in various crops to knockout mutations and identify the specific genes of interest for producing sustainable crops with abiotic and biotic stress resistance. There are various effectors built by the plants to manipulate the adaptation for defence reaction. This allowed not only discovering new biotechnological tools (Doron et al. 2018) but also to interrupt plant-pathogen/plant-insect interactions by target mutation to inhibit host influence and thereby accelerating the resistance. The latest study in tomato by (Guzman et al. 2020) demonstrated that TARK1 (tomato receptor kinase), required for pattern triggered immunity (PTI) when lost, inhibited bacteria and bacteria-associated induction of stomatal reopening. These types of phenomena are also seen in other crops like barley. Powdery mildew resistance in barley was developed by mutating MLO (Mildew resistant locus O) which does not confer resistance to the crop (Kusch and Panstruga 2017). The loss of MLO causes decline in growth and yield as this is a negative controller of plant defence (Brown and Rant 2013). Host-induced gene silencing (HIGS) involving miRNAs anticipated to resist away virus-related, mycological, pest, or nematode worm pathogens (Younis et al. 2014; Huang et al. 2019c; Iqbal et al. 2016) as pathogens employ RNAi mechanism including 'small noncoding sRNAs' which help in gene silencing (Weiberg et al. 2013). Natural mutations create variations which are undirected and may lead to off-target mechanisms. These may result in negative phenomenon like decline in growth and crop production. Advancements in genome editing approaches aim to establish mutations at specific loci depending on SSNs (Single Specific Nucleases) and can be useful for biotic stress resistance. Moreover, foreign gene insertion which is conferring the resistance at loci helps in inhibiting the negative effects caused by that introgression and increase the resistance. Association mapping identified QTLs or phenotypes by GWAS further aiding the prediction of the potential candidate genes (Alqudah et al. 2020). Due to the availability of genome-sequencing information for most of the crops, precise genome editing could be done to the target allele at specific locus in complex genomes including tetraploid crops such as potato (Johansen et al. 2019). Despite several challenges in using this technology, it has been successfully implemented in various vegetable crops. Classical breeding takes several years to incorporate resistance through backcrossing. Virus resistance is incorporated without any classical approach and transgenics in cucumber for the first time (*Cucumis sativus* L.) utilizing Cas9/sub genomic RNA (sgRNA) technique for knocking the role of recessive eIF4E (eukaryotic translation initiation factor 4E) factor (Chandrasekaran et al. 2016).

CRISPR/Cas has been effectively executed in producing resistant plant varieties against bacteria, fungi, and viral pathogens. Factors encoding susceptibility is extensively studied as knockout of genes can directly cause resistance (Zaidi et al. 2018). MLO genes in many crop species confer resistance to downy mildew when it is in recessive state. Mutation of MLO in tomato using CRISPR/Cas9 has rendered resistance against powdery mildew (Nekrasov et al. 2017). Similarly, in cucumber, mutation in the recessive eIF4E has induced resistance against *Cucumber Vein Yellowing Virus* (CVYV) (Chandrasekaran et al. 2016). Various other susceptibility genes like *DMR6* are also seen in *Arabidopsis* (Van Damme et al. 2008) and potato (Sun et al. 2016) for downy mildew resistance. The same gene in tomato was reported to bring about the resistance against various other pathogens like *Pseudomonas syringae* pv. tomato (Pto), *Phytophthora capsici*, and *Xanthomonas* spp (de Toledo Thomazella et al. 2016).

Manipulation of gene regulatory pathways in plant defence systems helps in increasing the resistance. Exploitation of these pathway, like SA-JA, indicate positive impact on agronomical traits and shows negative impact on defence. CRISPR/Cas9 system has made it possible to prevent defence potential trade-offs in tomato crop by inactivating JAZ2. This helped in avoiding the reopening of stomata in reaction to toxin (COR) coronatine generated by *Pseudomonas syringae*, with no adverse influence on the defence reaction in contrast to necrotroph *Botrytis cinerea* (Ortigosa et al. 2019). In a study related to conferring resistance to late blight of potato, the role of the StCCoAOMT (full form) gene has been examined via gene-editing technique. A SNP mutation of the StCCoAOMT gene was found to increase the resistance-related biosynthetic genes which in turn reduced the disease severity (Hegde et al. 2021). Through CRISPR/Cas technology, it is possible to introgress the resistance (R) genes in the host plants which may enhance the pathogenic resistance. But in hosts, R genes are less and need to be recognized by interrelating helper NLR (nucleotide-binding site leucine-rich repeat) proteins (Wang et al. 2019d). Moreover, dominant allele-mediated resistance lasts less in contrast to the recessive allele-mediated resistance and mostly confined for a less race-specific isolates, which is able to rapidly develop resistance through higher mutation rate (Pandolfi et al. 2017; Kourelis and Van Der Hoorn 2018). Therefore, gene pyramiding helps to attain high durability in resistance that could be further assisted by CRISPR/Cas9-generated HDR (Pandolfi et al. 2017). In addition to this, the transmission of pattern recognition receptors may also give significantly enhanced resistance that has been demonstrated successfully in crops like tomato and *Arabidopsis* (Karasov et al. 2017). These methodologies help plants to sense novel pests and counteract the defence reaction with pathogen accordingly by reduction in the yield potential penalty.

Some of the notable studies with CRISPR/Cas technology in enhancing the nature of resistance by viral genomes targeting directly have been studied in potato where the targeted mutagenesis in regions of P3 (potyviral membrane protein), CI (laminate cytoplasmic inclusion bodies), Nlb (viral replicase), CP (capsid protein) confers resistance to *Potato Virus Y* (PVY) through Cas13a effector (Zhan et al. 2019). Similarly, in tomato mutations in the target region CP and Rep (replication-associated protein) induce resistance to *Tomato Yellow Leaf Curl Virus* (TYLCV) via Cas9 (Tashkandi et al. 2018). Further, targeting a *CseIF4E* host gene in cucumber provide resistance to various viral diseases such as CVYV, *Zucchini Yellow Mosaic Virus* (ZYMV), *Papaya Ringspot Virus* (PRSV) (Chandrasekaran et al. 2016).

Tomato is one of the essential crops grown worldwide, and infestation by various pathogens and pests hampers the production, cultivation and shelf life. Initial report of mutagenesis through genome editing was done in tomato with the application of TALEN and CRISPR/Cas9 (Brooks et al. 2014; Lor et al. 2014). Mostly cas9 base editing (Shimatani et al. 2018), homologous recombination (Čermák et al. 2015) and prime editing (Lu et al. 2021) have been used to detect the variations through genome editing. Studies have suggested that an increase in salicylic acid confers resistance to many oomycetes and bacterial pathogens when there is a mutation in single gene *DMR6* in *Arabidopsis* (de Toledo Thomazella et al. 2016). CRISPR technology also has been exploited to increase the pathogen and pest resistance by targeting the gene *SlMLO1-Oidium neolycopersici* against powdery mildew (Nekrasov et al. 2017), *Pseudomonas syringae* pv. tomato for gene SlJAZ2 (Ortigosa et al. 2019), *P. syringae, Phytophthora capsica* for gene *SlDMR6* (de Toledo Thomazella et al. 2016), *Xanthomonas sp, Oidium neolycopersici* (powdery mildew) for gene *SlPMR4* Oidium *via* Cas9 (Martínez et al. 2020) with no negative effect on *Lycopersicon esculentum* development and growth.

Insects are also the main reason for lower yields in most of the essential crops grown worldwide. For control of these insects, farmers usually adopt chemical control causing detrimental effects to the environment and humans. These challenges encouraged researchers to create new and eco-friendly management strategies (Bisht et al. 2019). Resistance in insect or pest population can be obtained through the alterations in genetic material associated with receptor molecules responsible for distracting insects and release of toxin. Despite all these complex challenges, genome editing helps in combating the pests and serves as a new breeding strategy (Razzaq et al. 2019; Vats et al. 2019). Knockout of the targeted key regulatory sequences through genome editing may remain the long-term future strategy in tackling insect pests (Tyagi et al. 2020). Editing of the genomes is effectively utilized within a two-step approach including the alteration of target pests, as well as their consequent release in the environment (Bitew 2018). Knocking down the detoxification genetic factors in pests such as gossypol-producing cytochrome p450 ensued in susceptibility of the pest. Similarly, in polyphagous pests like *Helicoverpa*, knocking down of CYP6AE factor demonstrated the function of enzymes in decontamination of several toxins and phytochemicals (Maher et al. 2020). Other strategies of pest management via genome editing are by targeting the genes that will destroy the chemical signalling and detection of suitable partner for mating (Larsson et al. 2004). These strategies could revolutionize the establishment of effective plant insect interactions. Olfactory receptors in the insects are vital for recognizing the mating partner. In *Drosophila*, the preference of egg-laying site in the host is disrupted by causing the mutation in gene *Or83b*. Moreover, due to this, the cumulative effect on impaired olfactory detection was also affected. Correspondingly, *Orco* (olfactory receptor coreceptor) genetic factor in *Spodopthera* via CRISPR/Cas9 knockout resulted in interruption of the mating companion choice and leading to anosmia (Koutroumpa et al. 2016). Loss of function mutants through CRISPR/ Cas9 resulted in the generation of *abd-A* mutant phenotypes in agricultural pests like *Spodoptera litura* (Sun et al. 2017), *Spodoptera frugiperda* (Wu et al. 2018), and *Plutella xylostella* (Sun et al. 2017). Insects created in this way had deformed segments over the body, unarmed prolegs, abnormal sex glands, and embryonic fatality, showing that genome editing was effective (Belfort and Bonocora 2014) and can be extrapolated to other related insect species as well.

2.4 Biosafety and Legal Regulations

The apprehensions on the possible negative effects of transgenic plants/ GMOs/ LMOs on the environment have led to the requirement of biosafety and risk assessment for products developed through transgenic or genome editing technologies. These concerns arose because these genetic-engineering technologies may unintentionally interfere with well-structured physiological or biochemical pathways through alteration in the expression of various associated genes. The recombinant DNA evolved novel trait such as herbicide tolerance is a concern for the environment, food, animal feed and agricultural systems (Beckie et al. 2019). Besides these, the foreign DNA has also been reported to cause genomic irregularities through the alternation of secondary metabolites of plants (Aharoni and Galili 2011). Thus, traceability and tagging of genetically engineered products allow consumers choice as well as monitoring of ill effects on food, non-target and environment. Some of the countries have developed well-defined regulatory structures for GMOs; however, it is still in debate whether to consider GE products under the same legislation or to have a separate framework for their risk assessment and biosafety. The GE products also classified as NBTs (new breeding technologies) represent two different technologies viz. site-directed nucleases (SDN) and oligonucleotide-directed mutagenesis (ODM). Unlike GMOs, the GE technologies such as TALEN (transcription activator-like effector nucleases), MN (meganucleases), and CRISPR do not necessarily involve the integration of foreign DNA into the target genome. The SDN 1 and SDN 2 approaches are employed to deliver small insertions and deletions in the DNA, while SDN3 integrates a long chuck of DNA at the target sites (Podevin et al. 2013; Petolino and Kumar 2016). The traceability of SDN1- and SDN 2-based alteration remains a challenge due to few bases changes, which are indistinguishable from those obtained through naturally occurring mutations (Gao 2018). There is a larger opinion to make SDN1- and SDN2-based products available in the field in the shortest possible time owing to their high similarity with mutation breeding, while the GE products evolving from the SDN3 approach have to be considered as transgenic (Lassoued et al. 2019; Bhattacharya et al. 2021). Generally, any risk assessment procedure cannot identify as well as quantitate all aspects of the threats to the environment and non-targets because of lack of complete knowledge of intended and unintended effects on modified organisms, receiving environment, and their interactions. This is the reason that the risk assessment approach of well-known global agencies has been criticized because of their conjectures that every genetic change in GE organisms acts separately from any and all other genes and changes (Hilbeck et al. 2020). Thus, risk assessment considerations and methodologies for any type of genetically engineered organisms should be strengthened in terms of precautionary principles due to limited safety data and high scientific uncertainties. All GE applications either based on SDN1, SDN2 or SDN3 have the potential to cause genomic irregularities and off-target effects that may be related to their process or the trait (Norris et al. 2020; Modrzejewski et al. 2019). The unintended effects in GE crops can cause numerous unanticipated effects for example cleaving of an untargeted (off-target) gene (s) by nuclease, which may lead to change in some physiological processes. This will, in turn, have an impact on the biochemical, metabolic, and protein profile of the organism or the final product leading to its safety concerns to consumers or the environment. Before placing any GE product in the open environment for commercial use or even for evaluation, such effects need to trace as well as evaluated for their consequences on the environment. The risk of each genome-edited product has to be assessed individually focusing on genomic as well as epigenomic irregularities, on-site intended and off-target unintended alteration occurring beyond the recognition error of gRNA. Currently, the majority of GE studies use *in silico* tools for search for off-target hits as well as prediction of target sites (Modrzejewski et al. 2019); however, these tools are not fully reliable and may miss detecting unintended insertions and genomic irregularities. There is a need to identify protocols to address these issues through a combination of whole-genome sequencing and potent bioinformatics tools. The major challenge with the genome sequences will be to differentiate between the naturally occurring variations and unintended effects. The unintended effects of GE are not only restricted to the genome but can have an impact on the epigenome, transcriptome, proteome, and metabolome. Thus collective 'omics' approach can assist in the risk assessment of genome-edited products and processes (Kawall et al. 2020). The European Union Directive of 2001 and Regulation # 1829/2003 for first-generation GMOs seek information for both unintended and intended effects; however, the molecular data is restricted to insert and flanking regions of the insert site. For GE products these directives need to be revised as they cannot be restricted to a specific site but have to be evaluated in context to the earlier discussed 'omics' approach.

The regulation, release, and trade of GMOs were established in the Cartagena Protocol on Biosafety and having implications to this treaty the GE technologies fall under its dictate. However, a wide disparity has been observed in the regulation, production, and consumption of these genome-edited organisms/products, with some countries banning their production while others producing and consuming them (Turnbull et al. 2021). Similar to first-generation genetically modified organism regulations, the genome-edited products also fall under two substructures, i.e the regulation of the process and the final product. The regulation of GE products/organisms is not uniform across countries with some having established their biosafety norms while others are still in the deciding phase (El-Mounadi et al. 2020; Eckerstorfer et al. 2019). The hitches in developing the regulatory policies are also because the SDN1- or SDN2-based products are in the dilemma of being considered as GMO or non-transgenic as the latter is similar to the plant varieties created by naturally occurring genetic variations. Thus, the biosafety and commercialization of these products may escape the strict regulatory guidelines (Van Vu et al. 2019) as declared by the United States Department of Agriculture (USDA) in 2018 that for some specific cases the GE is equivalent to conventional breeding (Waltz 2016a). Genetically engineered mushroom for resisting browning has not been regulated and approved for commercialization with the basis that it didn't carry any transgene (Waltz 2016b). As mentioned earlier the regulatory framework for GE products/ organisms has not been laid uniformly across the globe with most countries/ unions still in the process of devising guidelines for these technologies. The European Union (EU) has maintained its stand for GE products similar to its earlier opinions for transgenic crops. In 2018, the Court of Justice of the European Union (ECJ) declared that the gene-edited crops must pass the same stringent regulations like first-generation transgenic plants (Waltz 2016a). Australia in its amendment to the schedule of the Gene Technology Act (GT Act) in 2019 has excluded organisms modified through SDN activity and CRISPR-Cas9. New Zealand, which has one of the most stringent regulatory frameworks for the development and field testing of GMOs under the aegis of Hazardous Substances and New Organisms Act 1996 (HSNO Act), has clarified its stand in 2016 that all GE products/processes will be under the same GMO regulations. In India, the evaluation of GMOs is carried out under rules and regulations of the Institutional Biosafety Committee, the Genetic Engineering Evaluation Committee, and the Genetic Engineering Review Committee. These regulations suggest that, for SDN1, extensive data demonstrating successful gene editing is required, while for SDN2, in addition to the data proving effectiveness of gene editing, field trial data demonstrating effectives of transformation are also required. The SDN3-based approach must pass through GMO-based regulations and safety guidelines (Bhattacharya et al. 2021). Likewise, the status of genome-editing based organisms and products in different countries is listed in Table 2. No matter the points discussed in this section give a general overview of the safety and regulatory framework of genome-edited organisms, the regulations are the same for food crops, vegetables, animals, or any other modified organism.

2.5   Conclusion

Conventional breeding techniques have taken major strides over the years for the improvement of agronomic traits such as yield, disease and pest resistance and abiotic stress tolerance as well as nutritional quality improvement. However, due to limited genetic germplasm and genetic diversity, conventional breeding has now been taken over by more modern technologies. Modern cultivars are sensitive to various stresses due to long-term artificial selection and domestication based on the excessive utilization of naturally occurring allelic variation. There is a need to develop vegetable cultivars with high yield and tolerance to various abiotic stresses.

Table 2 Country-wise status of regulatory framework of genome-edited organisms or products

Country	Status
United States of America	Non-regulated
European Union, Canada, India, Malaysia, New Zealand, South Africa, Thailand, Mexico	Regulated
Argentina, Chile, Brazil, Colombia	Case-by-case if no foreign DNA then not regulated
Paraguay, Uruguay, Bangladesh, Nigeria, Kenya	Likely case-by-case if no foreign DNA then not regulated as GMOs
Phillippines, Indonesia United Kingdom Norway	Discussion ongoing Discussion ongoing for post-Brexit Discussion ongoing and the proposal is if no foreign DNA then not regulated as GMO
Japan, Australia, Israel	If no foreign DNA then not regulated as GMO
China, Burma, Nepal, most of the African countries except Kenya and Nigera, Russia, and other European nations except those in EU, UK and Norway, Mediterranean nations, Latin American nations- Ecuador, Peru, Bolivia, Venezuela, Guyana, French Guyana, Cuba, Other nations not listed in above categories	No information on regulatory and biosafety of GE crops/organisms/products

Modified from Schmidt et al. (2020)

It is important to select such new cultivars to meet the global food supply and market demands owing to the availability of whole-genome sequences of various crops, CRISPR/Cas9 has now been utilized for obtaining germplasm of various cultivars with genetic diversity. While CRISPR/Cas9 has evidently achieved tremendously in terms of crop improvement, the selection of genes is vital in the process. Additionally, traits such as abiotic stress, disease resistance and nutritional improvement are highly complex traits which are more often than usual governed by multiple genes. Therefore, it is essential to target multiple genes for obtaining desired phenotypic changes. Additionally, the regeneration of edited plants in major crop systems such as rice, wheat, maize and relevant model crops have been achieved over the years. However, this does not hold true for vegetable crops where the embryogenic competency is low. Appropriate and effective transformation protocols with expression of regulatory genes are crucial for efficient regeneration of edited vegetable crops. While *Agrobacterium*-mediated, PEG and electroporation have been utilized for major crops, other deliver methods such as plant RNA and DNA virus vector systems might prove to be more effective in the case of vegetables.

Despite these challenges, CRISPR/Cas system remains one of the most promising tools in revolutionizing vegetable breeding pipelines as compared to its conventional counterpart. Genome editing will undoubtedly open newer avenues for exploration of several other crops which can eventually be extended for industrial benefits.

Acknowledgement Mr. Rahul Nitnavare acknowledges the grant (File no: 11015/137/2017-SCD-V) received from the Ministry of external affairs, Government of India. Ms. Joorie Bhattacharya acknowledges DST INSPIRE fellowship received from the Department of Science and Technology (DST/INSPIRE Fellowship/IF160547), Government of India. Rothamsted Research receives grant-aided support from the Biotechnology and Biological Sciences Research Council (BBSRC) of the Designing Future Wheat strategic program (BB/P016855/1).

References

Aharoni A, Galili G (2011) Metabolic engineering of the plant primary secondary metabolism interface. Curr Opin Biotechnol 22(2):239–244

Alqudah AM, Sallam A, Baenziger PS, Börner A (2020) GWAS: Fast-forwarding gene identification and characterization in temperate Cereals: lessons from Barley–a review. J Adv Res 22:119–135

Beckie HJ, Busi R, Bagavathiannan MV, Martin SL (2019) Herbicide resistance gene flow in weeds: under-estimated and under-appreciated. Agric Ecosyst Environ 283:106566

Belfort M, Bonocora RP (2014) Homing endonucleases: from genetic anomalies to programmable genomic clippers. In: Homing Endonucleases. Humana, pp 1–26

Bertier LD, Ron M, Huo H, Bradford KJ, Britt AB, Michelmore RW (2018) High-resolution analysis of the efficiency, heritability, and editing outcomes of CRISPR/Cas9-induced modifications of NCED4 in lettuce (*Lactuca sativa*). G3 (Bethesda) 8(5):1513–1521

Bhattacharya A, Parkhi V, Char B (2021) Genome editing for crop improvement: a perspective from India. In Vitro Cell Dev Biol Plant 57:565–573

Bisht DS, Bhatia V, Bhattacharya R (2019) Improving plant-resistance to insect-pests and pathogens: the new opportunities through targeted genome editing. In: Seminars in cell & developmental biology, vol 96. Academic, pp 65–76

Bitew MK (2018) Significant role of wild genotypes of tomato trichomes for *Tuta absoluta* resistance. Indian J Genet Plant Breed 2(1):104

Bouzroud S, Gasparini K, Hu G, Barbosa MAM, Rosa BL, Fahr M, Bendaou N, Bouzayen M, Zsögön A, Smouni A, Zouine M (2020) Down regulation and loss of Auxin response Factor 4 function using CRISPR/Cas9 alters plant growth, stomatal function and improves tomato tolerance to salinity and osmotic stress. Genes (Basel) 11(3):272

Brooks C, Nekrasov V, Lippman ZB, Van Eck J (2014) Efficient gene editing in tomato in the first generation using the clustered regularly interspaced short palindromic repeats/CRISPR-associated9 system. Plant Physiol 166(3):1292–1297

Brown JK, Rant JC (2013) Fitness costs and trade-offs of disease resistance and their consequences for breeding arable crops. Plant Pathol 62:83–95

Canter LW (2018) Environmental impact of agricultural production activities. CRC Press

Cardi T, Batelli G, Nicolia A (2017) Opportunities for genome editing in vegetable crops. Emerg Top Life Sci 1:193–207

Chai C, Fang J, Liu Y, Tong H, Gong Y, Wang Y, Liu M, Wang Y, Qian Q, Cheng Z, Chu C (2011) *ZEBRA2*, encoding a carotenoid isomerase, is involved in photoprotection in rice. Plant Mol Biol 75(3):211–221

Chandrasekaran J, Brumin M, Wolf D, Leibman D, Klap C, Pearlsman M, Sherman A, Arazi T, Gal-On A (2016) Development of broad virus resistance in non-transgenic cucumber using CRISPR/Cas9 technology. Mol Plant Pathol 17(7):1140–1153

Chen F, Song Y, Li X, Chen J, Mo L, Zhang X, Lin Z, Zhang L (2019) Genome sequences of horticultural plants: past, present, and future. Hortic Res 6(1):1–23

Chen Y, Mao W, Liu T, Feng Q, Li L, Li B (2020) Genome editing as a versatile tool to improve horticultural crop qualities. Hortic Plant J 6:372–384

Choudhury FK, Rivero RM, Blumwald E, Mittler R (2017) Reactive oxygen species, abiotic stress and stress combination. Plant J 90(5):856–867

Corte LED, Mahmoud LM, Moraes TS, Mou Z, Grosser JW, Dutt M (2019) Development of improved fruit, vegetable, and ornamental crops using the CRISPR/cas9 genome editing technique. Plants 8(12):601

Čermák T, Baltes NJ, Čegan R, Zhang Y, Voytas DF (2015) High-frequency, precise modification of the tomato genome. Genome Biol 16(1):1–5

de Toledo Thomazella DP, Brail Q, Dahlbeck D, Staskawicz B (2016) CRISPR-Cas9 mediated mutagenesis of a *DMR6* ortholog in tomato confers broad-spectrum disease resistance. BioRxiv. 064824

Doron S, Melamed S, Ofir G, Leavitt A, Lopatina A, Keren M, Amitai G, Sorek R (2018) Systematic discovery of antiphage defense systems in the microbial pangenome. Science 359(6379):eaar4120

Dos Reis SP, Lima AM, de Souza CRB (2012) Recent molecular advances on downstream plant responses to abiotic stress. Int J Mol Sci 13(7):8628–8647

Eckerstorfer MF, Engelhard M, Heissenberger A, Simon S, Teichmann H (2019) Plants developed by new genetic modification techniques—comparison of existing regulatory frameworks in the EU and non-EU countries. Front In Bioeng Biotechnol 7:26

El-Mounadi K, Morales-Floriano ML, Garcia-Ruiz H (2020) Principles, applications, and biosafety of plant genome editing using CRISPR-Cas9. Front Plant Sci 11:56

Fang H, Meng Q, Xu J, Tang H, Tang S, Zhang H, Huang J (2015) Knock-down of stress inducible OsSRFP1 encoding an E3 ubiquitin ligase with transcriptional activation activity confers abiotic stress tolerance through enhancing antioxidant protection in rice. Plant Mol Biol 87(4):441–458

Gao (2018) The future of CRISPR technologies in agriculture. Nat Rev Mol Cell Biol 19:275–276

Gao Z, Liu Q, Zhang Y, Chen D, Zhan X, Deng C, Cheng S, Cao L (2020) OsCUL3a-associated molecular switches have functions in cell metabolism, cell death, and disease resistance. J Agric Food Chem 68(19):5471–5482

Gao Y, Zhang Y-P, Fan Z-Q, Jing Y, Chen J-Y, Grierson D, Yang R, Fu D-Q (2021) Mutagenesis of SlNAC4 by CRISPR/Cas9 alters gene expression and softening of ripening tomato fruit. Veg Res 1(1):1–12

González MN, Massa GA, Andersson M, Turesson H, Olsson N, Fält AS, Storani L, Décima Oneto CA, Hofvander P, Feingold SE (2020) Reduced enzymatic browning in potato tubers by specific editing of a polyphenol oxidase gene via ribonucleoprotein complexes delivery of the CRISPR/Cas9 system. Front Plant Sci 10:1–12

Guzman AR, Kim JG, Taylor KW, Lanver D, Mudgett MB (2020) Tomato atypical receptor Kinase1 is involved in the regulation of pre-invasion defense. Plant Physiol 183(3):1306–1318

Hegde N, Joshi S, Soni N, Kushalappa AC (2021) The caffeoyl-CoA O-methyltransferase gene SNP replacement in Russet Burbank potato variety enhances late blight resistance through cell wall reinforcement. Plant Cell Rep 40(1):237–254

Hilbeck A, Meyer H, Wynne B, Millstone E (2020) GMO regulations and their interpretation: how EFSA's guidance on risk assessments of GMOs is bound to fail. Environ Sci Eur 32(1):1–15

Hua D, Ma M, Ge G, Suleman M, Li H (2020) The role of cyanide-resistant respiration in *Solanum tuberosum* L. against high light stress. Plant Biol 22(3):425–432

Huang Y, Xuan H, Yang C, Guo N, Wang H, Zhao J, Xing H (2019a) GmHsp90A2 is involved in soybean heat stress as a positive regulator. Plant Sci 285:26–33

Huang Y, Cao H, Yang L, Chen C, Shabala L, Xiong M, Niu M, Liu J, Zheng Z, Zhou L, Peng Z, Bie Z, Shabala S (2019b) Tissue-specific respiratory burst oxidase homolog-dependent $H_2 O_2$ signaling to the plasma membrane H + -ATPase confers potassium uptake and salinity tolerance in Cucurbitaceae. J Exp Bot 70(20):5879–5893

Huang CY, Wang H, Hu P, Hamby R, Jin H (2019c) Small RNAs–big players in plant-microbe interactions. Cell Host Microbe 26(2):173–182

Iorizzo M, Ellison S, Senalik D, Zeng P, Satapoomin P, Huang J, Bowman M, Iovene M, Sanseverino W, Cavagnaro P, Yildiz M (2016) A high-quality carrot genome assembly provides new insights into carotenoid accumulation and asterid genome evolution. Nat Genetic 48(6):657–666

Iqbal MS, Hafeez MN, Wattoo JI, Ali A, Sharif MN, Rashid B, Tabassum B, Nasir IA (2016) Prediction of host-derived miRNAs with the potential to target PVY in potato plants. Front Genet 14(7):159

Isaacson T, Ronen G, Zamir D, Hirschberg J (2002) Cloning of tangerine from tomato reveals a carotenoid isomerase essential for the production of β-carotene and xanthophylls in plants. Plant Cell 14(2):333–342

Javed T, Shabbir R, Ali A, Afzal I, Zaheer U, Gao S-J (2020) Transcription factors in plant stress responses: challenges and potential for sugarcane improvement. Plants (Basel, Switzerland) 9(4):491

Johansen IE, Liu Y, Jørgensen B, Bennett EP, Andreasson E, Nielsen KL, Blennow A, Petersen BL (2019) High efficacy full allelic CRISPR/Cas9 gene editing in tetraploid potato. Sci Rep 27;9(1):1–7

Kakeshpour T, Tamang TM, Motolai G, Fleming ZW, Park J-E, Wu Q, Park S (2021) CGFS-type glutaredoxin mutations reduce tolerance to multiple abiotic stresses in tomato. Physiol Plant 173(3):1263–1279

Kamburova VS, Nikitina EV, Shermatov SE, Buriev ZT, Kumpatla SP, Emani C, Abdurakhmonov IY (2017) Genome editing in plants: an overview of tools and applications. Int. J. Agron 2017:1–15

Karasov TL, Chae E, Herman JJ, Bergelson J (2017) Mechanisms to mitigate the trade-off between growth and defense. Plant Cell 29(4):666–680

Kawall K, Cotter J, Then C (2020) Broadening the GMO risk assessment in the EU for genome editing technologies in agriculture. Environ Sci Eur 32:106

Klap C, Yeshayahou E, Bolger AM, Arazi T, Gupta SK, Shabtai S, Usadel B, Salts Y, Barg R (2017) Tomato facultative parthenocarpy results from SlAGAMOUS-LIKE 6 loss of function. Plant Biotechnol J 15(5):634–647

Klimek-Chodacka M, Oleszkiewicz T, Lowder LG, Qi Y, Baranski R (2018) Efficient CRISPR/Cas9-based genome editing in carrot cells. Plant Cell Rep 37(4):575–586

Kourelis J, Van Der Hoorn RA (2018) Defended to the nines: 25 years of resistance gene cloning identifies nine mechanisms for R protein function. Plant Cell 30(2):285–299

Koutroumpa FA, Monsempes C, François MC, de Cian A, Royer C, Concordet JP, Jacquin-Joly E (2016) Heritable genome editing with CRISPR/Cas9 induces anosmia in a crop pest moth. Sci Rep 12;6(1):1–9

Kusano H, Ohnuma M, Mutsuro-Aoki H, Asahi T, Ichinosawa D, Onodera H, Asano K, Noda T, Horie T, Fukumoto K, Kihira M, Teramura H, Yazaki K, Umemoto N, Muranaka T, Shimada H (2018) Establishment of a modified CRISPR/Cas9 system with increased mutagenesis frequency using the translational enhancer dMac3 and multiple guide RNAs in potato. Sci Rep 8(1):1–9

Kusch S, Panstruga R (2017) mlo-based resistance: an apparently universal "weapon" to defeat powdery mildew disease. Mol Plant-Microbe Interact 30(3):179–189

Larsson MC, Domingos AI, Jones WD, Chiappe ME, Amrein H, Vosshall LB (2004) Or83b encodes a broadly expressed odorant receptor essential for *Drosophila* olfaction. Neuron 2;43(5):703–714

Lassoued R, Macall DM, Smyth SJ, Phillips PWB, Hesseln H (2019) Risk and safety considerations of genome edited crops: expert opinion. Curr Res Biotechnol 1:11–21

Li R, Li R, Li X, Fu D, Zhu B, Tian H, Luo Y, Zhu H (2018a) Multiplexed CRISPR/Cas9-mediated metabolic engineering of γ-aminobutyric acid levels in *Solanum lycopersicum*. Plant Biotechnol J 16(2):415–427

Li R, Zhang L, Wang L, Chen L, Zhao R, Sheng J, Shen L (2018b) Reduction of tomato-plant chilling tolerance by CRISPR–Cas9-mediated *SlCBF1* mutagenesis. J Agric Food Chem 66(34):9042–9051

Li R, Liu C, Zhao R, Wang L, Chen L, Yu W, Zhang S, Sheng J, Shen L (2019) CRISPR/Cas9-Mediated *SlNPR1* mutagenesis reduces tomato plant drought tolerance. BMC Plant Biol 19(1):38

Liu L, Zhang J, Xu J, Li Y, Guo L, Wang Z, Zhang X, Zhao B, Guo Y-D, Zhang N (2020) CRISPR/Cas9 targeted mutagenesis of *SlLBD40*, a lateral organ boundaries domain transcription factor, enhances drought tolerance in tomato. Plant Sci 301:110683

Lor VS, Starker CG, Voytas DF, Weiss D, Olszewski NE (2014) Targeted mutagenesis of the tomato PROCERA gene using transcription activator-like effector nucleases. Plant Physiol 166(3):1288–1291

Lu Y, Tian Y, Shen R, Yao Q, Zhong D, Zhang X, Zhu JK (2021) Precise genome modification in tomato using an improved prime editing system. Plant Biotechnol J 19(3):415

Maher MF, Nasti RA, Vollbrecht M, Starker CG, Clark MD, Voytas DF (2020) Plant gene editing through *de novo* induction of meristems. Nat Biotechnol 38(1):84–89

Makhotenko AV, Khromov AV, Snigir EA, Makarova SS, Makarov VV, Suprunova TP, Kalinina NO, Taliansky ME (2019) Functional analysis of Coilin in virus resistance and stress tolerance of Potato *Solanum tuberosum* using CRISPR-Cas9 editing. Dokl Biochem Biophys 484(1):88–91

Mantri N, Patade V, Penna S, Ford R, Pang E (2012) Abiotic stress responses in plants: present and future. In: Abiotic stress responses in plants. Springer, New York, pp 1–19

Martínez MI, Bracuto V, Koseoglou E, Appiano M, Jacobsen E, Visser RG, Wolters AM, Bai Y (2020) CRISPR/Cas9-targeted mutagenesis of the tomato susceptibility gene *PMR4* for resistance against powdery mildew. BMC Plant Biol 20(1):1–3

Modrzejewski D, Hartung F, Sprink T, Krause D, Kohl C, Wilhelm R (2019) What is the available evidence for the range of applications of genome editing as a new tool for plant trait modification and the potential occurrence of associated off-target effects: a systematic map. Environ Evid 8:27

Nakayasu M, Akiyama R, Lee HJ, Osakabe K, Osakabe Y, Watanabe B, Sugimoto Y, Umemoto N, Saito K, Muranaka T, Mizutani M (2018) Generation of α-solanine-free hairy roots of potato by CRISPR/Cas9 mediated genome editing of the *St16DOX* gene. Plant Physiol Biochem 131(February):70–77

Nekrasov V, Wang C, Win J, Lanz C, Weigel D, Kamoun S (2017) Rapid generation of a transgene-free powdery mildew resistant tomato by genome deletion. Sci Rep 28;7(1):1–6

Norris AL, Lee SS, Greenlees KJ, Tadesse DA, Miller MF, Lombardi HA (2020) Template plasmid integration in germline genome-edited cattle. Nat Biotechnol 38(2):163–164

Okuzaki A, Ogawa T, Koizuka C, Kaneko K, Inaba M, Imamura J, Koizuka N (2018) CRISPR/Cas9-mediated genome editing of the fatty acid desaturase 2 gene in *Brassica napus*. Plant Physiol Biochem 131:63–69

Ortigosa A, Gimenez-Ibanez S, Leonhardt N, Solano R (2019) Design of a bacterial speck resistant tomato by CRISPR/Cas9-mediated editing of *Sl JAZ 2*. Plant Biotechnol J 17(3):665–673

Pandolfi V, Ribamar Costa Ferreira Neto J, Daniel da Silva M, Lindinalva Barbosa Amorim L, Carolina Wanderley-Nogueira A, Lane de Oliveira Silva R, Akio Kido E, Crovella S, Maria Benko Iseppon A (2017) Resistance (R) genes: applications and prospects for plant biotechnology and breeding. Curr Protein Pept Sci 18(4):323–334

Parry MA, Madgwick PJ, Bayon C, Tearall K, Hernandez-Lopez A, Baudo M, Rakszegi M, Hamada W, Al-Yassin A, Ouabbou H, Labhilili M (2009) Mutation discovery for crop improvement. J Exp Bot 60(10):2817–2825

Petolino JF, Kumar S (2016) Transgenic trait deployment using designed nucleases. Plant Biotechnol J 14(2):503–509

Podevin N, Davies HV, Hartung F, Nogue F, Casacuberta JM (2013) Site-directed nucleases: a paradigm shift in predictable, knowledge-based plant breeding. Trends Biotechnol 31(6):375–383

Ramírez Gonzales L, Shi L, Bergonzi SB, Oortwijn M, Franco-Zorrilla JM, Solano-Tavira R, Visser RGF, Abelenda JA, Bachem CWB (2021) Potato CYCLING DOF FACTOR 1 and its lncRNA counterpart *StFLORE* link tuber development and drought response. Plant J 105(4):855–869

Razzaq A, Saleem F, Kanwal M, Mustafa G, Yousaf S, Imran Arshad HM, Hameed MK, Khan MS, Joyia FA (2019) Modern trends in plant genome editing: an inclusive review of the CRISPR/Cas9 toolbox. Int J Mol Sci 20(16):4045

Shimatani Z, Fujikura U, Ishii H, Terada R, Nishida K, Kondo A (2018) Herbicide tolerance-assisted multiplex targeted nucleotide substitution in rice. Data Brief 20:1325–1331

Su T, Yu S, Zhang JW, Yu Y, Zhang D, Zhao X, Wang W (2015) Loss of function of the carotenoid isomerase gene *BrCRTISO* confers orange color to the inner leaves of Chinese cabbage (*Brassica rapa* L. ssp. *pekinensis*). Plant Mol Biol Rep 33(3):648–659

Sun K, Wolters AM, Vossen JH, Rouwet ME, Loonen AE, Jacobsen E, Visser RG, Bai Y (2016) Silencing of six susceptibility genes results in potato late blight resistance. Transgenic Res 25(5):731–742

Sun D, Guo Z, Liu Y, Zhang Y (2017) Progress and prospects of CRISPR/Cas systems in insects and other arthropods. Front Physiol 6(8):608

Sun B, Jiang M, Zheng H, Jian Y, Huang WL, Yuan Q, Zheng AH, Chen Q, Zhang YT, Lin YX, Wang Y, Wang XR, Wang QM, Zhang F, Tang HR (2020) Color-related chlorophyll and carotenoid concentrations of Chinese kale can be altered through CRISPR/Cas9 targeted editing of the carotenoid isomerase gene BoaCRTISO. Hortic Res 7(1):161

Schmidt SM, Belisle M, Frommer WB (2020) The evolving landscape around genome editing in agriculture: Many countries have exempted or move to exempt forms of genome editing from GMO regulation of crop plants. EMBO reports 21(6):e50680.

Takeuchi A, Ohnuma M, Teramura H, Asano K, Noda T, Kusano H, Tamura K, Shimada H (2021) Creation of a potato mutant lacking the starch branching enzyme gene *stsbe3* that was generated by genome editing using the CRISPR/dMac3-Cas9 system. Plant Biotechnol 38(3):345–353

Tashkandi M, Ali Z, Aljedaani F, Shami A, Mahfouz MM (2018) Engineering resistance against tomato yellow leaf curl virus via the CRISPR/Cas9 system in tomato. Plant Signal Behav 13(10):e1525996

Tian S, Xing S, Xu Y (2021) Advances in CRISPR / Cas9-mediated genome editing on vegetable crops. In Vitro Cell Dev Biol Plant 57:672–682

Tran MT, Doan DTH, Kim J, Song YJ, Sung YW, Das S, Kim E, Son GH, Kim SH, Van Vu T, Kim J-Y (2021) CRISPR/Cas9-based precise excision of SlHyPRP1 domain(s) to obtain salt stress-tolerant tomato. Plant Cell Rep 40(6):999–1011

Tuncel A, Corbin KR, Ahn-Jarvis J, Harris S, Hawkins E, Smedley MA, Harwood W, Warren FJ, Patron NJ, Smith AM (2019) Cas9-mediated mutagenesis of potato starch-branching enzymes generates a range of tuber starch phenotypes. Plant Biotechnol J 17(12):2259–2271

Turnbull C, Lillemo M, Hvoslef-Eide TAK (2021) Global regulation of genetically modified crops amid the gene edited Crop Boom – a review. Front Plant Sci 12:630396

Tyagi S, Kesiraju K, Saakre M, Rathinam M, Raman V, Pattanayak D, Sreevathsa R (2020) Genome editing for resistance to insect pests: an emerging tool for crop improvement. ACS Omega 5(33):20674–20683

Van Butselaar T, Van den Ackerveken G (2020) Salicylic acid steers the growth–immunity trade-off. Trends Plant Sci 25(6):566–576

Van Damme M, Huibers RP, Elberse J, Van den Ackerveken G (2008) *Arabidopsis DMR6* encodes a putative 2OG-Fe (II) oxygenase that is defense-associated but required for susceptibility to downy mildew. Plant J 54(5):785–793

Van Vu T, Sung YW, Kim J, Doan DTH, Tran MT, Kim J-Y (2019) Challenges and perspectives in homology-directed gene targeting in monocot plants. Rice 12:95

Vats S, Kumawat S, Kumar V, Patil GB, Joshi T, Sonah H, Sharma TR, Deshmukh R (2019) Genome editing in plants: exploration of technological advancements and challenges. Cells 8(11):1386

Waltz E (2016a) CRISPR-edited crops free to enter market, skip regulation. Nat Biotechnol 34:582

Waltz E (2016b) Gene-edited CRISPR mushroom escapes US regulation. Nature 532:293

Wang L, Chen L, Li R, Zhao R, Yang M, Sheng J, Shen L (2017) Reduced drought tolerance by CRISPR/Cas9-mediated *SlMAPK3* mutagenesis in tomato plants. J Agric Food Chem 65(39):8674–8682

Wang D, Samsulrizal NH, Yan C, Allcock NS, Craigon J, Blanco-Ulate B, Ortega-Salazar I, Marcus SE, Bagheri HM, Perez-Fons L, Fraser PD, Foster T, Fray R, Paul Knox J, Seymour GB (2019a) Characterization of CRISPR mutants targeting genes modulating pectin degradation in ripening tomato. Plant Physiol 179(2):544–557

Wang H, Wu Y, Zhang Y, Yang J, Fan W, Zhang H, Zhao S, Yuan L, Zhang P (2019b) CRISPR/Cas9-based mutagenesis of starch biosynthetic genes in Sweet Potato (Ipomoea Batatas) for the improvement of starch quality. Int J Mol Sci 20(19):4702

Wang R, Tavano EC da R, Lammers M, Martinelli AP, Angenent GC, de Maagd RA (2019c) Re-evaluation of transcription factor function in tomato fruit development and ripening with CRISPR/Cas9-mutagenesis. Sci Rep 9(1):1–10

Wang L, Zhao L, Zhang X, Zhang Q, Jia Y, Wang G, Li S, Tian D, Li WH, Yang S (2019d) Large-scale identification and functional analysis of NLR genes in blast resistance in the Tetep rice genome sequence. PNAS 116(37):18479–18487

Wang Z, Hong Y, Zhu G, Li Y, Niu Q, Yao J, Hua K, Bai J, Zhu Y, Shi H, Huang S, Zhu J-K (2020) Loss of salt tolerance during tomato domestication conferred by variation in a Na+/K+ transporter. EMBO J 39(10):e103256

Weiberg A, Wang M, Lin FM, Zhao H, Zhang Z, Kaloshian I, Huang HD, Jin H (2013) Fungal small RNAs suppress plant immunity by hijacking host RNA interference pathways. Science 342(6154):118–123

Wu K, Shirk PD, Taylor CE, Furlong RB, Shirk BD, Pinheiro DH, Siegfried BD (2018) CRISPR/Cas9 mediated knockout of the abdominal-A homeotic gene in fall armyworm moth (*Spodoptera frugiperda*). PLoS One 13(12):e0208647

Xiong JS, Ding J, Li Y (2015) Genome-editing technologies and their potential application in horticultural crop breeding. Hortic Res 2(February):1–10

Xiong X, Liu W, Jiang J, Xu L, Huang L, Cao J (2019) Efficient genome editing of *Brassica campestris* based on the CRISPR/Cas9 system. Mol Genet Genomic 294(5):1251–1261

Yang C, Huang Y, Lv W, Zhang Y, Bhat JA, Kong J, Xing H, Zhao J, Zhao T (2020) *GmNAC8* acts as a positive regulator in soybean drought stress. Plant Sci 293:110442

Yin Y, Qin K, Song X, Zhang Q, Zhou Y, Xia X, Yu J (2018) BZR1 transcription factor regulates heat stress tolerance through FERONIA receptor-like kinase-mediated reactive oxygen species signaling in tomato. Plant Cell Physiol 59(11):2239–2254

Younis A, Siddique MI, Kim CK, Lim KB (2014) RNA interference (RNAi) induced gene silencing: a promising approach of hi-tech plant breeding. Int J Biol Sci 10(10):1150

Yu QH, Wang B, Li N, Tang Y, Yang S, Yang T, Xu J, Guo C, Yan P, Wang Q, Asmutola P (2017) CRISPR/Cas9-induced targeted mutagenesis and gene replacement to generate long-shelf life tomato lines. Sci Rep 7(1):1–9

Yuan L, Xie GZ, Zhang S, Li B, Wang X, Li Y, Liu T, Xu X (2021) GmLCLs negatively regulate ABA perception and signalling genes in soybean leaf dehydration response. Plant Cell Environ 44(2):412–424

Zafar SA, Zaidi SS-A, Gaba Y, Singla-Pareek SL, Dhankher OP, Li X, Mansoor S, Pareek A (2020) Engineering abiotic stress tolerance via CRISPR/ Cas-mediated genome editing. J Exp Bot 71(2):470–479

Zaidi SS, Mukhtar MS, Mansoor S (2018) Genome editing: targeting susceptibility genes for plant disease resistance. Trends Biotechnol 36(9):898–906

Zhan X, Zhang F, Zhong Z, Chen R, Wang Y, Chang L, Bock R, Nie B, Zhang J (2019) Generation of virus-resistant potato plants by RNA genome targeting. Plant Biotechnol J 17(9):1814–1822

Zhang H, Zhu J, Gong Z, Zhu J-K (2021) Abiotic stress responses in plants. Nat Rev Genet 23:104–119

Zhao P, You Q, Lei M (2019) A CRISPR/Cas9 deletion into the phosphate transporter *SlPHO1*;1 reveals its role in phosphate nutrition of tomato seedlings. Physiol Plant 167(4):556–563

Zhou X, Zha M, Huang J, Li L, Imran M, Zhang C (2017) StMYB44 negatively regulates phosphate transport by suppressing expression of PHOSPHATE1 in potato. J Exp Bot 68(5):1265–1281

Zhu J-K (2016) Abiotic stress signaling and responses in plants. Cell 167(2):313–324

CRISPR/Cas9-Mediated Targeted Mutagenesis in Medicinal Plants

Meghna Patial, Kiran Devi, and Rohit Joshi ⓘ

Abstract At present, traditional biotechnology techniques have usually been exploited for medicinal plant breeding. However, recent biotechnology-based breeding techniques, i.e., genome editing diversify the platform to develop custom-designed medicinal plants. The sequence-specific nucleases of TALENs, ZFNs, and Cas are advance genome editing tools which produce user-defined plants. CRISPR/Cas-based genome editing is an emerging technique that utilizes artificially engineered nucleases for digesting DNA at targeted locations in the genome for high-throughput biotechnology-based breeding of valuable medicinal plants. Due to its wide application in gene mutagenesis, transcriptional regulation, high efficiency and easy manipulation, several plant-specific CRISPR/Cas9 vector systems have been designed. This methodology is based upon the type II adaptive immunity response of prokaryotes, which comprises of a CRISPR-associated (Cas)9 protein and an engineered sgRNA that specifically targets the nucleic acid sequence to induce selective mutagenesis. In this chapter, various CRISPR/Cas-based approaches are discussed with emphasis on CRISPR/Cas9 vector platforms, multiplex editing strategies, analysis methods for induced mutations, and its applications in medicinal plants. This chapter provides the advancements in genome editing technologies and their associated strategies, giving an insight on the limitations of CRISPR/Cas9 technique and its future advances to improve the quality of traditional medicinal herbs. This new system will further open new arena to manage synthetic biology of medicinal plants for industrial purposes and to investigate the function of gene to accelerate basic plant research.

M. Patial · K. Devi
Division of Biotechnology, CSIR-Institute of Himalayan Bioresource Technology, Palampur, Himachal Pradesh, India

R. Joshi (✉)
Division of Biotechnology, CSIR-Institute of Himalayan Bioresource Technology, Palampur, Himachal Pradesh, India

Academy of Scientific and Innovative Research (AcSIR), Ghaziabad, Uttar Pradesh, India

© The Author(s), under exclusive license to Springer Nature Switzerland AG 2022
S. H. Wani, G. Hensel (eds.), *Genome Editing*,
https://doi.org/10.1007/978-3-031-08072-2_3

Keywords CRISPR/Cas9 · *Dendrobium officinale* · *Dioscorea zingiberensis* · Double strand DNA breaks · Genome engineering · Targeted mutagenesis · *Salvia miltiorrhiza*

1 Introduction

In the modern era, plants are defined as "green factories," having a broad range of chemical complexity to support food, feed, and biopharma industries (Niazian 2019). In the present context, plant-derived medicines are very much popularized due to their safe application (Upadhay et al. 2014; Niazian 2019). Despite their numerous applications and advantages, medicinal plants are constantly being deprived of domestication and breeders are constantly giving efforts to develop faster methods for their proliferation by utilizing modern biotechnology-based tools (Bahuguna et al. 2011). With the advancement of high-throughput omics, i.e., proteomics, transcriptomics, genomics, and metabolomics, a new area called "phytochemical genomics" has emerged, which investigates the evolution, function, and regulation of phytochemical metabolites. In addition, recent genome editing technology has emerged as a reliable molecular tool to precisely manipulate target metabolites (Fuller et al. 2015).

In a plant breeding program, a well-known method of producing novel traits is randomly introducing mutations into the genome, leading to an improvement in quality and yield. Over the past few years, a sharp increase in induced mutagenesis was observed which includes broad-range screening for characterization of superior traits in plants. Mutagens used for inducing mutations include physical agents like radiations (gamma rays and ion beams) as well as chemicals such as EMS (ethyl methane sulfonate). The double-stranded DNA break (DSB) repair mechanism, which occurs during early meiosis, is essential for inducing precise breaks in the genome (Osakabe et al. 2012; Vats et al. 2019). DSBs occur due to the production of reactive oxygen species by environmental exposure to radiation such as UV or chemical agents that cause damage to DNA bases and cause lesions that block DNA replication. Homologous recombination (HR)/ Homologous end joining (HEJ)/ Homology-directed repair (HDR) and non-homologous end joining (NHEJ) are two methods possessed by cells to repair DSBs (Negritto 2010). NHEJ is dominant in eukaryotes whereas HEJ dominates in bacteria and yeast.

HEJ involves a general recombination mechanism that uses homologous sequence to repair DNA. This process requires recombination proteins to identify and bring together regions of DNA sequence that match the corresponding chromosomes. The undamaged chromosome is then used as template to transfer genetic information to the broken chromosome and repair the chromosome without changing DNA sequence. NHEJ is an emergency solution to repair DSB, in which, broken ends are put together and rejoined by DNA ligation. Usually one or more nucleotides got lost at the joining site. Broken ends are detected via hetero-dimeric Ku70 and Ku80 proteins that allow Artemis and DNA-dependent protein kinase (DNA-PKcs) to act as nuclease. Artemis shows both endonuclease and exonuclease activity. Finally, DNA ligase IV in association with XRCC4 joins the double strand ends (Pannunzio et al. 2018). NHEJ is particularly common in the G0 and G1 phase whereas Homologous recombination predominates in the S and G2 phases (Lieber 2010). NHEJ is more subject to errors as it causes insertions, substitutions, and deletions of nucleotides. Double strand DNA breaks (DSBs) lay the concept of genome engineering. An efficient method for genetic engineering is genome editing with engineered nucleases (GEEN) which digest DNA at the specified site in genome of plants using artificially engineered nucleases (Osakabe and Osakabe 2015). An induced DSB introduced by the engineered nuclease at a particular site then undergoes repair through natural processes of HR and NHEJ.

There are three types of engineered nucleases: zinc finger nuclease (ZFN), transcription activator-like effector nuclease (TALEN), and CRISPR (clustered regularly inter-spaced short palindromic repeats)/Cas9 (CRISPR-associated9) (Joshi et al. 2019). For genome editing TALEN and CRISPR/Cas9 are widely used in plants. The ZFNs were the first genuinely targeted protein reagents to change the entirety of genome manipulation field (Chandrasegaran and Carroll 2016). ZFN was first identified in *Xenopus laevis* (African clawed frog). ZFNs have been commonly used in various plant species, such as Arabidopsis, tobacco, and maize, for targeted genome modifications (Vats et al. 2019; Iqbal et al. 2020). ZFNs structurally consists of the restriction enzyme FokI endonuclease domain and DNA binding zinc finger protein (ZFP; Kim et al. 1996). Type II restriction endonuclease FokI has been isolated from the N-terminal DNA-binding domain and C-terminal DNA cleavage domain of *Flavobacterium okeanokoites* (Durai et al. 2005). With FokI monomer, two DNA-binding ZFNs are attached to develop a 5–6 bp spacer, which allows FokI to be functional on dimerization and create DSBs (Bibikova et al. 2003). TALENs were first described in plant pathogenic bacteria (*Xanthomonas*). TALENs are highly specific as they target only one nucleotide at a target site (Joung and Sander 2013). TALENs have three domains: C-terminal domain including an activator domain, N-terminal domain, and DNA binding domain. DSB introducing activity of TALENs is influenced by length of N and C terminal. TALENs have been used successfully for genome editing of both angiosperms and bryophytes. CRISPR/*Cas*9 is based on acquired immune response of microbes that aids bacteria to defend themselves from invading phages, offering an excellent alternative to first generation site-directed nucleases (Terns and Terns 2011). Despite been identification of CRISPRs in bacterial DNA in 1987, their role in providing immunity with Cas protein was demonstrated in 2007. For gene editing, TALENS and ZFNs have been successfully used, but CRISPR/Cas has many advantages over other methods in terms of versatility, sensitivity, design, multiplexing, and expense (Boch and Bonas 2010). Gene editing using CRISPR/Cas was first done using cell-free system by Jinek et al. (2012). Later in 2013, gene editing via CRISPR/Cas9 was first shown in *Nicotiana benthamiana*, *Arabidopsis,* and *Oryza sativa* as model plants (Li et al. 2013; Nekrasov et al. 2013; Shan et al. 2013). In this chapter, we provide details of

CRISPR/Cas9 mechanism, different CRISPR/Cas9 vector system for plants and their efficiency, medicinal plants modified using CRISPR/Cas9 and applications of genome editing for further improvement of medicinal plants.

2   CRISPR/Cas9 Mechanism

The locus of CRISPRs includes tandem repeats and spacers, where the former consist of the same sequence and later consist of various exotic DNA-derived sequences. CRISPR repeats and spacers differ in size, respectively, from 23 to 47 bp and from 21 to 72 bp respectively (Fig. 1). CRISPR-associated (*Cas*) genes are required for providing immunity against invading foreign DNA. *Cas* genes encode proteins for nuclease, helicase, and polymerase. CRISPR/Cas9 comprises sgRNA (single guided RNA), CRISPR-RNA (crRna), trans-activating crRNA (tracrRNA) and nuclease protein Cas9.

When editing a gene, crRNA and trans-crRNA are linked together to form single guide RNA (sgRNA). Because of its complementarity to crRNA, the target sequences (DNA) are recognized by crRNA. The ribonucleoprotein invades the target gene sequence with the sgRNA sequence and forms a 20 bp hybrid of RNA/DNA. Cas9 is a nuclease protein consisting of two endonuclease domains: RuvC (named for an *E. coli* protein involved in DNA repair)-like nuclease domain which cleaves non-target DNA strand and HNH (named for characteristic histidine and asparagine residues) nuclease domain that cleaves the complementary strand of crRNA (target DNA). Cas9 binds to DNA-RNA hybrid and cleaves a RuvC-like nuclease domain with the displaced DNA strand and HNH nuclease domain with a complementary DNA strand (target DNA strand) (Nishimasu et al. 2014). After encountering the Protospacer adjacent motif (PAM), such as NGG sequence, it also displaces the opposite non-target DNA strand. Researchers have exploited this process by constructing synthetic guide RNAs (sgRNA) that direct the cas9 nuclease to genomic targets.

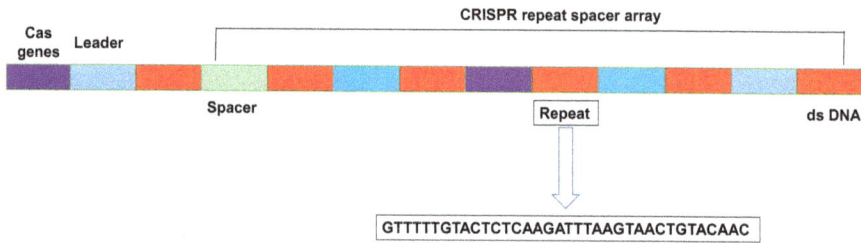

Fig. 1 Schematic diagram depicting CRISPR/*Cas*9 locus organization. CRISPR loci consist of 23–47 bp palindromic repeat sequences which are separated by 21–72 bp spacer sequences. *Cas* genes are located close to CRISPR loci

2.1 Cleavage Activity of Cas9

Three main components involved in CRISPR/Cas9 technology are: the Cas9 protein, sgRNA, and target sites of DNA upstream of PAM (Jinek et al. 2012; Doudna and Charpentier 2014). DNA sequences, which expresses crRNA having protospacer and tracrRNA, are linked to develop sgRNA; the target sequence does not include PAM rather the target sequence is present in the 5′ terminal position of sgRNA so that it can pair with the target site (Ma et al. 2016). The sgRNA precisely imitates the initial duplex of tracrRNA– crRNA and ease the manipulation of CRISPR/Cas9. A Cas9 nuclease complex forms the Cas9 protein and the sgRNA (Ma et al. 2016). Cas9/sgRNA complex search for the target sequence of DNA next to PAM that is required for effective detection of target (Sternberg et al. 2014). After identifying the target site, Cas9 cleaves the desired sequence allowing sgRNA to link with the complementary target strand (Fig. 2). The RucV and HNH domains cleave from three bases upstream of PAM in both strands of the target DNA, forming a blunt-ended DSB. For precise gene editing, DSB can be mended either via HDR in the presence of homologous donor DNA template to induce gene knock-in or by NHEJ to create point mutations in the desired location. Importantly, Cas9 can be guided to the correct locations inside the same cells by different sgRNA with specific target sequences (Cong et al. 2013; Ma et al. 2016).

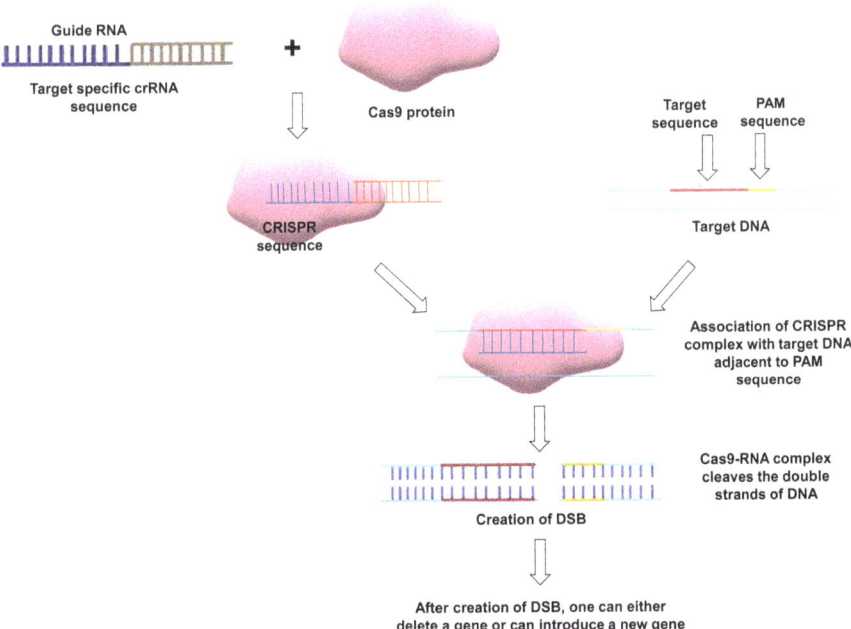

Fig. 2 Schematic diagram depicting the mechanism of CRISPR/Cas9 and cleavage activity of Cas9

3 CRISPR/Cas9 Vector System for Plants

The vector system of CRISPR/cas9 for plants comprises expression cassettes which are part of vector DNA. Expression cassettes have a regulating sequence which directs the cell for transcription and translation.

3.1 sgRNA Expression Cassettes

sgRNA is a result of fusions of critical regions of tracrRNA with the guide and PAM domains of crRNA (Jiang et al. 2013). The expression of sgRNAs is driven by U3/U6 small nuclear RNA promoters (snRNA) in plants. sgRNA is expressed by RNA polymerase III (Wang et al. 2018). sgRNA expression cassettes can be developed with target sequences by overlapping PCR or target adaptor ligation. The cassette developed is known as U3/U6 promoter: sgRNA expression cassette (Long et al. 2018). This cassette can be incorporated into CRISPR/cas9 binary vector through Gibson assembly or Golden gate cloning. This technique was previously reported to be more efficient and simpler to construct expression clones for CRISPR/Cas9-mediated genome editing in rice (Ma et al. 2020).

3.2 Cas9 Expression Cassettes

Cas9 has a coding sequence of length 4107 bp initially. Fusion of NLS (nuclear localization signal) with the coding sequences of cas9 results in nuclear localization of cas9 in eukaryotes (Wang et al. 2018). Cas9 genes can be modified to enhance the expression of Cas9 cassettes for effective genetic engineering. For example, in the case of rice at 5′ terminal region, which is known as plant codon optimized Cas9 gene (Cas9p), GC content has been increased to modify Cas9 gene (Zhang et al. 2014; Ma et al. 2016). For callus based transformations, promoters such as ubiquitin of *Arabidopsis, Zea mays* and *Oryza sativa* and of Cauliflower mosaic virus (CaMV) 35S can be applied to induce the Cas9 gene in plants for successful genome editing. In certain cases, using ubiquitin promoters has achieved better performance in editing than the CaMV 35S promoter (Ren et al. 2021). Cas9 and sgRNA vectors are induced in plant cells through *Agrobacterium*-mediated transformation as it is the most successful method of transformation in plants. Other methods to induce vectors in plants include PEG mediated transformation, protoplast fusion, and particle bombardment (Liu et al. 2019).

4 CRISPR/dCas9 and Epigenome Editing in Plants

dCas9 is nuclease dead Cas9. In dCas9, activity of endonuclease is removed through point mutations in its endonuclease domain. The CRISPR/dCas9 system consists of three major components: a nuclease-dead Cas9 (also known as dcas9 null mutant), sgRNA, and transcriptional activators.

4.1 Nuclease-Dead Cas9

The cas9 nuclease domains were mutated in *S. pyogens* by introducing D10A mutation in RuvC domain and H840A mutation in HNH domain (Jinek et al. 2012). This dead version of the Cas9 cannot bind DNA anymore, but it still targets and connects DNA to sgRNA with the same precision. Rather than permanently changing the genome, linking of dCas9 disrupts transcription at the target location, resulting in temporary gene silencing. Instead, standard CRISPR/Cas9 relies on DSBs by means of Cas9 endonuclease action, which is then accompanied by manipulating gene editing DNA repair mechanisms, dCas9 activation systems integrate transcriptional N and C-terminal protein enabler (Piatek et al. 2015; Polstein and Gersbach 2015; Li et al. 2017). Thus, dCas9 can modulate gene expression without permanently modifying the DNA.

4.2 sgRNA

Fusion of tracrRNA and crRNA results in the formation of gRNA (Jinek et al. 2012). The gRNA comprises of scaffolding and spacer regions. The spacer region includes nucleotides complementary to the sequences present in the promoter region of target gene. Scaffolding region forms a complex with dcas9. Both regions bind with cas9 and guide it to target gene. CRISPR systems are extremely versatile as any genes or nucleotides with a complementary base to the spacer region can be used as targets. Due to spacer region, CRISPR systems are extremely versatile as the gRNA in former region can be changed for any genes or nucleotides that is complementary to the spacer region. The Cas9 binding sequence within the sgRNA is required for the formation of the dCas9 complex (Ivanov et al. 2020).

4.3 Transcriptional Effectors

Transcriptional regulators are fusion proteins, whose functional domain is connected to DNA binding domain (Piatek et al. 2015). Transcriptional effectors include domains of proteins or proteins that are linked to dcas9 or sgRNA. They also include transcriptional activators or repressors. Transcription effectors have two functions; firstly, the gene that is regulated by its DNA-binding domains is defined and secondly, they are capable of stimulating transcription. A gene sequence with active or repressive domains can be recruited for the purpose to improve transcription of RNA polymerases. In eukaryotes DNA nucleosome interactions can be loosen or modifications can be made in histones for easier gene transcription (Ma et al. 2015). For CRISPR activation and CRISPR interference (CRISPRi), sgRNAs are necessary, as enhanced or suppressed transcription of the target gene can be achieved by dCas9 repressors or activators directed by gRNA(s). To activate or suppress the function of gene with dCas9 activators or suppressors, the gene of interest (GOI) promoter area should also be targeted by sgRNAs (Dahlman et al. 2015). The plant transcription regulatory experiments based on CRISPR-dCas9 also involve a single T-DNA expression of individual gRNAs and dCas9-effector fusion proteins (Lowder et al. 2017). Synthetic transcriptional regulators may include a zinc-finger protein trans-activation region, TALE transcription activators, Herpes simplex viral protein 16 (VP16) (Adli 2018) and other unrelated sequence transcription activation domains (TADs). Studies have suggested that these simple fusion proteins trigger or suppress endogenous genes, although their efficiency in the transcriptional activation is limited by a single gRNA to an average two to five-fold. Transcriptional effectors can be incorporated to CRISPR/dCas9 system by fusing them with sgRNA or dCas9. As artificial guide RNAs can be modified easily to achieve specific new target, dCas9-TAD are simple and have multiplex-ability in contrast to ZFP-TAD and TALE-TAD (Moradpour and Abdulah 2020).

5 Analysis and Efficiency of Targeted Mutations

To validate the use of CRISPR/Cas9 in genome editing, it is necessary to confirm and determine the target mutations. This can be validated through reporter genes, single strand conformation polymorphism, high resolution melting, high throughput sequencing, and sanger sequencing. The details are discussed below:

5.1 Reporter Genes

Reporter genes can be utilized as an indicator of edited mutations to monitor the functionality of the designed CRISPR/Cas9 vector framework. For instance, reporter gene can be modified to have an objective position that causes a frame shift; Cas9/sgRNA complex mutation of the target can repair the gene's reading frame for restoring its expression. Also in Arabidopsis, green fluorescent protein (GFP) was used to quantify the targeted mutation via CRISPR/Cas9 (LeBlanc et al. 2018).

5.2 Single-Strand Conformation Polymorphism (SSCP)

ssDNA with nucleotide variations can modify the components of the single-strand conformation polymorphism (SSCP). When dsDNA is denatured, it takes its primary conformation; if there is mutation, then its primary conformation will differ from the original conformation. This can be displayed as different migration rates into a non-denatured PAGE gel.

5.3 High-Resolution Melting (HRM)

HRM is performed on double-strand DNA. First, the target DNA with mutation is amplified using polymerase chain reaction. After amplification, HRM analysis starts that employs real-time PCR which is achieved by fluorescent dyes which bind specifically with dsDNA. During HRM analysis, as the temperature rises separation of dsDNA strands begin (Li et al. 2020). Initially, fluorescence is brighter as dye has many copies of amplified dsDNA to bind with, but gradually the fluorescence decreases because DNA strands separate due to melting at higher temperature. Melting curve of different DNA can be compared as slight change in base pair or a single base cause's change in melting temperature (Denbow et al. 2017). Thus, the two DNA strands one with mutation and other one with original strand, will have different melting temperature and therefore the mutations can be verified (Zischewski et al. 2017).

5.4 High-Throughput Sequencing (HTS or Deep Sequencing)

NGS technique is used for detection of rare mutations and complex chimeric mutations specifically for finding potential off-target mutations across the entire genome either by using whole genome or single or multiple PCR amplicons. However, the technique is expensive (Fauser et al. 2014; Feng et al. 2014).

5.5 Sanger Sequencing

Sanger sequencing or chain terminator method is used to determine the sequence of nucleotides in a given fragment of DNA. First, the target DNA strand is amplified in the presence of dNTPs and a complementary DNA strand which serve as a primer. To determine any mutation, dideoxynucleotide triphosphates (ddNTPs) are incorporated. These ddNTPs are labelled with fluorescent dyes which are observed by capillary electrophoresis technique after which the sequences are determined (Xie et al. 2019).

6 Medicinal Plants Modified Using CRISPR/Cas9

Medicinal plants are a vital component of herbal remedies, cosmetics, beauty products, and any natural product. They are utilized all around the world because of the secondary metabolites. These secondary metabolites are nevertheless produced in small quantities but have a high demand in the pharmaceutical industry, so the CRISPR-Cas9 mediated strategy, which greatly improves metabolism and pathway technology due to its simplicity and effectiveness, will be used to combat this and improve their quality and quantity. Genetic engineering enables to play with the genes involved in various pathways of metabolite synthesis. This technology has been successfully used in several plant species, of those few medicinal plants such as *Salvia miltiorrhiza*, *Dendrobium officinale*, and *Dioscorea zingiberensis* are described below.

6.1 Salvia militorrhiza

The first medicinal plant whose genome was edited successfully via CRISPR/Cas9 is *Salvia miltiorrhiza* (Li et al. 2017). It is a medicinal herb used for the treatment of cardio-vascular and cerebrovascular conditions. The presence of phenolic compounds like lithospermic acid, salvianolic acid, and rosmarinc acid and diterpenes such as tanshinones means that these plants have their medicinal features. In the *Salvia miltiorrhiza* roots, tanshinones are most prevalent. Mevalonic acid pathway and methylerythritol phosphate pathway are biosynthesis pathways of tanshinones (Yang et al. 2013). The gene for diterpenes synthase (*SmCPS1*) in *Salvia milltiorrhiza* is involved in tanshinone biosynthesis. Rosmarinic acid synthesis is through Shikimic acid and phenylpropanoid pathway (Chen et al. 2018). Rosmarinic acid synthase gene (*SmRAS*) is involved in biosynthesis of rosmarinic acid. CRISPR/Cas9 has been used to knockout *SmCPS1* gene and edit *SmRAS* gene. By *Agrobacterium rhizogenes*-mediated transformation, eight chimeric and three homozygous mutants were produced from 26 separate transgenic hairy root lines while knocking out the *SmCPS1* gene. Metabolomic analyses showed that homozygotic mutants are completely missing from tanshinones. Tanshinones that resemble a previously reported RNAi SmCPS1 study were lowered but can be detected in chimeric mutants. These results demonstrate the simple and effective genome editing tool used by the *Agrobacterium rhizogenes* transformation using CRISPR/Cas9. To improve quality and output of this valuable traditional Chinese herb, it is necessary to elucidate its secondary metabolites (Li et al. 2017). *SmRAS* was edited using CRISPR/cas9. In order to accurately edit the main *SmRAS* gene from the 11 family members, sgRNA was selected by bioinformatics. The sequencing results showed that 50 percent genomes have been successfully edited from the transgenic regenerated hair radicals. Expression and metabolic analysis subsequently showed by successfully editing hairy-root lines, specifically in the homozygotic mutants, the contents of phenolic acids, including Rosmarinic acid (RA) and RAS expression levels decreased. In addition, 3, 4-dihydroxyphenyllactic acid was substantially increased in the RA precursor. These findings show that with bio informatics studies CRISPR/Cas9 could be used to classify key genes in a several set of similar genes (Zhou et al. 2018).

6.2   Dendrobium officinale

Dendrobium officinale is a pharmacologically patented orchid. Presence of high content of lignocellulose affects its taste and popularity. So, CRISPR/Cas9 was employed to knock out the genes involved in the biosynthesis of lignocellulose. In *D. officinale* CaMV 35S promoter, superfold green fluorescence protein (SGP) and β-glucuronidase (GUS) were introduced as reporter genes in plants through *Agrobacterium rhizogenes*. Five genes involved in biosynthesis of ligno-cellulose are *IRREGULAR XYLEM5 (IRX)*, *CINNAMOYL COENZYME A REDUCTASE (CCR)*, *CINNAMATE 4-HYDROXYLASE (C4H)*, *4-COUMARATE: COENZYME A LIGASE (4CL)*, and *COUMARATE 3-HYDROXYLASE (C3H)* were targeted. Fluorescence emission from transformed plants indicated that this technique can produce modifications (insertions, deletions or substitutions) ranging from 10% to 100%. Gene activities with different promoters was compared and it was found that the 35S promoter for the transformation was as successful as *MMV* (Mirabilis mosaic virus), *CVMV* (Cassava vein mosaic virus) and *PCISV* (Peanut Chlorotic Streak Caulimovirus) promoters (Kui et al. 2017).

6.3   Dioscorea zingiberensis

It has high content of diosgenin in its rhizome. Diosgenin serves as a raw material for various steroid hormones. Due to asexual reproduction, diosgenin content has been decreased in this plant. The precursor molecule of diosgenin is farnesyl pyrophosphate (FPP) which is synthesized through mevalonic acid pathway and methylerythritol phosphate pathway (Dhar et al. 2013). The key enzyme involved in the synthesis of FPP is farnesyl pyrophosphate synthase (FPS) which is the product of gene *Dzfps*. Through CRISPR/Cas9, *Dzfps* gene was targeted and sgRNA expression cassette was used as a CRISPR/Cas9 vector. Transformation was done through *Agrobacterium tumefaciens* with high mutation frequency. The squalene concentration was decreased in the mutant plants when compared to wild plants which showed that the activity of FPS was reduced. Hence, CRISPR/Cas9 has successfully targeted the *Dzfps* which lead to decrease in its function (Feng et al. 2018).

7 Applications of Genome Editing in Medicinal Plants

Genome editing via CRISPR/Cas9 has many uses in studying the metabolic pathways and genes involved in various pathways. With CRISPR/Cas9 plants can be improved genetically for their yield, disease resistance, and stress tolerance. Through CRISPR/Cas9 metabolic pathways can be altered to enhance plant's quality and metabolite production by transcriptional regulation, gene knockout, and gene mutation (Joshi et al. 2018). With CRISPR/Cas9, the following techniques can be performed:

- *Gene knock out*: Using CRISPR/Cas9 genes can be inactivated for a particular function. Genome editing through CRISPR/Cas9 can be used to eliminate undesirable trait from plants.
- *DNA free gene silencing*: In this DNA vectors are not used, instead it requires RNA or protein components. It has the advantage that undesirable genetic alterations occurring due to plasmid DNA at the cut site can be avoided.
- *Gene knock in/gene replacement*: CRISPR/Cas9 method uses cells homology-directed repair to insert a gene.
- *Transient gene silencing*: For this Cas9 protein is modified so that it doesn't cut DNA. To reduce transcriptional activity and gene expression, promoter region of the gene is targeted by altered Cas9 guided by RNA.
- The main practical advantage of CRISPR/Cas9 is the ease of multiplexing. The DSBs can be introduced at various sites simultaneously and thus several genes can be edited at the very same time (Li et al. 2013; Mao et al. 2013).
- For development of synthetic biology system and to enhance the secondary metabolite production in medicinal plants, an artificial polyploidy is induced, subsequent to CRISPR/Cas9-armed *Agrobacterium rhizogenes*-mediated hairy root culture.

8 Conclusion

In both developed and developing countries, long-term consumption of synthetic medicines were reported to cause various health problems. In this condition, secondary metabolites of medicinal plants provide valuable and promising material for quality health. Increasing evidence shows that genes regulating biosynthetic pathways form clusters, which can provide in-depth insights into evolution and function of specialized metabolisms. With the evolution of genomics and metabolomics, medicinal plants emerged as an untapped reservoir for novel bioactive secondary metabolites. However, conventional methods of cultivation and breeding are not enough to remove the technological hurdles of industrial-scale production of these wild-crafted and low-accessible medicinal plants. High throughput omic techniques provide promising avenues to overexpress the precursors, regulatory enzymes and rate limiting enzymes for biosynthesis of these metabolites. CRISPR/Cas9 has emerged as a flexible, versatile and highly efficient system with broad applicability for targeted mutation in the genome of medicinal plants and subsequent alteration in their metabolism (Wani et al. 2018). CRISPR/Cas9 is different from ZFNs and TALENs as in former case target sequences can simply be inserted into sgRNA, and after identification of target site via Cas9/sgRNA nuclease complex desired site can be engineered whereas ZFNs and TALENs rely on protein domains binding to DNA. At present, the use of CRISPR/Cas9 is limited in medicinal plants because in medicinal plants whole genome sequence information along with regeneration and genetic transformation protocols are lacking. Till date, it is premature to proclaim that CRISPR/Cas9 technology is in a golden era for manipulation of medicinal plants for drugs. Still, this advance tool can be used to regulate the biosynthesis pathway of many plants to increase the production of various metabolites. In conclusion, newly emerged gene editing technologies of medicinal plants will be quite promising to improve global future by cooperation to tackle the sustainable developmental goals.

Acknowledgments MP, KD and RJ gratefully acknowledge the director, CSIR-Institute of Himalayan Bioresource Technology, Palampur, for providing the facilities to carry out this work. CSIR support in the form of project MLP0201, MLP0165, MLP0170, and MLP0172 and agriculture department, Shimla in the form of project SSP-0125 and SSP-0126 for this study is highly acknowledged. This manuscript represents CSIR-IHBT communication no. 4891.

References

Adli M (2018) The CRISPR tool kit for genome editing and beyond. Nat Commun 9:1911

Bahuguna RN, Joshi R, Singh G, Shukla A, Gupta R, Bains G (2011) Micropropagation and total alkaloid extraction of Indian snake root (*Rauwolfia serpentina*). Indian J Agric Sci 81:1124–1129

Bibikova M, Beumer K, Trautman JK, Carroll D (2003) Enhancing gene targeting with designed zinc finger nucleases. Science 300(5620):764

Boch J, Bonas U (2010) Xanthomonas AvrBs3 family-type III effectors: discovery and function. Annu Rev Phytopathol 48:419–436

Chandrasegaran S, Carroll D (2016) Origins of programmable nucleases for genome engineering. J Mol Biol 428(5):963–989

Chen Z, Wu J, Ma Y, Wang P, Gu Z, Yang R (2018) Biosynthesis, metabolic regulation and bioactivity of phenolic acids in plant food materials. Food Sci 39(7):321–328

Cong L, Ran FA, Cox D, Lin S, Barretto R, Habib N, Hsu PD, Wu X, Jiang W, Marraffini LA, Zhang F (2013) Multiplex genome engineering using CRISPR/Cas systems. Science 339(6121):819–823

Dahlman JE, Abudayyeh OO, Joung J, Gootenberg JS, Zhang F, Konermann S (2015) Orthogonal gene knockout and activation with a catalytically active Cas9 nuclease. Nat Biotechnol 33(11):1159–1161

Denbow CJ, Lapins S, Dietz N, Scherer R, Nimchuk ZL, Okumoto S (2017) Gateway-compatible CRISPR-Cas9 vectors and a rapid detection by high-resolution melting curve analysis. Front Plant Sci 8:1171

Dhar MK, Koul A, Kaul S (2013) Farnesyl pyrophosphate synthase: a key enzyme in isoprenoid biosynthetic pathway and potential molecular target for drug development. New Biotechnol 30(2):114–123

Doudna JA, Charpentier E (2014) The new frontier of genome engineering with CRISPR-Cas9. Science 346:12580

Durai S, Mani M, Kandavelou K, Wu J, Porteus MH, Chandrasegaran S (2005) Zinc finger nucleases: custom-designed molecular scissors for genome engineering of plant and mammalian cells. Nucleic Acids Res 33(18):5978–5990

Fauser F, Schiml S, Puchta H (2014) Both CRISPR/Cas-based nucleases and nickases can be used efficiently for genome engineering in *Arabidopsis thaliana*. Plant J 79:348–359

Feng Z, Mao Y, Xu N, Zhang B, Wei P, Yang DL, Wang Z, Zhang Z, Zheng R, Yang L, Zeng L (2014) Multigeneration analysis reveals the inheritance, specificity, and patterns of CRISPR/Cas-induced gene modifications in Arabidopsis. Proc Natl Acad Sci 111(12):4632–4637

Feng S, Song W, Fu R, Zhang H, Xu A, Li J (2018) Application of the CRISPR/Cas9 system in Dioscorea zingiberensis. Plant Cell Tissue Organ Cult 135(1):133–141

Fuller KK, Chen S, Loros JJ, Dunlap JC (2015) Development of the CRISPR/Cas9 system for targeted gene disruption in *Aspergillus fumigatus*. Eukaryot Cell 14(11):1073–1080

Iqbal Z, Iqbal MS, Ahmad A, Memon AG, Ansari MI (2020) New prospects on the horizon: genome editing to engineer plants for desirable traits. Curr Plant Biol 20:100171

Ivanov IE, Wright AV, Cofsky JC, Aris KD, Doudna JA, Bryant Z (2020) Cas9 interrogates DNA in discrete steps modulated by mismatches and supercoiling. Proc Natl Acad Sci 117(11):5853–5860

Jinek M, Chylinski K, Fonfara I, Hauer M, Doudna JA, Charpentier E (2012) A programmable dual-RNA–guided DNA endonuclease in adaptive bacterial immunity. Science 337(6096):816–821

Joshi R, Singh B, Chinnusamy V (2018) Genetically engineering cold stress-tolerant crops: approaches and challenges. In: Cold tolerance in plants. Springer, Cham, pp 179–195

Joshi R, Gupta BK, Pareek A, Singh MB, Singla-Pareek SL (2019) Functional genomics approach towards dissecting out abiotic stress tolerance trait in plants. In: Genetic enhancement of crops for tolerance to abiotic stress: mechanisms and approaches. Springer, Cham, pp 1–24

Joung JK, Sander JD (2013) TALENs: a widely applicable technology for targeted genome editing. Nat Rev Mol Cell Biol 14(1):49–55

Jiang W, Zhou H, Bi H, Fromm M, Yang B, Weeks DP (2013) Demonstration of CRISPR/Cas9/sgRNA-mediated targeted gene modification in Arabidopsis, tobacco, sorghum and rice. Nucl Acids Res 41(20): e188.

Kim YG, Cha J, Chandrasegaran S (1996) Hybrid restriction enzymes: zinc finger fusions to Fok I cleavage domain. Proc Natl Acad Sci 93(3):1156–1160

Kui L, Chen H, Zhang W, He S, Xiong Z, Zhang Y, Yan L, Zhong C, He F, Chen J, Zeng P (2017) Building a genetic manipulation tool box for orchid biology: identification of constitutive promoters and application of CRISPR/Cas9 in the orchid, *Dendrobium officinale*. Front Plant Sci 7:2036

LeBlanc C, Zhang F, Mendez J, Lozano Y, Chatpar K, Irish VF, Jacob Y (2018) Increased efficiency of targeted mutagenesis by CRISPR/Cas9 in plants using heat stress. Plant J 93(2):377–386

Li JF, Norville JE, Aach J, McCormack M, Zhang D, Bush J, Church GM, Sheen J (2013) Multiplex and homologous recombination–mediated genome editing in *Arabidopsis* and *Nicotiana benthamiana* using guide RNA and Cas9. Nat Biotechnol 31(8):688–691

Li B, Cui G, Shen G, Zhan Z, Huang L, Chen J, Qi X (2017) Targeted mutagenesis in the medicinal plant Salvia miltiorrhiza. Sci Rep 7(1):43320

Li R, Ba Y, Song Y, Cui J, Zhang X, Zhang D, Yuan Z, Yang L (2020) Rapid and sensitive screening and identification of CRISPR/Cas9 edited rice plants using quantitative real-time PCR coupled with high resolution melting analysis. Food Control 112:107088

Lieber MR (2010) The mechanism of double-strand DNA break repair by the nonhomologous DNA end-joining pathway. Annu Rev Biochem 79:181–211

Liu J, Gunapati S, Mihelich NT, Stec AO, Michno JM, Stupar RM (2019) Genome editing in soybean with CRISPR/Cas9. In: Plant genome editing with CRISPR systems. Humana Press, New York, pp 217–234

Long L, Guo DD, Gao W, Yang WW, Hou LP, Ma XN, Miao YC, Botella JR, Song CP (2018) Optimization of CRISPR/Cas9 genome editing in cotton by improved sgRNA expression. Plant Methods 14(1):85

Lowder L, Malzahn A, Qi Y (2017) Rapid construction of multiplexed CRISPR-Cas9 systems for plant genome editing. In: Plant pattern recognition receptors. Humana Press, New York, pp 291–307

Ma X, Zhang Q, Zhu Q, Liu W, Chen Y, Qiu R, Wang B, Yang Z, Li H, Lin Y, Xie Y (2015) A robust CRISPR/Cas9 system for convenient, high-efficiency multiplex genome editing in monocot and dicot plants. Mol Plant 8(8):1274–1284

Ma X, Zhu Q, Chen Y, Liu YG (2016) CRISPR/Cas9 platforms for genome editing in plants: developments and applications. Mol Plant 9(7):961–974

Ma X, Zhang X, Liu H, Li Z (2020) Highly efficient DNA-free plant genome editing using virally delivered CRISPR–Cas9. Nat Plants 6(7):773–779

Mao Y, Zhang H, Xu N, Zhang B, Gou F, Zhu JK (2013) Application of the CRISPR–Cas system for efficient genome engineering in plants. Mol Plant 6(6):2008–2011

Moradpour M, Abdulah SN (2020) CRISPR/dC as9 platforms in plants: strategies and applications beyond genome editing. Plant Biotechnol J 18(1):32–44

Negritto MC (2010) Repairing double-strand DNA breaks. Nat Educ 3:26

Nekrasov V, Staskawicz B, Weigel D, Jones JD, Kamoun S (2013) Targeted mutagenesis in the model plant *Nicotiana benthamiana* using Cas9 RNA-guided endonuclease. Nat Biotechnol 31(8):691–693

Niazian M (2019) Application of genetics and biotechnology for improving medicinal plants. Planta 249(4):953–973

Nishimasu H, Ran FA, Hsu PD, Konermann S, Shehata SI, Dohmae N, Ishitani R, Zhang F, Nureki O (2014) Crystal structure of Cas9 in complex with guide RNA and target DNA. Cell 156(5):935–949

Osakabe Y, Osakabe K (2015) Genome editing with engineered nucleases in plants. Plant Cell Physiol 56(3):389–400

Osakabe K, Saika H, Okuzaki A, Toki S (2012) Site-directed mutagenesis in higher plants. In: Shu QY, Forster BP, Nakagawa H (eds) Plant mutation breeding and biotechnology. CAB eBooks, Wallingford, pp 523–533

Pannunzio NR, Watanabe G, Lieber MR (2018) Nonhomologous DNA end-joining for repair of DNA double-strand breaks. J Biol Chem 293(27):10512–10523

Piatek A, Ali Z, Baazim H, Li L, Abulfaraj A, Al-Shareef S, Aouida M, Mahfouz MM (2015) RNA-guided transcriptional regulation in planta via synthetic dC as9-based transcription factors. Plant Biotechnol J 13(4):578–589

Polstein LR, Gersbach CA (2015) A light-inducible CRISPR-Cas9 system for control of endogenous gene activation. Nat Chem Biol 11(3):198–200

Ren C, Liu Y, Guo Y, Duan W, Fan P, Li S, Liang Z (2021) Optimizing the CRISPR/Cas9 system for genome editing in grape by using grape promoters. Hortic Res 8(1):52

Shan Q, Wang Y, Li J, Zhang Y, Chen K, Liang Z, Zhang K, Liu J, Xi JJ, Qiu JL, Gao C (2013) Targeted genome modification of crop plants using a CRISPR-Cas system. Nat Biotechnol 31(8):686–688

Sternberg SH, Redding S, Jinek M, Greene EC, Doudna JA (2014) DNA interrogation by the CRISPR RNA-guided endonuclease Cas9. Nature 507(7490):62–67

Terns MP, Terns RM (2011) CRISPR-based adaptive immune systems. Curr Opin Microbiol 14(3):321–327

Upadhyay J, Upadhyay G, Joshi R, Juyal V (2014) Effect of rhododendron flower juice on the bioavailability of amlodipine in rats. Int J Bioassays 3(2):1734–1737

Vats S, Kumawat S, Kumar V, Patil GB, Joshi T, Sonah H, Sharma TR, Deshmukh R (2019) Genome editing in plants: exploration of technological advancements and challenges. Cell 8(11):1386

Wang M, Mao Y, Lu Y, Wang Z, Tao X, Zhu JK (2018) Multiplex gene editing in rice with simplified CRISPR-Cpf1 and CRISPR-Cas9 systems. J Intgr Plant Biol 60(8):626–631

Wani SH, Tripathi P, Zaid A, Challa GS, Kumar A, Kumar V, Upadhyay J, Joshi R, Bhatt M (2018)
Transcriptional regulation of osmotic stress tolerance in wheat (*Triticum aestivum* L.). Plant
Mol Biol 97(6):469–487

Xie X, Ma X, Liu YG (2019) Decoding sanger sequencing chromatograms from CRISPR-induced
mutations. In: Plant genome editing with CRISPR systems. Humana Press, New York, pp 33–43

Yang L, Ding G, Lin H, Cheng H, Kong Y, Wei Y, Fang X, Liu R, Wang L, Chen X, Yang C (2013)
Transcriptome analysis of medicinal plant Salvia miltiorrhiza and identification of genes
related to tanshinone biosynthesis. PLoS One 8(11):e80464

Zhang H, Zhang J, Wei P, Zhang B, Gou F, Feng Z, Mao Y, Yang L, Zhang H, Xu N, Zhu JK (2014)
The CRISPR/C as9 system produces specific and homozygous targeted gene editing in rice in
one generation. Plant Biotechnol J 12(6):797–807

Zhou Z, Tan H, Li Q, Chen J, Gao S, Wang Y, Chen W, Zhang L (2018) CRISPR/Cas9-mediated
efficient targeted mutagenesis of RAS in *Salvia miltiorrhiza*. Phytochemistry 148:63–70

Zischewski J, Fischer R, Bortesi L (2017) Detection of on-target and off-target mutations generated by CRISPR/Cas9 and other sequence-specific nucleases. Biotechnol Adv 35(1):95–104

Genome Editing: A Review
of the Challenges and Approaches

**Dimple Sharma, Harmanpreet Kaur, Harsimran Kaur Kapoor,
Rajat Sharma, Harpreet Kaur, and Mohd Kyum**

Abstract Genome editing is a recent technological advancement in life sciences
that is being used to create novel genetic changes in the genome across different
species, including plants, bacteria, and animals. Site-directed nucleases were earlier
used for genome editing, and nowadays, CRISPR/Cas (clustered regularly interspaced short palindromic repeats)-based genome editing technology is popular
among scientists due to its simplicity, flexibility, and ease of access. In this review,
mechanisms such as repairing double-stranded breaks through non-homologous
end joining and homologous recombination as well as history of genome editing
and different genome-editing approaches, including ZFNs, TALENs, meganucleases, base editing, prime editing, and CRISPR/Cas, are discussed. CRISPR/Cas
have been successfully used for treating human diseases and crop improvement. But
despite numerous advantages of using CRISR/Cas as a tool of gene modification, it
is also facing major hurdles. The review highlights complex designing, inefficient

---

Authors Dimple Sharma, Harmanpreet Kaur, Harsimran Kaur Kapoor, Rajat Sharma, Harpreet
Kaur, and Mohd Kyum have been equally contributed to this chapter.

D. Sharma
Department of Food Science and Human Nutrition, Michigan State University,
East Lansing, MI, USA

H. Kaur
Department of Plant and Environmental Sciences, New Mexico, State University,
Las Cruces, NM, USA

H. K. Kapoor
Department of Plant and Soil Sciences, Texas Tech University, Lubbock, TX, USA

R. Sharma (✉)
Department of Biology, University of Texas Rio Grande Valley, Edinburg, TX, USA
e-mail: rajat.sharma01@utrgv.edu

H. Kaur
Department of Agriculture, Guru Kashi University, Bathinda, Punjab, India

M. Kyum
Department of Plant Breeding and Genetics, Punjab Agricultural University,
Ludhiana, Punjab, India

S. H. Wani, G. Hensel (eds.), *Genome Editing*,
https://doi.org/10.1007/978-3-031-08072-2_4

71

delivery systems, selection of target sites, design of guide RNA (gRNA), occurrence of off-targets, weak efficiency of repair of eukaryotes, endonuclease activity, and cytotoxicity of Cas9 as the major challenges. The information provided in this review will facilitate in understanding genome editing and approaches that it includes along with their technological advancements and challenges.

Keywords Genome editing · CRISPR/Cas9 · Base editing · Prime editing · gRNA

1 Introduction

Plant breeders have tried to improve crop varieties in order to satisfy the hunger of exponentially growing human population. Furthermore, steady changes in climatic conditions and reduction in natural resources gave birth to new problems which limited the scientists to achieve desirable outcomes within time. Traditional breeding methods utilize already available genetic variation in natural population to produce a new variety which takes around 8–10 years, lacking behind in the race to feed the ever-growing population. In addition, it leads to degradation of genomic diversity, ultimately resulting in the generation of vulnerable genetic stock (Haroon et al. 2020). Genome-editing (GE) technology creates fundamental insight into biology of crop plants ultimately revolutionizing the agriculture sector at commercial scale (Chen et al. 2019). The GE techniques including the custom-based site-specific nucleases (SSNs), such as meganucleases, zinc finger nucleases (ZFNs), and transcription activator like effector nucleases (TALENs), come under the traditional techniques. However, GE came into the limelight after CRISPR/Cas9 was developed and was considered as a modern technique (Chen et al. 2020). CRISPR/Cas9 technique is mostly utilized in plant breeding programs for the sake of its high efficiency, easy to perform, and high flexibility in comparison to conventional GE techniques. Plants generated through these techniques are almost similar to their wild types except the corresponding trait allowing them to be separate from the genetically modified organisms (GMOs) legislation (Ran et al. 2017). The GE techniques are now known as new breeding techniques (NBT) which influenced the academic institutions, legislation authorities, and government bodies to rewrite the regulation document. Together with conventional plant breeding methods, NBT have shown immense potential in future of trait improvement of elite cultivars. SSNs directed generation of double-stranded breaks (DSBs) are simultaneously repaired by two natural mechanisms either non-homologous end joining (NHEJ) or homologous recombination (HR), resulting into a loss of function or replacement of gene, respectively (Yin et al. 2017). Using the idea of natural mechanism, plant breeders are exploiting NHEJ pathway for the production of knock-out mutants and synthetically deriving homology-directed repair (HDR) pathway for development of knock-in mutants. However, GE offers great opportunities to basic and applied research areas, but on the other hand it is also arousing many dynamic challenges. Independent responses of a particular plant species or cultivar against in vitro processes, transformations, and survival rate are its major drawbacks. Furthermore, off-targets and unintended modification due to integration of cassettes in plant genomic background affect the productivity of GE techniques (Ellison et al. 2020). Since the beginning of CRISPR in plant GE, there has been tremendous improvement in this technique with the introduction of novel tools. These include DNA free editing, base editing, prime editing, epigenome editing, CRISPRa (gene activation by CRISPR), CRISPRi (gene induction by CRISPR), etc. (Zhang et al. 2019, 2020). The future of plant science is looking promising, although substitution of every novel technique needs to be simultaneously addressed to avoid any delay for betterment of agricultural sciences. This chapter highlights the current scenario of GE techniques along with their challenges and new approaches with future perspective.

2   Mechanisms of Repairing Double-Stranded Breaks

2.1   Non-Homologous End Joining (NHEJ)

NHEJ is a type of DSB repair which is not a result of immense homology. There are two types of NHEJ – classical NHEJ and alternative NHEJ. The former one needs a lot of factors such as Ligase 4, KU70/80, XRCC4, etc. (Burma et al. 2006), whereas the latter one will lead to the least of the DSB repair that is required, and it doesn't need any of the mentioned factors that are needed in classical NHEJ. Alternative NHEJ usually results in minimum homology and a deletion at the repair junction. It is not clear up to what extent is alternative-NHEJ different from homology directed repair (Guirouilh-Barbat et al. 2004). NHEJ can lead to mutations and the error rates can be as high as 50% (Paris et al. 2015).

The DNA repair mechanisms either NHEJ or HDR play a role in genome editing. In bacteria, the DSB can be repaired by either HDR or NHEJ. In eukaryotes, breaks by CRISPR/Cas can be most effectively repaired by NHEJ, which leads to indel mutations (Bernheim et al. 2017). For example, different pathways can affect how CRISPR/Cas will perform. These pathways lead to the regulation of the DSB, if it will be available or if it will compete with the CRISPR/Cas machinery for DNA substrate. Also, as the DNA substrate becomes available for the CRISPR/Cas mechanism, it might inhibit the DNA repair mechanism pathways to work. It has been observed that after a DSB, spacers from CRISPR/Cas have been obtained from a DNA repair mechanism called RecBCD pathway (Levy et al. 2015).

2.2 Genome editing and Homologous Recombination

Homologous recombination refers to the exchange of identical DNA sequences. This mechanism makes sure that the precise replacement and joining of DNA molecules happen; however, the exchange might not be possible if there is less homology. This process is very helpful when certain mutations are required to be brought into the organism's system or when certain mutations are needed out of the same system. Different kinds of mutations can be introduced into the DNA with the help of certain nucleases called SSNs (sequence-specific nucleases). As the name suggests, SSNs are very specific in cutting the double-stranded DNA at a particular targeted sequence. The natural DSB repair mechanisms of the host come into play afterwards and have been studied in yeast and bacteria (Doudna and Charpentier 2014). This indispensable mechanism has a lot of applications in the biological systems. This is an efficient and simple method for gene deletion as a minimal level of gene homology would also lead to targeting a specific gene. Gene targeting is very successful in mouse model system. Mammalian cells were not targeted that often but as such techniques improved, it leads to manipulating the non-selectable genes more frequently and with much higher efficiency (Müller et al. 1999; Sedivy and Dutriaux 1999). Once damage is done to the DNA, the DSB can be repaired by SDSA (synthesis dependent strand annealing) pathway, or by the formation of a DSBR (double-stranded break repair) which follows either a non-crossover or a crossover approach, or SDSA, which follows only the non-crossover approach (San Filippo et al. 2008).

2.2.1   History of Genome Editing

Gene targeting was first done in animal cells, which were earlier considered hard to work with. But shortly after the creation of first knockouts in animal cells, it was seen that the same can be done in plant cells as well. *Nicotiana* spp. was used as a transformation system using PEG (polyethylene glycol)-mediated transformation, which had led to low transformation efficiency and is also time consuming (Paszkowski et al. 1988); however, *Agrobacterium*-mediated transformation has been later proved to be more efficient (Offringa et al. 1990). *Agrobacterium*-mediated transformation method is more efficient in terms of transformation efficiencies but less efficient in terms of targeting efficiencies, and it is also considered to be less labor intensive. With *Agrobacterium*-mediated transformation, transformation efficiency increased and the targeting efficiency decreased (Offringa et al. 1990). False positives can be a problem. Some gene targeting products that were thought to be positive proved to be random integrations. It may have happened because of the cell's repair mechanism that resulted in the integration of a random sequence in place of the target sequence. *Agrobacterium*-mediated transformation has also shown low transformation efficiencies as PEG (Hrouda and Paszkowski 1994). Lower organisms such as *Chlamydomonas* also show low transformation efficiencies as higher organisms such as tobacco or *Arabidopsis* (Smart and Selman 1991; Sodeinde and Kindle 1993; Gumpel et al. 1994).

Most of the time, the tissues used were from the mesophyll protoplasts of leaf from *Arabidopsis* or tobacco, and sometimes the root tissue from *Arabidopsis* was also used (Miao and Lam 1995). Vacuum infiltration is another method that made use of inflorescence of *Arabidopsis* (Bechtold 1993). It was hypothesized that the positive-negative selection as well as the endogenous genes might also have caused the efficiency to be low. By using a negative selection system, a very high efficiency has been seen in rice (Terada et al. 2002).

2.2.2   Homologous Recombination in *E. coli*

It is common knowledge that prokaryotic systems are easier to understand as compared to eukaryotes (Roca and Cox 1997). In eukaryotes, as well as prokaryotes, there are different kinds of enzymes involved in HR. DSBs are identified and repaired by the RecBCD pathway (Kowalczykowski et al. 1994). The heterotrimer of RecB, RecC, and RecD proteins recognizes the break and uses its exonuclease and helicase activity, which also requires Mg^{2+} ions. Recombination hot spots are created by Chi(χ)-site sequences, with which the said heterotrimer complex interacts and the enzyme degrades the 3′ terminal strand, followed by 5′-terminal degradation. Single-stranded DNA on the 3′ terminal is produced by RecBCD complex, followed by RecQ helicase providing a substrate for RecA to produce a nucleoprotein filament by coating the tail of 3′ssDNA (Bianco and Kowalczykowski 1997; Arnold and Kowalczykowski 2000). RecBCD displaces single-stranded binding proteins from the ssDNA, which is stabilized with the binding of RecBCD (Meyer and Laine 1990).

Branch migration is promoted by proteins such as RuvA, RuvB, and RecG. RuvA identifies the holliday junction, RuvB is important for the migration of branch, and RecG is also important for branch migration, but at a smaller scale (West 1996; Whitby and Lloyd 1998). Mutants of RecA proteins show no recombination events (Cox 1999).

2.2.3   HR in *Saccharomyces cerevisiae*

HR is more important when it comes to *S. cerevisiae* as compared to NHEJ. The cells that were competent for HR didn't show any DSB. NHEJ activity could only be seen when the HR system is disabled, acting as a backup system (Siede et al. 1996). *RAD* genes such as *RAD54, RAD55, RAD52, RAD50, RAD50, RAD 57, RAD55*, and *MRE11, XRS2* are involved in HR. The function of these genes is written in Table 1. The mutants of these genes are not affected by infrared but by ultraviolet light (Petes 1991; Pâques and Haber 1999; Pastink et al. 2001; Symington 2002; van den Bosch et al. 2002). *RAD52* is the most important of all the said genes for HR. Mutant of *rad52* shows IR (Infrared) sensitivity. But when a double mutant

Table 1 List of factors involved in homologous recombination

S.N.	HR factor in *S. cerevisiae*	Gene function
1	Rad50	ATPase activity, DNA binding activity
2	Mre11	ssDNA endonuclease, dsDNA exonuclease, DNA duplex unwinding, and DNA binding
3	Rad51	DNA strand exchange, DNA binding, homologous pairing
4	Rad52	DNA strand exchange, ssDNA annealing, homologous pairing, DNA binding
5	Rdh54/Tid1	DNA binding, ATPase activity
6	Rad55	ATPase, DNA binding, homologous pairing, creation of filamentous structures with Rad51D
7	Rad57	ATPase, homologous pairing, DNA binding, creation of filamentous structures with Rad51C
8	Rad58	ssDNA annealing, DNA binding activity

Source: Dudáš and Chovanec (2004)

is made with *rad52* and one of *rad52*, *rad54*, *rad55*, or *rad57* genes, the phenotype is consistent with each other. Other than *RAD52*, *RAD51* is also needed for some HR events. The mutants of other genes involved in HR in *S. cerevisiae* such as *xrs2*, *mre11*, and *rad50* also show similar phenotypes as described before (Game and Mortimer 1974).

Different genes in *S. cerevisiae* that are involved in HR are connected by networks.

2.2.4 HR in Higher Organisms

Genes involved in homologous recombination in higher organisms are a little different than the ones found in lower organisms. Gene targeting in mice helped scientists generate thousands of mutations, which lead to loss of function of a protein (Smithies 1987). One integration event in a hundred could be seen in embryonic stem cells (Jasin et al. 1996). The same has not been seen in plants, where the transformation frequencies are much lower (Paszkowski et al. 1988). Generally, transformation methods such as using polyethylene glycol, electroporation, and *Agrobacterium*-mediated transformation are employed; however, the gene transformation efficiencies are always as low as 1 in 10,000 or 1 in 100,000 (Zupan et al. 2000; Potrykus and Spangenberg 2013). Up to 22 kb DNA has been transferred (Thykjær et al. 1997). Recombination can be increased in plants by induction of DSBs (Pâques and Haber 1999).

Transformation mediated by *Agrobacterium* in *Arabidopsis* to target *TGA* (TGA1A-related gene 3) locus has been used. The number of calluses that were used was 2580, whereas only one of them showed the targeted *TGA* locus (Miao and Lam 1995). In *Arabidopsis*, MADS-box gene *AGL5* (Agamous-like 5) was knocked out. Out of 750 events, one showed to have the said gene actually targeted, when vacuum infiltration was used (Kempin et al. 1997). Out of the two models of recombination double-stranded break repair (DSBR) and SDSA, it is observed that chromosomal rearrangements, specifically translocations can occur, according to the DSBR model, whereas translocations are completely avoided according to the SDSA model (Gorbunova and Levy 1999; Puchta 1999).

Gene targeting has also been done in moss, *Physcomitrella patens*, where DNA was very efficiently integrated into the organism using homologous recombination (Schaefer 2001). The reason for more efficient gene transfer in moss compared to plants can be given to the fact that gene transfer occurs at a particular stage in the life cycle of a moss, which is the G_2/M phase (Reski 1999).

3 Different Genome Editing Techniques

3.1 Meganucleases (MNs)

Meganucleases or homing endonucleases are the type of endonucleases that cleaves DNA at a larger recognition site of around 14–40 bp (Iqbal et al. 2020). They are naturally occurring restriction enzymes that are found in prokaryotic and unicellular eukaryotic organisms (Carroll 2017). The recognition site of MNs is bigger than normal type II restriction enzymes and can alter the target sequence in a highly efficient manner. MNs are encoded by mobile genetic elements and are composed of both DNA binding and DNA cleavage domains. The double-stranded breaks formed by MNs are repaired by NHEJ or HDR process (Silva et al. 2011).

Based on their structural and sequence motifs, MNs have been characterized into five families: HNH, His-Cys box, GIG-YIG, PD-(D/E) XK, and LAGLIDADG (Zhao et al. 2007). Among all these families, the LAGLIDADG family is well characterized and is highly used for genome modification purposes. *I-SceI* (*Saccharomyces cerevisiae*), *I-CreI* (*Chlamydomonas reinhardtii*), and *I-DmoI* (*Desulfurococcus mobilis*) are widely used meganucleases of the LAGLIDADG family (Khandagale and Nadaf 2016). They can withstand site-specific polymorphism without loss of binding and cleavage activity. This technology has been successful with *I-SceI*-mediated transformation in prokaryotes and eukaryotes, but the structure of *I-SceI* is quite complex which makes it difficult for re-engineering to target genes of interest (Zaman et al. 2019). The structure of I-*CreI* is less complex and has been widely used to knockout genes in several organisms (Arnould et al. 2007). For example, The *Cre*-I-based meganuclease was used to target two maize loci, namely, *liguless* 1 and *ms26,* which upon treatment induced the insertion and deletion mutations at target loci (Gao et al. 2010; Djukanovic et al. 2013). Furthermore, it has been shown that the DSB repair by NHEJ led to gene knockout in *Arabidopsis* and tobacco (Kirik et al. 2000). Thus, overall meganucleases are easy to use and can be used to edit the genomes of plants and animals. They also possess a small size (40 kD) which makes them compatible with viral vectors with shorter coding sequences (Iqbal et al. 2020). Despite these advantages, they have not been commonly used in genome engineering as other genome-editing tools due to certain limitations. The first limitation is that the DNA binding domain and catalytic domain are overlapping. To edit the target gene, one has to engineer the DNA recognition sites of MNs but as both domains overlap each other, it is really hard to re-engineer the MNs compared to other genome-editing tools (Khandagale and Nadaf 2016; Iqbal et al. 2020). Second, meganucleases are prone to sequence degeneracy which can highly result in off-target binding and cleavage (Argast et al. 1998). So, in order to overcome the limitations of MNs, researchers were focusing on other simple and efficient methods of gene editing which gave rise to ZFN, TALENs, and CRISPR.

3.2 Zinc Finger Nucleases (ZFNs)

ZFNs are the proteins, designed to cut the DNA at specific sites known as DSBs, which subsequently leads to induction of HR or NHEJ. These repair mechanisms result in deletions, insertions, and base mutations at the site of cleavage (Carroll 2011). Hence, this technology has been employed for the editing of plant and mammalian genomes. ZFNs have different DNA-binding and DNA-cleavage domains (Li et al. 1992).

Fok1 is a type IIS restriction enzyme (Kim and Chandrasegaran 1994). It consists of N-terminal DNA-binding domain and non-specific DNA cleavage domain at the C-terminal end. The cleavage domain has no sequence specificity and hence, can be redirected by substituting with the alternative recognition domains and the most useful for these were Cys_2His_2 zinc fingers (Kim and Chandrasegaran 1994). Various sequences can be attacked by using novel assemblies of ZFNs. When both sets of ZFNs bind to their recognition sequences on the DNA, dimerization and cleavage is achieved. Short linkers of 5–6 bp (base pair) are generally used between the domains of the protein and binding sites (Bibikova et al. 2001; Händel et al. 2009; Shimizu et al. 2009).

Kim et al. (1996) created first ZFNs as chimeric restriction endonucleases, and the first success was achieved using a ZFN pair that targeted the genome of *Drosophila*. However, frequency of target modification varies, and ZFN pairs have been successfully used in a wide range of organisms and cell types (Carroll 2011). The success of this technique of genome editing depends on the delivery method used to deliver ZFNs into the host cell. Earlier experiments were dependent on the genomic integration of ZFN-coding sequences and donor DNA via P-element-mediated transformation, (Bibikova et al. 2002, 2003; Beumer et al. 2006) which used to require elaborate and complex construction of delivery system of the ZFNs. A major breakthrough in this technology occurred when it was demonstrated that both DSB repair mechanisms could be obtained through injecting ZFN mRNAs and donor DNA into the host embryo (Beumer et al. 2006). This method is well-established in zebrafish, rat (Geurts et al. 2009; Mashimo et al. 2010), frog

(Young et al. 2011) as well as sea urchin (Ochiai et al. 2010), and ZFN-induced mutagenesis had been achieved in a number of genes (Doyon et al. 2008; Meng et al. 2008; Foley et al. 2009). In the higher organisms such as plants, including *Arabidopsis thaliana* and several crop species, *Agrobacterium*-mediated transformation had been used through the delivery of coding sequences which are under the control of viral promoter (Lloyd et al. 2005; Cai et al. 2009; De Pater et al. 2009; Osakabe et al. 2010; Zhang et al. 2010). In addition to this, direct transfer of DNA (Wright et al. 2005; Cai et al. 2009; Shukla et al. 2009; Townsend et al. 2009) and viral delivery has been a success in the plants as well (Ira et al. 2010).

The design of ZFNs to target genetic modifications is smooth; however, substantial proportion of ZFN pairs fail (Ramirez et al. 2008; Joung et al. 2010; Carroll 2011). This is the reason that scientists at Sigma-Aldrich and Sangamo Biosciences always practice making multiple pairs for sequences within a single target gene to do extensive testing. There are various methods to select three sets of ZFNs from partially randomized libraries, which can be quite time consuming (Meng et al. 2007). ZFNs for some DNA triplets derived by ToolGen explain the individual finger in their collection that behaves best in modular assembly (Kim et al. 2011).

The modular structure of ZF motifs and recognition by ZF domains allows designing of the artificial DNA binding domains to facilitate genetic modification at the specific target site in the genome (Pabo et al. 2001; Beerli and Barbas 2002). The ZF motifs binds the DNA through insertion of its α-helix into the major groove of helical structure of DNA (Pavletich and Pabo 1991). Key amino acids are present at -1, $+1$, $+2$, $+3$, $+4$, $+5$, and $+6$ positions (relative to the start of α -helix) of ZF motif (Pavletich and Pabo 1991; Shi and Berg 1995; Elrod-Erickson and Pabo 1999). In order to bind long sequences of DNA, several ZF motifs are linked in a tandem fashion which forms zinc finger proteins (ZFPs) (Kim et al. 1996; Liu et al. 1997; Beerli et al. 1998). Zinc finger activators (ZFAs), zinc finger transcription repressors (ZFRs), and zinc finger methylases (ZFMs) make the whole ZFP platform (Xu and Bestor 1997; Bartsevich and Juliano 2000; Zhang et al. 2000; Liu et al. 2001; McNamara et al. 2002; Rebar et al. 2002; Ren et al. 2002; Bartsevich et al. 2003; Snowden et al. 2003; Dai et al. 2004; Rebar 2004).

ZF recognition depends upon the match to the target DNA sequence as well as mechanisms being employed for DSB repair. This ability of ZFNs to modify specific sequences of genes in order to create variants with loss-in-function is a powerful tool for investigating the function of genes as well as development of new products (Osakabe et al. 2010). Gene knockouts have been prepared by the scientists in zebrafish. Similarly, mutagenesis and gene replacement had been achieved in mice (Carbery et al. 2010). Alterations in the genomic loci had been also achieved in the crop plants. Tobacco (Townsend et al. 2009) and maize (Shukla et al. 2009) can be regrown again from the callus, modified in culture by ZFNs. ZFN knockout of *CCR5* gene is one of the therapeutic application in humans, which provide resistance to HIV with an improved immune system (Urnov et al. 2005). Similarly, efforts are being made to knock out the targeted genes in order to treat the neurodegenerative diseases in humans such as Huntington's, Parkinson's, Schizophrenia, and amyotrophic lateral necrosis (Swarthout et al. 2011).

This technology is broadly applicable and versatile in the genetic engineering of plants. Before the emergence of this technology, targeted gene modification was difficult in plants, engineering of plant traits had always been laborious, time consuming, and unpredictable (Puchta 2002). Zinc finger consortium (ZFC) had been established to ensure the development of ZFNs technology through creation of software, resources, and other required tools for engineering zinc fingers for performing genome editing (Wright et al. 2005; Maeder et al. 2008). For instance, using this publicly available consortium, ZFNs were engineered to recognize *SuR* loci to achieve high-frequency modification of plant genes (Townsend et al. 2009). Jeffrey A. Townsend et al. (2009) used ZFN to target acetolactate synthase genes (*ALS, SuRA, SuRB*) in tobacco, which resulted in herbicide-resistance mutations. ZFC had developed a method, called oligomerized pool engineering (OPEN), that uses genetic selections in bacteria for identifying variants of zinc finger arrays (ZFAs) that recognize specific target sequences in the genome (Maeder et al. 2008). These ZFAs function as ZFNs. Similarly, other researchers also used ZFNs to modify endogenous loci in plants, for instance, Shukla et al. (2009), described the use of this technology for genome editing in the crop species of *Zea mays*. Furthermore, ZFNs is a powerful tool for genome modification of animals as well (Rémy et al. 2010). Initially, genetic manipulations of embryonic cells were done by cloning through nuclear transfer which was limited to only some species and modification at specific loci started with emergence of ZFNs. It has been used to modify Drosophila, zebra fish, and rats (Beumer et al. 2008; Ekker 2008; Geurts et al. 2009). Mammalian cells, including the human genome had been modified permanently via HR of targeted DSB (Durai et al. 2005).

Scientists have also faced several challenges with delivery of the targeting materials. High level of somatic mutagenesis in the targets of genomes as well as extra chromosomal arrays was achieved using heat shock promoter for driving ZFN expression from a DNA template in *Caenorhabditis elegans*. Due to RNA interference, parallel expression in the germline was undetectable (Morton et al. 2006). Apart from the use of ZFN for mutagenesis, there have been other studies reporting gene replacement using ZFNs. Some genomic regions and sequences within a single gene are sometimes inaccessible due to compact chromatin structure or modifications in the DNA. For instance, chromatin structure prevents cleavage of intact recognition sites during mating-type switching in *Saccharomyces cerevisiae* (Rusche et al. 2003). DSB repair mechanisms differ with cell types and their developmental stages; hence, understanding of the biological system of every organism is essential to overcome these limitations. Another challenge is specificity of ZF binding, as some bind equally well to triplets other than their supposed preference. The addition of fingers can improve both, specificity and affinity; however, it might lead to binding at off-target sites. The separation of two-finger modules with a short linker was shown to improve specificity (Moore et al. 2001). Death of the host cells is the ultimate result due to off-target cleavage, as the number of breaks outstrips the DSB repair capacity of the DNA (Bibikova et al. 2002; Porteus and Baltimore 2003; Alwin et al. 2005).

3.3   Transcription Activator-Like Effector Nucleases (TALENs)

TALENs are the restriction enzymes which can be engineered to cut the DNA at specific sequences; hence, they are being used as site specific nucleases for targeted genome editing. TALENs are the fusions between non-specific DNA cleavage domain and a custom-designed DNA binding domain (Miller et al. 2011; Wood et al. 2011). DSBs are induced at the desired site on the DNA, which can be repaired by HDR or NHEJ to create small insertions or deletions at the cleavage sites. This technology emerged as an alternative to one of the similar genome-editing technologies, i.e., ZFNs. It contains DNA-binding domains which contain highly conserved repeats, derived from TALENs. These are proteins secreted by *Xanthomonas* spp. (Boch and Bonas 2010). Both ZFNs and TALENs can cleave the DNA at similar efficiency (Hockemeyer et al. 2011; Tesson et al. 2011; Reyon et al. 2012). The difference between them is that, TALENs more site specific with lesser off-target effects as compared to ZFNs (Chandrasegaran and Carroll 2016).

TALENs can be easily and rapidly designed using "protein-DNA code," relating DNA-binding TALE repeat domains to the target-binding site. One TAL effector repeatedly binds to one base pair of DNA (Boch et al. 2009; Moscou and Bogdanove 2009). These TAL effector repeats can also be joined together to develop extended arrays that can recognize new targets in DNA (Boch et al. 2009; Morbitzer et al. 2010; Miller et al. 2011; Weber et al. 2011). Most of the methods that are used to construct TALENs, use golden gate cloning method with some variations (Morbitzer et al. 2010; Cermak et al. 2011; Huang et al. 2011; Li et al. 2011; Sander et al. 2011; Weber et al. 2011). However, none of them are adaptable for automated high-throughput production (Maeder et al. 2008; Morbitzer et al. 2010; Cermak et al. 2011; Li et al. 2011; Sander et al. 2011; Weber et al. 2011). DNA-binding domains with highly conserved repeats of 33–35 amino acids are transferred into the host cells via Type III secretion system of the bacteria, *Xanthomonas* spp., hence, facilitating bacterial colonization to alter the transcription of genomic DNA of the host cells. The two hypervariable residues identify the site at the DNA where the TALE repeats bind. Hypervariable residues, *viz.*, NN, NI, HD, and NG, in nearly all engineered TALE repeats recognize guanine, adenine, cytosine, and thymine respectively (Joung and Sander 2013).

Fast ligation-based automatable solid-phase high throughput (FLASH) is a rapid and cost-effective technology for assembly of TALENs at large scale (Reyon et al. 2012). This technology had been used previously to construct 48 TALEN pairs which were targeted to diverse range of gene sequences in human beings and 100% of nucleases were active in human cells, similarly FLASH TALEN pairs had been targeted to 96 genes involved in epigenetic regulation in humans, out of which targeted alterations arose in 84 genes (Reyon et al. 2012).

Genes in a wide range of cell types or organisms can be engineered using this technology; thus, this technology has significant effects on biological research and has a potential to treat genetic diseases as well as has applications in crop improvement (Joung and Sander 2013). For reflecting its wide importance, it was named the

2011 "Method of the Year" by the journal "Nature Methods" (Baker 2011). This technique had been employed in a variety of organisms, including yeast (Li et al. 2011), zebrafish (Sander et al. 2011), frog (Lei et al. 2012), roundworm (Wood et al. 2011), rat (Tesson et al. 2011), cow, pig (Carlson et al. 2012), rice (Li et al. 2012), thale cress (Cermak et al. 2011), silkworm (Ma et al. 2012), cricket (Watanabe et al. 2012), fruit fly (Liu et al. 2012), and somatic and pluripotent stem cells of human beings (Cermak et al. 2011; Hockemeyer et al. 2011; Miller et al. 2011; Reyon et al. 2012).

Construction of TALE repeat arrays can be challenging due to the need for assembling multiple, identical repeat sequences. Different platforms have been designed, including "Golden Gate" cloning; solid phase assemble; standard restriction enzyme; and ligation-based cloning to facilitate the assembly of plasmids that encode TALE repeat arrays (Joung and Sander 2013). Usually, TALENs are built to bind 18-bp sequences or even longer than that; however, recent studies have suggested that use of larger TALENs may result in less specificity (Guilinger et al. 2014). Off-target effects are also one of the major concerns regarding TALENs, as, in one of the studies where this technology was used in human pluripotent stem cells, mutagenesis at 19 possible off-target sites was reported (Hockemeyer et al. 2011). The size of cDNA encoding TALEN is approximately 3 kb. This large size is also one of the disadvantages of TALENs, which makes it harder to deliver and express TALENs into the host cells. The ability of the TALENs to get delivered by some of the viral vectors is often impaired due to their highly repetitive nature (Holkers et al. 2013); however, this limitation can be overcome through diversification of the coding sequences of the TALE repeats (Yang et al. 2013).

3.4 Clustered Regularly Interspaced Short Palindromic Repeats/CRISPR Associated Protein

TALENs came as an alternative approach to the less accurate and error prone ZFNs. CRISPR is even easier to be prosecuted and is more efficient compared to TALENs and ZFNs. CRISPR and different types of Cas have been widely adapted as a gene-editing technology, which shows a lot of applications and many promising results. Emmanuelle Charpentier and Jennifer Doudna won the 2020 Noble prize in chemistry for CRISPR discovery (NoblePrize.org). Different types of Cas proteins are: Cas1, Cas2, Cas3, Cas5, Cas6, Cas7, Cas9, Cas10, etc. (Makarova et al. 2011). CRISPR/Cas is part of the immune system of bacteria to provide resistance against viruses. Once the bacteria encounter the same virus again, it attacks it with the memory from before (Horvath and Barrangou 2010). It consists of Cas endonuclease, a variety of which, i.e., Cas9 is derived from *Streptococcus pyogenes* and sgRNA (Heler et al. 2015). sgRNA consists of a tracrRNA and a crRNA (Deltcheva et al. 2011), also known as trans-activating crRNA and CRISPR RNA respectively. The sgRNA has a part that binds to the target sequence of the host, and another part that loops on itself (Cui et al. 2018), the former can be made specific for any target sequence (Cong and Zhang 2015). The target sequence consists of the region to be targeted by sgRNA, and a PAM (protospacer adjacent motif) sequence, which usually is NGG, where N pertains to any nucleotide, and G is guanosine. The target sequence is usually 20 nucleotides. If the PAM site is absent, the Cas9 endonuclease will not be able to recognize and hence cleave the target sequence. If the PAM site is present right before the sgRNA, the Cas9 endonuclease can bind to it and leads to the creation of a DSB exactly the size of the target sequence. Once that sequence is cut, the natural mechanisms of the host cells come into play to repair the DSB, by the mechanisms such as NHEJ and HDR (Wyman and Kanaar 2006). The latter will not, but the former one might lead to the creation of insertions and deletions, which leads to the loss of gene function.

Figure 1 shows different parts of CRISPR. CRISPR cannot only lead to loss of gene function, i.e., gene knockouts (CRISPRko) (Mali et al. 2013) but is also helpful in inducing the gain of function of a gene. CRISPRi (Qi et al. 2013) includes a non- functional Cas9 endonuclease, which doesn't cleave the target sequence. CRISPRa (Gilbert et al. 2014) can be used for the activation of the gene when non-functional Cas9 is attached to a transcriptional activation domain. The sgRNA can be designed in various ways. Along with binding to the target sequence, sgRNA also

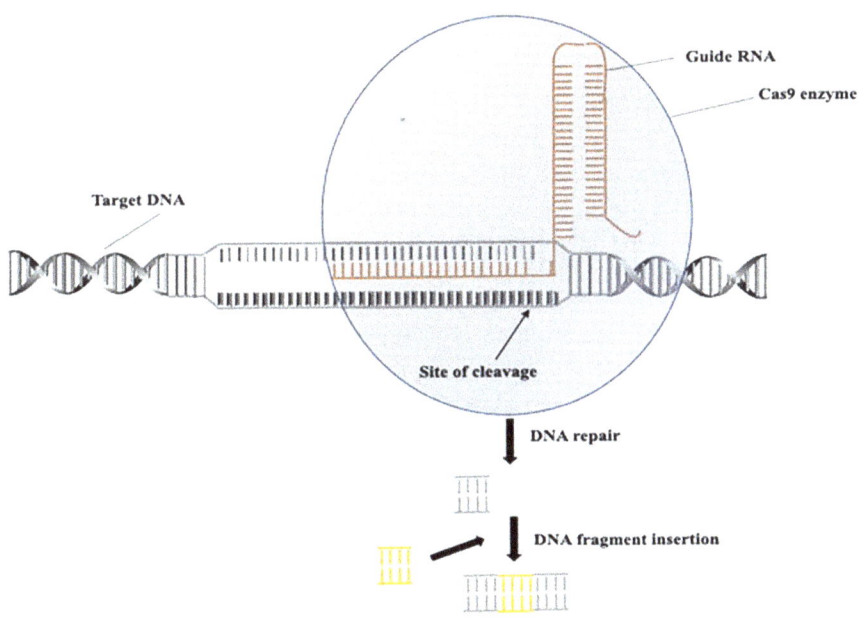

Modified from https://www.labiotech.eu/in-depth/crispr-cas9-drug-discovery/

Fig. 1 Mechanism of CRISPR/Cas. (Modified from https://www.labiotech.eu/in-depth/crispr-cas9-drug-discovery/)

works in recognizing the target sequence in the whole genome. A good sgRNA will have specificity as well as efficacy.

The mutations that are created by NHEJ can be identified by Sanger sequencing, AFLP (amplified fragment length polymorphism), and restriction enzyme assays (Belhaj et al. 2013; Belhaj et al. 2015). Instead of targeting only one gene, multiple genes can be targeted together, which is called target gene multiplexing (Cong et al. 2013; Čermák et al. 2017). This can be used either to target different genes or different sequences in the same gene to increase efficiency, by having multiple sgRNAs either under different promoters or under the same promoter (Čermák et al. 2017). The other versions include Cys4-gRNA or tRNA-gRNA (Xie et al. 2015). The former is an endoribonuclease from *Pseudomonas aeruginosa*. It has been applied in plants such as wheat, *Arabidopsis*, tomato, potato, rice, banana, and tobacco (Upadhyay et al. 2013; Andersson et al. 2018; Kaur et al. 2018; Castel et al. 2019).

4   Challenges in the CRISPR/cas9 System

Despite having several advantages, CRISPR/Cas9 system in gene modification faces major hurdles, lowering the efficiency or complete failure of genome editing.

4.1   Complex Designing

Despite of being highly specific, the conventional techniques are restricted to only research laboratories due to the inefficient understanding of their complex nature. The customized protein engineering varies from species to species and is more difficult in polyploidy genomes. Use of meganucleases is limited because of a rare homing site present in particular genome which requires extensive research to design DNA-binding domains ultimately narrowing their application in plant science (Wright et al. 2014). Additionally, it is also time consuming and costly in comparison to other GE techniques (Aglawe et al. 2018). ZFNs overcome few of the problems of meganucleases and broaden the scope of plant genome editing in various agriculturally important crops. However, the selection-based fabrication of large sized libraries for different traits make it inadequate to be used in diverse laboratories, since it requires high technical expertise (Maeder et al. 2008; Nelson and Gersbach 2016). The most specific technique, i.e., TALENs, requires monomer DNA-binding domains; perhaps, comprehensive knowledge is needed to achieve significant target efficiency. Various methods of TALEs construction have been developed using 20–30 monomers involving ligation, cloning, to generate dimers library, and subsequently, golden gate assembly is used which is very tedious and time-consuming process (Abdallah et al. 2015). Currently, CRISPR/Cas9 technique is a preferred approach, bypassing the meganuclease, ZFNs, and TALENS due to its simple RNA-dependent DNA binding followed by cleavage of target site via single protein (Zhang et al. 2014; Sharma et al. 2017).

4.2   Inefficient Delivery

To modify plant genome, the construct of GE machinery needs to be effectively introduced in plant cell, making it a very crucial step to achieve beneficial outcome. The methods available for delivery of cassettes into plants are based on *Agrobacterium*, viral vector, gene gun mediated, lipofection, etc. (Nelson and Gersbach 2016, Yin et al. 2017, Liu et al. 2020). The target tissues for transformation being utilized are callus, immature embryos, protoplast, shoot apical meristem, etc. (Altpeter et al. 2016, Ran et al. 2017). Delivery of construct is very challenging because of the specific requirement of independent genotypes, explants, and type of in vitro culture conditions (Ran et al. 2017). Additionally, the requirement of transgene free altered plants at commercial scale implies restriction on the most frequently used methods of delivery of GE cassettes, creating a tough task for the researchers. This emerges as a principal obstacle toward the generation of novel traits in plants (Baltes et al. 2014). Although considerable achievements have been obtained for transformation or delivery of GE reagent in various crops, their low regeneration ability and unstable integration are hindering the scientific endeavors (Altpeter et al. 2016).

4.3   Selection of Target Site and gRNA Design

The major advantage of CRISPR machinery is its ability to target ~21–23 base pair (bp) DNA sequence containing a PAM sequence on forward or reverse strand. The PAM motif on average occurs every 8bps which provides higher flexibility in target site selection (Ramakrishna et al. 2014). In addition, Cas9 proteins from several other organisms have been studied and different PAM sites have been identified. This diversity in PAM sites have increased choices for choosing target sequences. However, recent reports have shown that the target site selection and designing of guide RNA is not as simple as it was previously assumed. Due to post-transcriptional modifications of mRNA transcribed by RNA polymerase II, it is not possible to use RNA polymerase II for sgRNA construction (Zhang et al. 2014). Currently, RNA polymerase III along with Ubiquitin3 (U3) and U6 small nuclear RNA (snRNA) promoters are used for guide RNA production. The Ubiquitin genes are housekeeping genes and cannot be used to generate tissue specific or cell specific gRNAs (Gao and Zhao 2014). As the RNA polymerase III is not commercially available, this also limits the production U3- and U6-based guide RNA. Therefore, various approaches should be developed to overcome the application of ubiquitin promoters-based gRNA construction.

4.4   Off-Target Effect

In additional to the rational design of sgRNA, off-target DNA cleavage by Cas9 endonuclease is a major challenge, reducing the efficiency of the machinery. During the process of CRISPR/Cas9 gene editing, the interaction of endonuclease and target strand initiated by the recognition of PAM motif, which denatures DNA upstream of the motif, therefore allowing the binding of target sequence and sgRNA to form R-loop (Ebrahimi and Hashemi 2020). The Cas9 enzyme then cleaves the target DNA having less number of mismatches, ultimately resulting in the off-target cleavage (Herai 2019; Newton et al. 2019). This off-target cleavage came from natural combat between bacteria and virus where the DNA sequence of virus mutates itself to escape from the chopping effect of Cas9 nuclease, but Cas9 in return is capable to bind and chop target virus with minimum number of mismatches (Li et al. 2019). Based on these accepted number of mismatches, various algorithms have been developed to design guide RNA and to check off-target sequences. In general, highly specific sequence with zero mismatch is required for cleavage of DNA at 7–9 PAM proximal bases, and 3–4 mismatches away from PAM site is admissible for attachment of Cas9 but not cleavage (Dagdas et al. 2017; Singh et al. 2017).

The off-target effect is solely not based on the Cas9 endonuclease, but also depends highly on sgRNA design. Therefore, modifications in Cas9 protein and sgRNA can limit these off-target effects respectively. One approach is the modification of sgRNA backbone chemically. The introduction of 2′O-methyl-3′-phosphonoacetate at particular sites of sgRNA or partial substitution of ribonucleotides with deoxyribonucleotides can reduce off-target effects respectively (Ryan et al. 2018). Another approach includes creating mutant Cas9 systems where ZFNs and TALENS serve as an inspiration for improved target precision. This mutant Cas9 system involves the fusion of dcas9 (deactivated) and a Fok1 nuclease dimer. Here, the 20–21bp guide RNAs binds the target strand. After binding of sgRNA, the fok1 nuclease dimers become functional Fok1 and cause double-strand breaks (Guilinger et al. 2014; Tsai et al. 2014). This system significantly reduces off-target cleavage but increases the size of genome-editing tool, thereby reducing the efficiency of transformation in vivo. Another mutant approach involves the fusion of Cas9 protein with ZFPs or TALEs, which can target the genomic loci with better precision (Bolukbasi et al. 2015).

4.5   Weak Repair Efficiency of HDR in Eukaryotes

In eukaryotic cells, the double-strand breaks are repaired by NHEJ mechanism with higher efficiency. This mechanism commonly repairs the DNA without the use of template DNA resulting in indel mutations, which causes the initiation of frameshift resulting in establishment of gene knockout or knockdown (Shalem et al. 2014). On the other hand, the efficiency of HDR-mediated repair from double-strand break is very low in mammalian cells. It has been reported that the HDR repair in mice after cas9-based genome editing is 0.5–20% whereas the repair efficiency through NHEJ is ~20–60% (Maruyama et al. 2015).

Various strategies have been developed to increase the HDR repair process and reduce NHEJ efficacy. The use of tiny molecules known as inhibitors (for NHEJ) or inducers (for HDR) is one of the approach for improving recombination frequency (Aird et al. 2018). The application of SCR-7 (NHEJ inhibitor) or RS-1(HDR enhancer) have been reported to increase the HDR efficiency by several folds (Yu et al. 2015; Vartak and Raghavan 2015; Song et al. 2016). In addition, the other approaches include the gene silencing, use of cell lines deficit in NHEJ machinery (Weinstock and Jasin 2006), cell cycle synchronization, or controlled cas9 expression (Weber et al. 2015). The controlled delivery of CRISPR/Cas9 machinery along with cell cycle has increased the CRISPR/Cas9 mediated DNA repair. The synchronization of Cas9 protein with cell cycle progression can be made by fusing Cas9 protein with human DNA replication inhibitor (geminin), which modifies Cas9 endonuclease post-translationally (Gutschner et al. 2016). While the use of NHEJ inhibitors or HDR inducers have resulted in increased HDR mediated genome editing, they are really toxic to host cells and cause numerous problems for sufficient cell growth. To overcome these problems, another approach has been developed which rely on covalent tethering of repair template to ribonucleoprotein complex. The utilization of this approach led to the increase in HDR repair by ~30 folds, and it was proved that this strategy can be applied in various other organisms and target loci (Aird et al. 2018).

4.6 Cas9 Endonuclease Activity and Cytotoxicity

Several Cas9 proteins from different species have been found and used in genome editing, including *Staphylococcus aureus* (SaCas9) (Ran et al. 2013a, b), *S. thermophiles* (StCas9) (Kleinstiver et al. 2016), and *Streptococcus pyogenes* (SpCas9) (Vento et al. 2019). Among all, SpCas9 is an endonuclease enzyme commonly used for genome editing in prokaryotic bacteria. SpCas9 is a well-characterized endonuclease with simple PAM site and has high expression rate in various prokaryotes. On the other hand, different Cas9 proteins have been used in eukaryotes and the selection of specific Cas9 ortholog for each organism showed improved efficiency of editing for a specific sequence. In addition to the selection, many factors have shown to affect the activity of Cas9 protein. For gene editing to occur in eukaryotes, the Cas9 protein must translocate in nucleus, and the nuclear location signal (NLS) should be associated with Cas9 protein. It has been reported that by reducing the proximity of NLS and Cas9 using 32 amino acid spacer, the DNA cleavage activity increased to a higher extent (Shen et al. 2013). In addition, by increasing the guide RNA:Cas9 ratio was shown to increase on target chopping activity, by ensuring all Cas9 protein form active R loop with the sgRNA and DNA complex (Kim et al. 2014). Unlike other endonucleases, the activity of Cas9 protein is significantly less with a single turnover rate of ~0.4–1.0 per min (Jinek et al. 2012). Also, when bind to the target DNA, the displacement rate of Cas9 protein is quite challenging. It has been reported that even after DSB formation, the 1 nm Cas9 enzyme cleaved ~2 nm plasmid DNA after 2 h (Jinek et al. 2012). This shows Cas9 enzyme works more like actuator rather than a catalytic enzyme.

In addition to enzyme activity, the cytotoxic effect of Cas9 protein can be considered as a crucial obstacle in gene editing (Vento et al. 2019). Various attempts have been made to reduce this toxic effect of Cas9 protein for efficient genome editing using programmable DNA cleavage. The first approach includes the usage of inducible expression system for Cas9, where the activity of Cas9 protein is highly reduced when no inducer is present (Reisch and Prather 2015). Another approach includes the use of toxin-free endonucleases or variant nucleases. The variant of cas9 protein: cas9n has shown reduced toxicity because it targets only single strand of DNA (Standage-Beier et al. 2015).

5 Approaches

5.1 Base Editing

Base editing allows nucleotide substitutions in the genome by using modified Cas effectors without the requirement of producing DSBs (Komor et al. 2016). There are two types of base-editing tools, cytidine base editor (CBE) and adenine base-editor (ABE), that enable cytosine-guanine to thymine-adenine and adenine-thymine to guanine-cytosine transitions (Chen et al. 2019). These CBE and ABE tools make use of cytosine and adenosine deaminases for cytosine and adenine base editing (Komor et al. 2016; Nishida et al. 2016; Gaudelli et al. 2017; Ren et al. 2018; Wang et al. 2018). In the CBE system, nick/dead Cas protein fuses to cytidine deaminase, catalyzing conversion of cytidine (C) to uracil (U). In the ABE system, engineered *Escherichia coli* RNA adenosine deaminase (TadA) fuses to nick/dead Cas protein, leading to conversion of adenine (A) to inosine (I), which is recognized as guanine by DNA polymerase during replication of DNA. Plasmid transfection and viral delivery are the common methods of delivery of the base editors in the living cells (122,124). Hence, they introduce targeted substitutions in the genes, and it had been intensively employed in model plants and crops for the improvement of agricultural traits, including flowering, plant height, disease, and herbicide resistance (Chen et al. 2017; Shimatani et al. 2017; Kang et al. 2018; Tian et al. 2018; Chen et al. 2019; Wu et al. 2020). Other major applications are the study or treatment of disease-associated point mutations, use of base editors as the recorder of cellular events in biomedical research, and introduction of premature stop codons by CBA to disrupt genes in homogenous manner (Landrum et al. 2014, 2016; Farzadfard and Lu 2018). However, in mammalian cells, challenge is to circumvent DNA repair processes that oppose target base pair conversion (Rees and Liu 2018). Human cells undergo effective cellular repair of U.G intermediate through base excision repair of U.G in DNA, in which uracil N-glycosylate (UNG) recognizes U.G mismatch and cleaves glycosidic bond between uracil and backbone of deoxyribose (Kunz et al. 2009). Generation of indels, targeting limitations and off-target editing DNA base pairs, had been also reported (Rees and Liu 2018), and various scientists are doing efforts worldwide to overcome these limitations. Base editing in RNA is also possible that provide powerful capabilities to life sciences and medicine. To date, deamination of A to I has been only reported (Vogel and Stafforst 2019).

5.2 Prime Editing

Prime editors are being employed for precise editing, employs same mechanism as conventional CRISPR/Cas systems but does not require DSBs (Anzalone et al. 2019). It involves longer-than-usual guide-RNA, commonly known as pegRNA. Fusion of Moloney murine leukemia virus (M-MLV) reverse transcriptase (RT) and Cas9 nickase (nCas9) is the major component of prime editor. Complex of nCas/M-MLV/pegRNA mediates site-specific nicking by nCas9, which is then served as a template for RT and then at the end reverse transcriptions leads to production of stable edited DNA (Zhan et al. 2020). Base conversions and small insertions can be achieved through the use of prime editors (Anzalone et al. 2019). There are three prime editors: first is PE1 that is a combination of Cas9 H840A nickase and wild type (WT) M-MLV RT enzyme; second is PE2, improved thermostability, processivity, and DNA-RNA substrate affinity of the RT component; and third is PE3, in which second gRNA was introduced in addition to pegRNA (Anzalone et al. 2019). Prime-editing systems have the capability of performing precise genome editing in human cells, and scientists also tried to create mutations in the genome for treating rare genetic diseases, including SCD, Tay-Sach, and prion diseases in humans (Matsoukas 2020). Recently, Xu et al. (2020) has developed plant prime-editing system, plant prime editor 2 which was tested through targeted mutation on an HPT[ATG] reporter in rice. Its development is an essential addition to the genome-editing technologies and also addresses CRISPR/Cas limitations. However, this technology also has some challenges, it may not be able to create large DNA insertions or deletions as compared to conventional CRISPR/Cas systems, possibility of addition of cDNAs due to presence of RT, and large protein constructs may affect the delivery of full-length therapeutic protein (Matsoukas 2020). Hence, further research is required for optimizing prime editors and maximizing their efficiency in different cell types.

6 Conclusion and Future Prospects

The steady and undesirable variations in climatic conditions, supplemented with depletion of the natural resources and biodiversity, are creating new challenges toward sustainable crop production. The rapid and constantly evolving nature of GE techniques assists plant scientists via numerous applications in crop improvement.

These advances widen the scope of trait refinement in the diverse genetic background, irrespective of their natural mechanism. However, the efficient utilization of GE techniques seeks a deep understanding of the interaction between the genotype and their respective phenotype under the different environmental conditions as most of the agronomically important plant features are controlled through a complex genetic mechanism. Conventional GE techniques initially gained the attention of researchers due to their ability to induce directed DSBs followed by the natural repair mechanism. But, the natural weapon of the bacterial immune system transformed the era of GE and emerged as a principle genome modifying tool. The necessity of short sgRNA and single unit Cas9 nuclease protein makes it the first preference over the conventional methods. CRISPR opens up tremendous opportunities in the living world beyond the DSB mediated SDNs, including the study of regulatory elements, complex genetic mechanisms, cell signaling, chromatin modeling, etc. These developments broadened the way of assessment in various agriculturally important monocot and dicot species as well as the plants having multiple copies of genomes. The adequate information of functional genomics is a prerequisite to target any specific genotype. GE techniques, especially CRISPR, have shown the ability of precise mutation through knock-out, knock-in, and knock-down mutation resulting in loss of function, the gain of function, and specific transient modulation, respectively. The rapid advancement in CRISPR techniques accompanied by dynamic natural variations display enormous opportunities for the betterment of plant sciences. On the other hand, the technical difficulties and various shortcomings such as off-targets, genotype dependency, construct delivery, unintended effects, etc. take time to translate these techniques from basic research to applied studies. The techniques are already standardized in model plants such as Arabidopsis and Rice, but their exploitation in other important plant species needs to be simultaneously addressed. The exponential growth in recent years in the field of GE promises to plant scientists in providing customized and flexible solutions of their beneficial thoughts with regards to the nourishment of agricultural sciences. Conclusively, this handy approach has shown the potential to counter the upcoming challenges of agricultural, environmental, social, and geological issues ultimately hindering the fate of the food crisis in the scenario of climate alterations.

References

Abdallah NA et al (2015) Genome editing for crop improvement: Challenges and opportunities. GM Crops Food 6(4):183–205. https://doi.org/10.1080/21645698.2015.1129937

Aglawe SB et al (2018) New breeding technique "genome editing" for crop improvement: applications, potentials and challenges. 3. Biotech 8(8):1–20. https://doi.org/10.1007/s13205-018-1355-3

Aird EJ, Lovendahl KN, Martin AS, Harris RS, Gordon WR (2018) Increasing Cas9-mediated homology-directed repair efficiency through covalent tethering of DNA repair template. Commun Biol 1:1–6. https://doi.org/10.1038/s42003-018-0054-2

Altpeter F et al (2016) Advancing crop transformation in the era of genome editing. Plant Cell 28(7):1510–1520. https://doi.org/10.1105/tpc.16.00196

Alwin S, Gere MB, Guhl E, Effertz K, Barbas CF III, Segal DJ et al (2005) Custom zinc-finger nucleases for use in human cells. Mol Ther 12:610–617. https://doi.org/10.1016/j. ymthe.2005.06.094

Andersson M, Turesson H, Olsson N, Fält AS, Ohlsson P, Gonzalez MN et al (2018) Genome editing in potato via CRISPR-Cas9 ribonucleoprotein delivery. Physiol Plant 164:378–384. https://doi.org/10.1111/ppl.12731

Anzalone AV et al (2019) Search-and-replace genome editing without double-strand breaks or donor DNA. Nature 576(7785):149–157. https://doi.org/10.1038/s41586-019-1711-4

Argast GM et al (1998) I-PpoI and I-CreI homing site sequence degeneracy determined by random mutagenesis and sequential in vitro enrichment. J Mol Biol 280(3):345–353. https://doi.org/10.1093/nar/gkl645

Arnold DA, Kowalczykowski SC (2000) Facilitated loading of RecA protein is essential to recombination by RecBCD enzyme. J Biol Chem 275:12261–12265. https://doi.org/10.1074/jbc.275.16.12261

Arnould S et al (2007) Engineered I-CreI derivatives cleaving sequences from the human XPC gene can induce highly efficient gene correction in mammalian cells. J Mol Biol 371(1):49–65. https://doi.org/10.1016/j.jmb.2007.04.079

Baker M (2011) Gene-editing nucleases. In: Nature Publishing Group. V8 (12). https://doi.org/10.1101/cshperspect.a023754

Baltes NJ et al (2014) DNA replicons for plant genome engineering. Plant Cell 26(1):151–163. https://doi.org/10.1105/tpc.113.119792

Bartsevich VV, Juliano R (2000) Regulation of the *MDR1* gene by transcriptional repressors selected using peptide combinatorial libraries. Mol Pharmacol 58:1–10. https://doi.org/10.1124/mol.58.1.1

Bartsevich VV, Miller JC, Case CC, Pabo CO (2003) Engineered zinc finger proteins for controlling stem cell fate. Stem Cells 21:632–637. https://doi.org/10.1634/stemcells.21-6-632

Bechtold N (1993) In planta *Agrobacterium*-mediated gene transfer by infiltration of adult *Arabidopsis* plants. CR Acad Sci Ser III Sci Vie 316:1194–1199

Beerli RR, Barbas CF (2002) Engineering polydactyl zinc-finger transcription factors. Nat Biotechnol 20:135–141. https://doi.org/10.1038/nbt0202-135

Beerli RR, Segal DJ, Dreier B, Barbas CF (1998) Toward controlling gene expression at will: specific regulation of the *erbB-2/HER-2* promoter by using polydactyl zinc finger proteins constructed from modular building blocks. Proc Natl Acad Sci 95:14628–14633

Belhaj K, Chaparro-Garcia A, Kamoun S, Nekrasov V (2013) Plant genome editing made easy: targeted mutagenesis in model and crop plants using the CRISPR/Cas system. Plant Methods 9:1–10. https://doi.org/10.1186/1746-4811-9-39

Belhaj K, Chaparro-Garcia A, Kamoun S, Patron NJ, Nekrasov V (2015) Editing plant genomes with CRISPR/Cas9. Curr Opin Biotechnol 32:76–84. https://doi.org/10.1016/j.copbio.2014.11.007

Bernheim A et al (2017) Inhibition of NHEJ repair by type II-A CRISPR-Cas systems in bacteria. Nat Commun 8(1):1–9. https://doi.org/10.1038/s41467-017-02350-1

Beumer K, Bhattacharyya G, Bibikova M, Trautman JK, Carroll D (2006) Efficient gene targeting in *Drosophila* with zinc-finger nucleases. Genetics 172:2391–2403. https://doi.org/10.1016/j.copbio.2014.11.007

Beumer KJ, Trautman JK, Bozas A, Liu J-L, Rutter J, Gall JG, Carroll D (2008) Efficient gene targeting in *Drosophila* by direct embryo injection with zinc-finger nucleases. Proc Natl Acad Sci 105:19821–19826. https://doi.org/10.1073/pnas.0810475105

Bianco PR, Kowalczykowski SC (1997) The recombination hotspot Chi is recognized by the translocating RecBCD enzyme as the single strand of DNA containing the sequence 5′-GCTGGTGG-3′. Proc Natl Acad Sci 94:6706–6711. https://doi.org/10.1073/pnas.94.13.6706

Bibikova M, Carroll D, Segal DJ, Trautman JK, Smith J, Kim Y-G, Chandrasegaran S (2001) Stimulation of homologous recombination through targeted cleavage by chimeric nucleases. Mol Cell Biol 21:289–297. https://doi.org/10.1128/MCB.21.1.289-297.2001

Bibikova M, Golic M, Golic KG, Carroll D (2002) Targeted chromosomal cleavage and mutagenesis in *Drosophila* using zinc-finger nucleases. Genetics 161:1169–1175. https://doi.org/10.1093/genetics/161.3.1169

Bibikova M, Beumer K, Trautman JK, Carroll D (2003) Enhancing gene targeting with designed zinc finger nucleases. Science 300:764–764. https://doi.org/10.1126/science.1079512

Boch J, Bonas U (2010) *Xanthomonas* AvrBs3 family-type III effectors: discovery and function. Annu Rev Phytopathol 48:419–436. https://doi.org/10.1146/annurev-phyto-080508-081936

Boch J, Scholze H, Schornack S, Landgraf A, Hahn S, Kay S et al (2009) Breaking the code of DNA binding specificity of TAL-type III effectors. Science 326:1509–1512. https://doi.org/10.1126/science.1178811

Bolukbasi, M. F., Gupta, A., Oikemus, S., Derr, A. G., Garber, M., Brodsky, M. H., ... & Wolfe, S. A. (2015). DNA-binding-domain fusions enhance the targeting range and precision of Cas9. Nat Methods 12(12):1150–1156.

Burma S et al (2006) Role of non-homologous end joining (NHEJ) in maintaining genomic integrity. DNA Repair 5(9-10):1042–1048. https://doi.org/10.1016/j.dnarep.2006.05.026

Cai CQ, Doyon Y, Ainley WM, Miller JC, DeKelver RC, Moehle EA et al (2009) Targeted transgene integration in plant cells using designed zinc finger nucleases. Plant Mol Biol 69:699–709. https://doi.org/10.1007/s11103-008-9449-7

Carbery ID, Ji D, Harrington A, Brown V, Weinstein EJ, Liaw L, Cui X (2010) Targeted genome modification in mice using zinc-finger nucleases. Genetics 186:451–459. https://doi.org/10.1093/genetics/186.2.NP

Carlson DF, Tan W, Lillico SG, Stverakova D, Proudfoot C, Christian M et al (2012) Efficient TALEN-mediated gene knockout in livestock. Proc Natl Acad Sci 109:17382–17387. https://doi.org/10.1073/pnas.1211446109

Carroll D (2011) Genome engineering with zinc-finger nucleases. Genetics 188:773–782. https://doi.org/10.1534/genetics.111.131433

Carroll, D. (2017). Focus: Genome editing: genome editing: past, present, and future. Yale J Biol Med 90(4): 653.

Castel B, Tomlinson L, Locci F, Yang Y, Jones JD (2019) Optimization of T-DNA architecture for Cas9-mediated mutagenesis in *Arabidopsis*. PLoS One 14. https://doi.org/10.1371/journal.pone.0204778

Cermak T, Doyle EL, Christian M, Wang L, Zhang Y, Schmidt C et al (2011) Efficient design and assembly of custom TALEN and other TAL effector-based constructs for DNA targeting. Nucleic Acids Res 39:e82–e82. https://doi.org/10.1093/nar/gkr218

Čermák T, Curtin SJ, Gil-Humanes J, Čegan R, Kono TJ, Konečná E et al (2017) A multipurpose toolkit to enable advanced genome engineering in plants. Plant Cell 29:1196–1217. https://doi.org/10.1105/tpc.16.00922

Chandrasegaran S, Carroll D (2016) Origins of programmable nucleases for genome engineering. J Mol Biol 428:963–989. https://doi.org/10.1016/j.jmb.2015.10.014

Chen Y et al (2017) CRISPR/Cas9-mediated base-editing system efficiently generates gain-of-function mutations in Arabidopsis. Sci China Life Sci 60(5):520–523. https://doi.org/10.1007/s11427-017-9021-5

Chen K et al (2019) CRISPR/Cas genome editing and precision plant breeding in agriculture. Annu Rev Plant Biol 70:667–697. https://doi.org/10.1146/annurev-arplant-050718-100049

Chen G et al (2020) Gene editing to facilitate hybrid crop production. Biotechnol Adv 107676. https://doi.org/10.1016/j.biotechadv.2020.107676

Cong L, Zhang F (2015) Genome engineering using CRISPR-Cas9 system. In: Chromosomal mutagenesis, vol 8. Humana Press, Totowa, pp 2281–2308. https://doi.org/10.1038/nprot.2013.143

Cong L, Ran FA, Cox D, Lin S, Barretto R, Habib N et al (2013) Multiplex genome engineering using CRISPR/Cas systems. Science 339:819–823. https://doi.org/10.1126/science.1231143

Cox MM (1999) Recombinational DNA repair in bacteria and the RecA protein. In: Progress in nucleic acid research and molecular biology, vol 63. Academic, New York, pp 311–366. https://doi.org/10.1016/S0079-6603(08)60726-6

Cui Y, Xu J, Cheng M, Liao X, Peng S (2018) Review of CRISPR/Cas9 sgRNA design tools. Interdiscip Sci Comput Life Sci 10:455–465. https://doi.org/10.1007/s12539-018-0298-z

Dagdas YS, Chen JS, Sternberg SH, Doudna JA, Yildiz A (2017) A conformational checkpoint between DNA binding and cleavage by CRISPR-Cas9. Sci Adv 3:eaao0027. https://doi.org/10.1126/sciadv.aao0027

Dai Q, Huang J, Klitzman B, Dong C, Goldschmidt-Clermont PJ, March KL et al (2004) Engineered zinc finger–activating vascular endothelial growth factor transcription factor plasmid DNA induces therapeutic angiogenesis in rabbits with hindlimb ischemia. Circulation 110:2467–2475. https://doi.org/10.1161/01.CIR.0000145139.53840.49

De Pater S, Neuteboom LW, Pinas JE, Hooykaas PJ, Van Der Zaal BJ (2009) ZFN-induced mutagenesis and gene-targeting in *Arabidopsis* through *Agrobacterium*-mediated floral dip transformation. Plant Biotechnol J 7:821–835. https://doi.org/10.1111/j.1467-7652.2009.00446.x

Deltcheva E, Chylinski K, Sharma CM, Gonzales K, Chao Y, Pirzada ZA et al (2011) CRISPR RNA maturation by trans-encoded small RNA and host factor RNase III. Nature 471:602–607. https://doi.org/10.1038/nature09886

Djukanovic V et al (2013) Male-sterile maize plants produced by targeted mutagenesis of the cytochrome P 450-like gene (MS 26) using a re-designed I–C reI homing endonuclease. Plant J 76(5):888–899. https://doi.org/10.1111/tpj.12335

Doudna JA, Charpentier E (2014) The new frontier of genome engineering with CRISPR-Cas9. Science 346:6213. https://doi.org/10.1126/science.1258096

Doyon Y, McCammon JM, Miller JC, Faraji F, Ngo C, Katibah GE et al (2008) Heritable targeted gene disruption in zebrafish using designed zinc-finger nucleases. Nat Biotechnol 26:702–708. https://doi.org/10.1038/nbt1409

Dudáš A, Chovanec M (2004) DNA double-strand break repair by homologous recombination. Mutat Res Rev Mutat Res 566:131–167. https://doi.org/10.1016/j.mrrev.2003.07.001

Durai S, Mani M, Kandavelou K, Wu J, Porteus MH, Chandrasegaran S (2005) Zinc finger nucleases: custom-designed molecular scissors for genome engineering of plant and mammalian cells. Nucleic Acids Res 33:5978–5990. https://doi.org/10.1093/nar/gki912

Ebrahimi V, Hashemi A (2020) Challenges of in vitro genome editing with CRISPR/Cas9 and possible solutions: A review. Gene 144813. https://doi.org/10.1016/j.gene.2020.144813

Ekker SC (2008) Zinc Finger–based knockout punches for Zebrafish genes. Zebrafish 5:121–123. https://doi.org/10.1089/zeb.2008.9988

Ellison EE et al (2020) Multiplexed heritable gene editing using RNA viruses and mobile single guide RNAs. Nat Plant 6(6):620–624. https://doi.org/10.1038/s41477-020-0670-y

Elrod-Erickson M, Pabo CO (1999) Binding studies with Mutants of Zif268: contribution of individual side chains to binding affinity and specificity in the Zif268 zinc finger-DNA complex. J Biol Chem 274:19281–19285. https://doi.org/10.1074/jbc.274.27.19281

Farzadfard F, Lu TK (2018) Emerging applications for DNA writers and molecular recorders. Science 361(6405):870–875. https://doi.org/10.1126/science.aat9249

Foley JE, Yeh J-RJ, Maeder ML, Reyon D, Sander JD, Peterson RT, Joung JK (2009) Rapid mutation of endogenous zebrafish genes using zinc finger nucleases made by Oligomerized Pool ENgineering (OPEN). PLoS One 4:e4348. https://doi.org/10.1371/journal.pone.0004348

Game J, Mortimer R (1974) A genetic study of X-ray sensitive mutants in yeast. Mutat Res Fundam Mol Mech Mutagen 24:281–292. https://doi.org/10.1016/0027-5107(74)90176-6

Gao Y, Zhao Y (2014) Self-processing of ribozyme-flanked RNAs into guide RNAs in vitro and in vivo for CRISPR-mediated genome editing. J Integr Plant Biol 56:343–349. https://doi.org/10.1111/jipb.12152

Gao H et al (2010) Heritable targeted mutagenesis in maize using a designed endonuclease. Plant J 61(1):176–187. https://doi.org/10.1111/j.1365-313X.2009.04041.x

Gaudelli NM et al (2017) Programmable base editing of A• T to G• C in genomic DNA without DNA cleavage. Nature 551(7681):464–471. https://doi.org/10.1038/nature24644

Geurts AM, Cost GJ, Freyvert Y, Zeitler B, Miller JC, Choi VM et al (2009) Knockout rats via embryo microinjection of zinc-finger nucleases. Science 325:433–433. https://doi.org/10.1126/science.1172447

Gilbert LA, Horlbeck MA, Adamson B, Villalta JE, Chen Y, Whitehead EH et al (2014) Genome-scale CRISPR-mediated control of gene repression and activation. Cell 159:647–661. https://doi.org/10.1016/j.cell.2014.09.029

Gorbunova V, Levy AA (1999) How plants make ends meet: DNA double-strand break repair. Trends Plant Sci 4:263–269. https://doi.org/10.1016/S1360-1385(99)01430-2

Guilinger JP, Thompson DB, Liu DR (2014) Fusion of catalytically inactive Cas9 to FokI nuclease improves the specificity of genome modification. Nat Biotechnol 32:577. https://doi.org/10.1038/nbt.2909

Guirouilh-Barbat J et al (2004) Impact of the KU80 pathway on NHEJ-induced genome rearrangements in mammalian cells. Mol Cell 14(5):611–623. https://doi.org/10.1016/j.molcel.2004.05.008

Gumpel NJ, Rochaix J-D, Purton S (1994) Studies on homologous recombination in the green alga *Chlamydomonas reinhardtii*. Curr Genet 26:438–442. https://doi.org/10.1007/BF00309931

Gutschner T, Haemmerle M, Genovese G, Draetta GF, Chin L (2016) Post-translational regulation of Cas9 during G1 enhances homology-directed repair. Cell Rep 14:1555–1566. https://doi.org/10.1016/j.celrep.2016.01.019

Händel E-M, Alwin S, Cathomen T (2009) Expanding or restricting the target site repertoire of zinc-finger nucleases: the inter-domain linker as a major determinant of target site selectivity. Mol Ther 17:104–111. https://doi.org/10.1038/mt.2008.233

Haroon M, et al (2020) Conventional breeding, molecular breeding and speed breeding; brave approaches to Revamp the production of cereal crops. https://doi.org/10.20944/preprints202011.0667.v1

Heler R, Samai P, Modell JW, Weiner C, Goldberg GW, Bikard D, Marraffini LA (2015) Cas9 specifies functional viral targets during CRISPR–Cas adaptation. Nature 519:199–202. https://doi.org/10.1038/nature14245

Herai RH (2019) Avoiding the off-target effects of CRISPR/cas9 system is still a challenging accomplishment for genetic transformation. Gene 700:176–178. https://doi.org/10.1016/j.gene.2019.03.019

Hockemeyer D, Wang H, Kiani S, Lai CS, Gao Q, Cassady JP et al (2011) Genetic engineering of human pluripotent cells using TALE nucleases. Nat Biotechnol 29:731–734. https://doi.org/10.1038/nbt.1927

Holkers M, Maggio I, Liu J, Janssen JM, Miselli F, Mussolino C et al (2013) Differential integrity of TALE nuclease genes following adenoviral and lentiviral vector gene transfer into human cells. Nucleic Acids Res 41:e63–e63. https://doi.org/10.1093/nar/gks1446

Horvath P, Barrangou R (2010) CRISPR/Cas, the immune system of bacteria and archaea. Science 327:167–170. https://doi.org/10.1126/science.1179555

Hrouda M, Paszkowski J (1994) High fidelity extrachromosomal recombination and gene targeting in plants. Mol Gen Genet MGG 243:106–111. https://doi.org/10.1007/BF00283882

Huang P, Xiao A, Zhou M, Zhu Z, Lin S, Zhang B (2011) Heritable gene targeting in zebrafish using customized TALENs. Nat Biotechnol 29:699–700. https://doi.org/10.1038/nbt.1939

Iqbal Z et al (2020) New prospects on the horizon: genome editing to engineer plants for desirable traits. Curr Plant Biol 100171. https://doi.org/10.1016/j.cpb.2020.100171

Ira M, Zuker A, Shklarman E, Zeevi V, Tovkach A, Roffe S et al (2010) Non-transgenic genome modification in plant cells. Plant Physiol. https://doi.org/10.1104/pp.110.164806

Jasin M, Moynahan ME, Richardson C (1996) Targeted transgenesis. Proc Natl Acad Sci U S A 93:8804. https://doi.org/10.1073/pnas.93.17.8804

Jinek M, Chylinski K, Fonfara I, Hauer M, Doudna JA, Charpentier E (2012) A programmable dual-RNA–guided DNA endonuclease in adaptive bacterial immunity. Science 337:816–821. https://doi.org/10.1126/science.1225829

Joung JK, Sander JD (2013) TALENs: a widely applicable technology for targeted genome editing. Nat Rev Mol Cell Biol 14:49–55. https://doi.org/10.1038/nrm3486

Joung JK, Voytas DF, Cathomen T (2010) Reply to "Genome editing with modularly assembled zinc-finger nucleases". Nat Methods 7:91–92. https://doi.org/10.1038/nmeth0210-91b

Kang B-C et al (2018) Precision genome engineering through adenine base editing in plants. Nat Plant 4(7):427–431. https://doi.org/10.1038/s41477-018-0178-x

Kaur N, Alok A, Kaur N, Pandey P, Awasthi P, Tiwari S (2018) CRISPR/Cas9-mediated efficient editing in phytoene desaturase (PDS) demonstrates precise manipulation in banana cv. Rasthali genome. Funct Integr Genomic 18:89–99. https://doi.org/10.1007/s10142-017-0577-5

Kempin SA, Liljegren SJ, Block LM, Rounsley SD, Yanofsky MF, Lam E (1997) Targeted disruption in *Arabidopsis*. Nature 389:802–803. https://doi.org/10.1038/39770

Khandagale K, Nadaf A (2016) Genome editing for targeted improvement of plants. Plant Biotechnol Rep 10(6):327–343. https://doi.org/10.1007/s11816-016-0417-4

Kim Y-G, Chandrasegaran S (1994) Chimeric restriction endonuclease. Proc Natl Acad Sci 91:883–887. https://doi.org/10.1073/pnas.91.3.883

Kim H, Kim J-S (2014) A guide to genome engineering with programmable nucleases. Nat Rev Genet 15:321–334. https://doi.org/10.1038/nrg3686

Kim Y-G, Cha J, Chandrasegaran S (1996) Hybrid restriction enzymes: zinc finger fusions to Fok I cleavage domain. Proc Natl Acad Sci 93:1156–1160. https://doi.org/10.1073/pnas.93.3.1156

Kim S, Lee MJ, Kim H, Kang M, Kim J-S (2011) Preassembled zinc-finger arrays for rapid construction of ZFNs. Nat Methods 8:7. https://doi.org/10.1038/nmeth0111-7a

Kim S, Kim D, Cho SW, Kim J, Kim J-S (2014) Highly efficient RNA-guided genome editing in human cells via delivery of purified Cas9 ribonucleoproteins. Genome Res 24:1012–1019. https://doi.org/10.1101/gr.171322.113

Kirik A et al (2000) Species-specific double-strand break repair and genome evolution in plants. EMBO J 19(20):5562–5566. https://doi.org/10.1093/emboj/19.20.5562

Kleinstiver BP, Pattanayak V, Prew MS, Tsai SQ, Nguyen NT, Zheng Z, Joung JK (2016) High-fidelity CRISPR–Cas9 nucleases with no detectable genome-wide off-target effects. Nature 529:490–495. https://doi.org/10.1038/nature16526

Komor AC et al (2016) Programmable editing of a target base in genomic DNA without double-stranded DNA cleavage. Nature 533(7603):420–424. https://doi.org/10.1038/nature24644

Kowalczykowski SC, Dixon DA, Eggleston AK, Lauder SD, Rehrauer WM (1994) Biochemistry of homologous recombination in *Escherichia coli*. Microbiol Rev 58:401–465

Kunz C et al (2009) DNA repair in mammalian cells. Cell Mol Life Sci 66(6):1021–1038. https://doi.org/10.1007/s00018-009-8739-9

Landrum MJ et al (2014) ClinVar: public archive of relationships among sequence variation and human phenotype. Nucleic Acids Res 42(D1):D980–D985. https://doi.org/10.1093/nar/gkt1113

Landrum MJ et al (2016) ClinVar: public archive of interpretations of clinically relevant variants. Nucleic Acids Res 44(D1):D862–D868. https://doi.org/10.1093/nar/gkv1222

Lei Y, Guo X, Liu Y, Cao Y, Deng Y, Chen X et al (2012) Efficient targeted gene disruption in Xenopus embryos using engineered transcription activator-like effector nucleases (TALENs). Proc Natl Acad Sci 109:17484–17489. https://doi.org/10.1073/pnas.1215421109

Levy A et al (2015) CRISPR adaptation biases explain preference for acquisition of foreign DNA. Nature 520(7548):505–510. https://doi.org/10.1038/nature14302

Li L, Wu LP, Chandrasegaran S (1992) Functional domains in Fok I restriction endonuclease. Proc Natl Acad Sci 89:4275–4279. https://doi.org/10.1073/pnas.89.10.4275

Li T, Huang S, Zhao X, Wright DA, Carpenter S, Spalding MH et al (2011) Modularly assembled designer TAL effector nucleases for targeted gene knockout and gene replacement in eukaryotes. Nucleic Acids Res 39:6315–6325. https://doi.org/10.1093/nar/gkr188

Li T, Liu B, Spalding MH, Weeks DP, Yang B (2012) High-efficiency TALEN-based gene editing produces disease-resistant rice. Nat Biotechnol 30:390. https://doi.org/10.1038/nbt.2199

Li D, Zhou H, Zeng X (2019) Battling CRISPR-Cas9 off-target genome editing. Springer. https://doi.org/10.1007/s10565-019-09485-5

Liu Q, Segal DJ, Ghiara JB, Barbas CF (1997) Design of polydactyl zinc-finger proteins for unique addressing within complex genomes. Proc Natl Acad Sci 94:5525–5530. https://doi.org/10.1073/pnas.94.11.5525

Liu P-Q, Rebar EJ, Zhang L, Liu Q, Jamieson AC, Liang Y et al (2001) Regulation of an endogenous locus using a panel of designed zinc finger proteins targeted to accessible chromatin regions: activation of vascular endothelial growth factor A. J Biol Chem 276:11323–11334. https://doi.org/10.1074/jbc.M011172200

Liu J, Li C, Yu Z, Huang P, Wu H, Wei C et al (2012) Efficient and specific modifications of the *Drosophila* genome by means of an easy TALEN strategy. J Genetic Genomic 39:209–215. https://doi.org/10.1016/j.jgg.2012.04.003

Liu, W., Rudis, M. R., Cheplick, M. H., Millwood, R. J., Yang, J. P., Ondzighi-Assoume, C. A., ... & Stewart, C. N. (2020). Lipofection-mediated genome editing using DNA-free delivery of the Cas9/gRNA ribonucleoprotein into plant cells. Plant Cell Rep 39(2):245–257.

Lloyd A, Plaisier CL, Carroll D, Drews GN (2005) Targeted mutagenesis using zinc-finger nucleases in *Arabidopsis*. Proc Natl Acad Sci 102:2232–2237. https://doi.org/10.1073/pnas.0409339102

Ma S, Zhang S, Wang F, Liu Y, Liu Y, Xu H et al (2012) Highly efficient and specific genome editing in silkworm using custom TALENs. PLoS One 7:e45035. https://doi.org/10.1371/journal.pone.0045035

Maeder ML, Thibodeau-Beganny S, Osiak A, Wright DA, Anthony RM, Eichtinger M et al (2008) Rapid "open-source" engineering of customized zinc-finger nucleases for highly efficient gene modification. Mol Cell 31:294–301. https://doi.org/10.1016/j.molcel.2008.06.016

Makarova KS, Haft DH, Barrangou R, Brouns SJ, Charpentier E, Horvath P et al (2011) Evolution and classification of the CRISPR–Cas systems. Nat Rev Microbiol 9:467–477. https://doi.org/10.1038/nrmicro2577

Mali P, Aach J, Stranges PB, Esvelt KM, Moosburner M, Kosuri S et al (2013) CAS9 transcriptional activators for target specificity screening and paired nickases for cooperative genome engineering. Nat Biotechnol 31:833–838. https://doi.org/10.1038/nbt.2675

Maruyama T, Dougan SK, Truttmann MC, Bilate AM, Ingram JR, Ploegh HL (2015) Increasing the efficiency of precise genome editing with CRISPR-Cas9 by inhibition of nonhomologous end joining. Nat Biotechnol 33:538–542. https://doi.org/10.1038/nbt.3190

Mashimo T, Takizawa A, Voigt B, Yoshimi K, Hiai H, Kuramoto T, Serikawa T (2010) Generation of knockout rats with X-linked severe combined immunodeficiency (X-SCID) using zinc-finger nucleases. PLoS One 5:e8870. https://doi.org/10.1371/journal.pone.0008870

Matsoukas IG (2020) Prime editing: genome editing for rare genetic diseases without double-strand breaks or donor DNA. Front Genet 11:528. https://doi.org/10.3389/fgene.2020.00528

McNamara AR, Hurd PJ, Smith AE, Ford KG (2002) Characterisation of site-biased DNA methyltransferases: specificity, affinity and subsite relationships. Nucleic Acids Res 30:3818–3830. https://doi.org/10.1093/nar/gkf501

Meng X, Thibodeau-Beganny S, Jiang T, Joung JK, Wolfe SA (2007) Profiling the DNA-binding specificities of engineered Cys2His2 zinc finger domains using a rapid cell-based method. Nucleic Acids Res 35:e81. https://doi.org/10.1093/nar/gkm385

Meng X, Noyes MB, Zhu LJ, Lawson ND, Wolfe SA (2008) Targeted gene inactivation in zebrafish using engineered zinc-finger nucleases. Nat Biotechnol 26:695–701. https://doi.org/10.1038/nbt1398

Meyer RR, Laine PS (1990) The single-stranded DNA-binding protein of Escherichia coli. Microbiol Mol Biol Rev 54:342–380

Miao ZH, Lam E (1995) Targeted disruption of the TGA3 locus in *Arabidopsis thaliana*. Plant J 7:359–365. https://doi.org/10.1046/j.1365-313X.1995.7020359.x

Miller JC, Tan S, Qiao G, Barlow KA, Wang J, Xia DF et al (2011) A TALE nuclease architecture for efficient genome editing. Nat Biotechnol 29:143–148. https://doi.org/10.1038/nbt.1755

Moore M, Klug A, Choo Y (2001) Improved DNA binding specificity from polyzinc finger peptides by using strings of two-finger units. Proc Natl Acad Sci 98:1437–1441. https://doi.org/10.1073/pnas.98.4.1437

Morbitzer R, Römer P, Boch J, Lahaye T (2010) Regulation of selected genome loci using de novo-engineered transcription activator-like effector (TALE)-type transcription factors. Proc Natl Acad Sci 107:21617–21622. https://doi.org/10.1073/pnas.1013133107

Morton J, Davis MW, Jorgensen EM, Carroll D (2006) Induction and repair of zinc-finger nuclease-targeted double-strand breaks in Caenorhabditis elegans somatic cells. Proc Natl Acad Sci 103:16370–16375. https://doi.org/10.1073/pnas.0605633103

Moscou MJ, Bogdanove AJ (2009) A simple cipher governs DNA recognition by TAL effectors. Science 326:1501–1501. https://doi.org/10.1126/science.1178817

Müller AE, Kamisugi Y, Grüneberg R, Niedenhof I, Hörold RJ, Meyer P (1999) Palindromic sequences and A+T-rich DNA elements promote illegitimate recombination in *Nicotiana tabacum*. J Mol Biol 291:29–46. https://doi.org/10.1006/jmbi.1999.2957

Nelson CE, Gersbach CA (2016) Engineering delivery vehicles for genome editing. Ann Rev Chem Biomol Eng 7:637–662. https://doi.org/10.1146/annurev-chembioeng-080615-034711

Newton MD, Taylor BJ, Driessen RP, Roos L, Cvetesic N, Allyjaun S et al (2019) DNA stretching induces Cas9 off-target activity. Nat Struct Mol Biol 26:185–192. https://doi.org/10.1038/s41594-019-0188-z

Nishida K, et al (2016) Targeted nucleotide editing using hybrid prokaryotic and vertebrate adaptive immune systems. Science 353(6305). See all authors and affiliations. https://doi.org/10.1126/science.aaf8729

Ochiai H, Fujita K, Suzuki KI, Nishikawa M, Shibata T, Sakamoto N, Yamamoto T (2010) Targeted mutagenesis in the sea urchin embryo using zinc-finger nucleases. Genes Cells 15:875–885. https://doi.org/10.1111/j.1365-2443.2010.01425.x

Offringa R, de Groot MJ, Haagsman HJ, Does MP, van den Elzen PJ, Hooykaas PJ (1990) Extrachromosomal homologous recombination and gene targeting in plant cells after *Agrobacterium* mediated transformation. EMBO J 9:3077–3084. https://doi.org/10.1002/j.1460-2075.1990.tb07504.x

Osakabe K, Osakabe Y, Toki S (2010) Site-directed mutagenesis in *Arabidopsis* using custom-designed zinc finger nucleases. Proc Natl Acad Sci 107:12034–12039. https://doi.org/10.1073/pnas.1000234107

Pabo CO, Peisach E, Grant RA (2001) Design and selection of novel Cys2His2 zinc finger proteins. Annu Rev Biochem 70:313–340. https://doi.org/10.1146/annurev.biochem.70.1.313

Pâques F, Haber JE (1999) Multiple pathways of recombination induced by double-strand breaks in *Saccharomyces cerevisiae*. Microbiol Mol Biol Rev 63:349–404. https://doi.org/10.1128/MMBR.63.2.349-404.1999

Paris Ü et al (2015) NHEJ enzymes LigD and Ku participate in stationary-phase mutagenesis in Pseudomonas putida. DNA Repair 31:11–18. https://doi.org/10.1016/j.dnarep.2015.04.005

Pastink A, Eeken JC, Lohman PH (2001) Genomic integrity and the repair of double-strand DNA breaks. Mutat Res Fundam Mol Mech Mutagen 480:37–50. https://doi.org/10.1016/S0027-5107(01)00167-1

Paszkowski J, Baur M, Bogucki A, Potrykus I (1988) Gene targeting in plants. Embo J 7:4021–4026. https://doi.org/10.1002/j.1460-2075.1988.tb03295.x

Pavletich NP, Pabo CO (1991) Zinc finger-DNA recognition: crystal structure of a Zif268-DNA complex at 2.1 A. Science 252:809–817. https://doi.org/10.1126/science.2028256

Petes TD (1991) Recombination in yeast. In: The molecular and cellular biology of the yeast Saccharomyces: genome dynamics, protein synthesis and energetics. Cold Spring Harbor Laboratory Press, New York, pp 407–521

Porteus MH, Baltimore D (2003) Chimeric nucleases stimulate gene targeting in human cells. Science 300:763. https://doi.org/10.1126/science.1078395

Potrykus I, Spangenberg G (2013) Gene transfer to plants. Springer, Tokyo

Puchta H (1999) Double-strand break-induced recombination between ectopic homologous sequences in somatic plant cells. Genetics 152:1173–1181. https://doi.org/10.1093/genetics/152.3.1173

Puchta H (2002) Gene replacement by homologous recombination in plants. In: Functional genomics. Springer, Dordrecht, pp 173–182. https://doi.org/10.1007/978-94-010-0448-0_12

Qi LS, Larson MH, Gilbert LA, Doudna JA, Weissman JS, Arkin AP, Lim WA (2013) Repurposing CRISPR as an RNA-guided platform for sequence-specific control of gene expression. Cell 152:1173–1183. https://doi.org/10.1016/j.cell.2013.02.022

Ramakrishna S, Dad A-BK, Beloor J, Gopalappa R, Lee S-K, Kim H (2014) Gene disruption by cell-penetrating peptide-mediated delivery of Cas9 protein and guide RNA. Genome Res 24:1020–1027. https://doi.org/10.1101/gr.171264.113

Ramirez CL, Foley JE, Wright DA, Müller-Lerch F, Rahman SH, Cornu TI et al (2008) Unexpected failure rates for modular assembly of engineered zinc fingers. Nat Methods 5:374–375. https://doi.org/10.1038/nmeth0508-374

Ran FA, Hsu PD, Lin C-Y, Gootenberg JS, Konermann S, Trevino AE et al (2013a) Double nicking by RNA-guided CRISPR Cas9 for enhanced genome editing specificity. Cell 154:1380–1389. https://doi.org/10.1016/j.cell.2013.08.021

Ran FA, Hsu PD, Wright J, Agarwala V, Scott DA, Zhang F (2013b) Genome engineering using the CRISPR-Cas9 system. Nat Protoc 8:2281–2308. https://doi.org/10.1038/nprot.2013.143

Ran, Y., Liang, Z., & Gao, C. (2017). Current and future editing reagent delivery systems for plant genome editing. Sci China Life Sci 60(5):490–505.

Rebar EJ (2004) Development of pro-angiogenic engineered transcription factors for the treatment of cardiovascular disease. Expert Opin Investig Drugs 13:829–839. https://doi.org/10.1517/13543784.13.7.829

Rebar EJ, Huang Y, Hickey R, Nath AK, Meoli D, Nath S et al (2002) Induction of angiogenesis in a mouse model using engineered transcription factors. Nat Med 8:1427–1432. https://doi.org/10.1038/nm1202-795

Rees HA, Liu DR (2018) Base editing: precision chemistry on the genome and transcriptome of living cells. Nat Rev Genet 19(12):770–788. https://doi.org/10.1038/s41576-018-0059-1

Reisch CR, Prather KL (2015) The no-SCAR (S carless C as9 A ssisted R ecombineering) system for genome editing in *Escherichia coli*. Sci Rep 5:1–12. https://doi.org/10.1038/srep15096

Rémy S, Tesson L, Ménoret S, Usal C, Scharenberg AM, Anegon I (2010) Zinc-finger nucleases: a powerful tool for genetic engineering of animals. Transgenic Res 19:363–371. https://doi.org/10.1007/s11248-009-9323-7

Ren D, Collingwood TN, Rebar EJ, Wolffe AP, Camp HS (2002) PPARγ knockdown by engineered transcription factors: exogenous PPARγ2 but not PPARγ1 reactivates adipogenesis. Genes Dev 16:27–32. https://doi.org/10.1101/gad.953802

Ren, B., Yan, F., Kuang, Y., Li, N., Zhang, D., Zhou, X., ... & Zhou, H. (2018). Improved base editor for efficiently inducing genetic variations in rice with CRISPR/Cas9-guided hyperactive hAID mutant. Mol Plant 11(4):623–626.

Reski R (1999) Molecular genetics of Physcomitrella. Planta 208:301–309. https://doi.org/10.1007/s004250050563

Reyon D, Tsai SQ, Khayter C, Foden JA, Sander JD, Joung JK (2012) FLASH assembly of TALENs for high-throughput genome editing. Nat Biotechnol 30:460–465. https://doi.org/10.1038/nbt.2170

Roca AI, Cox MM (1997) RecA protein: structure, function, and role in recombinational DNA repair. Prog Nucleic Acid Res Mol Biol 56:129–223. https://doi.org/10.1016/s0079-6603(08)61005-3

Rusche LN, Kirchmaier AL, Rine J (2003) The establishment, inheritance, and function of silenced chromatin in *Saccharomyces cerevisiae*. Annu Rev Biochem 72:481–516. https://doi.org/10.1146/annurev.biochem.72.121801.161547

Ryan DE, Taussig D, Steinfeld I, Phadnis SM, Lunstad BD, Singh M et al (2018) Improving CRISPR–Cas specificity with chemical modifications in single-guide RNAs. Nucleic Acids Res 46:792–803. https://doi.org/10.1093/nar/gkx1199

San Filippo J, Sung P, Klein H (2008) Mechanism of eukaryotic homologous recombination. Annu Rev Biochem 77:229–257. https://doi.org/10.1146/annurev.biochem.77.061306.125255

Sander JD, Cade L, Khayter C, Reyon D, Peterson RT, Joung JK, Yeh J-RJ (2011) Targeted gene disruption in somatic zebrafish cells using engineered TALENs. Nat Biotechnol 29:697–698. https://doi.org/10.1038/nbt.1934

Schaefer DG (2001) Gene targeting in Physcomitrella patens. Curr Opin Plant Biol 4:143–150. https://doi.org/10.1016/S1369-5266(00)00150-3

Sedivy JM, Dutriaux A (1999) Gene targeting and somatic cell genetics–a rebirth or a coming of age? Trends Genet 15:88–90. https://doi.org/10.1016/s0168-9525(98)01689-8

Shalem O, Sanjana NE, Hartenian E, Shi X, Scott DA, Mikkelsen TS et al (2014) Genome-scale CRISPR-Cas9 knockout screening in human cells. Science 343:84–87. https://doi.org/10.1126/science.1247005

Sharma S et al (2017) Recent advances in CRISPR/Cas mediated genome editing for crop improvement. Plant Biotechnol Rep 11(4):193–207. https://doi.org/10.1007/s11816-017-0446-7

Shen B, Zhang J, Wu H, Wang J, Ma K, Li Z et al (2013) Generation of gene-modified mice via Cas9/RNA-mediated gene targeting. Cell Res 23:720–723. https://doi.org/10.1038/cr.2013.46

Shi Y, Berg JM (1995) A direct comparison of the properties of natural and designed zinc-finger proteins. Chem Biol 2:83–89. https://doi.org/10.1016/1074-5521(95)90280-5

Shimatani Z et al (2017) Targeted base editing in rice and tomato using a CRISPR-Cas9 cytidine deaminase fusion. Nat Biotechnol 35(5):441–443. https://doi.org/10.1038/nbt.3833

Shimizu Y, Bhakta MS, Segal DJ (2009) Restricted spacer tolerance of a zinc finger nuclease with a six amino acid linker. Bioorg Med Chem Lett 19:3970–3972. https://doi.org/10.1016/j.bmcl.2009.02.109

Shukla VK, Doyon Y, Miller JC, DeKelver RC, Moehle EA, Worden SE et al (2009) Precise genome modification in the crop species *Zea mays* using zinc-finger nucleases. Nature 459:437–441. https://doi.org/10.1038/nature07992

Siede W, Friedl AA, Dianova I, Eckardt-Schupp F, Friedberg EC (1996) The *Saccharomyces cerevisiae* Ku autoantigen homologue affects radiosensitivity only in the absence of homologous recombination. Genetics 142:91–102. https://doi.org/10.1093/genetics/142.1.91

Silva G et al (2011) Meganucleases and other tools for targeted genome engineering: perspectives and challenges for gene therapy. Curr Gene Ther 11(1):11–27. https://doi.org/10.2174/156652311794520111

Singh V, Braddick D, Dhar PK (2017) Exploring the potential of genome editing CRISPR-Cas9 technology. Gene 599:1–18. https://doi.org/10.1016/j.gene.2016.11.008

Smart EJ, Selman BR (1991) Isolation and characterization of a *Chlamydomonas reinhardtii* mutant lacking the gamma-subunit of chloroplast coupling factor 1 (CF1). Mol Cell Biol 11:5053–5058. https://doi.org/10.1128/mcb.11.10.5053

Smithies O (1987) Targeted correction of a mutant HPRT gene in mouse ES cells. Nature 330:576–578. https://doi.org/10.1038/330576a0

Snowden AW, Zhang L, Urnov F, Dent C, Jouvenot Y, Zhong X et al (2003) Repression of vascular endothelial growth factor A in glioblastoma cells using engineered zinc finger transcription factors. Cancer Res 63:8968–8976

Sodeinde OA, Kindle KL (1993) Homologous recombination in the nuclear genome of *Chlamydomonas reinhardtii*. Proc Natl Acad Sci 90:9199–9203. https://doi.org/10.1073/pnas.90.19.9199

Song J, Yang D, Xu J, Zhu T, Chen YE, Zhang J (2016) RS-1 enhances CRISPR/Cas9-and TALEN-mediated knock-in efficiency. Nat Commun 7:1–7. https://doi.org/10.1038/ncomms10548

Standage-Beier K, Zhang Q, Wang X (2015) Targeted large-scale deletion of bacterial genomes using CRISPR-nickases. ACS Synth Biol 4:1217–1225. https://doi.org/10.1021/acssynbio.5b00132

Swarthout JT, Raisinghani M, Cui X (2011) Zinc Finger nucleases: a new era for transgenic animals. Ann Neurosci 18:25–28. https://doi.org/10.5214/ans.0972.7531.1118109

Symington LS (2002) Role of RAD52 epistasis group genes in homologous recombination and double-strand break repair. Microbiol Mol Biol Rev 66:630–670. https://doi.org/10.1128/mmbr.66.4.630-670.2002

Terada N, Hamazaki T, Oka M, Hoki M, Mastalerz DM, Nakano Y et al (2002) Bone marrow cells adopt the phenotype of other cells by spontaneous cell fusion. Nature 416:542–545. https://doi.org/10.1038/nature730

Tesson L, Usal C, Ménoret S, Leung E, Niles BJ, Remy S et al (2011) Knockout rats generated by embryo microinjection of TALENs. Nat Biotechnol 29:695–696. https://doi.org/10.1038/nbt.1940

Thykjær T, Finnemann J, Schauser L, Christensen L, Poulsen C, Stougaard J (1997) Gene targeting approaches using positive-negative selection and large flanking regions. Plant Mol Biol 35:523–530. https://doi.org/10.1023/A:1005865600319

Tian S et al (2018) Engineering herbicide-resistant watermelon variety through CRISPR/Cas9-mediated base-editing. Plant Cell Rep 37(9):1353–1356. https://doi.org/10.1007/s00299-018-2299-0

Townsend JA, Wright DA, Winfrey RJ, Fu F, Maeder ML, Joung JK, Voytas DF (2009) High-frequency modification of plant genes using engineered zinc-finger nucleases. Nature 459:442–445. https://doi.org/10.1038/nature07845

Tsai SQ, Wyvekens N, Khayter C, Foden JA, Thapar V, Reyon D et al (2014) Dimeric CRISPR RNA-guided FokI nucleases for highly specific genome editing. Nat Biotechnol 32:569–576. https://doi.org/10.1038/nbt.2908

Upadhyay SK, Kumar J, Alok A, Tuli R (2013) RNA-guided genome editing for target gene mutations in wheat. G3: Genes. Genomes Genet 3:2233–2238. https://doi.org/10.1534/g3.113.008847

Urnov FD, Miller JC, Lee Y-L, Beausejour CM, Rock JM, Augustus S et al (2005) Highly efficient endogenous human gene correction using designed zinc-finger nucleases. Nature 435:646–651. https://doi.org/10.1038/nature03556

van den Bosch M, Lohman PH, Pastink A (2002) DNA double-strand break repair by homologous recombination. Biol Chem 383:873–892. https://doi.org/10.1515/bc.2002.095

Vartak SV, Raghavan SC (2015) Inhibition of nonhomologous end joining to increase the specificity of CRISPR/Cas9 genome editing. FEBS J 282:4289–4294. https://doi.org/10.1111/febs.13416

Vento JM, Crook N, Beisel CL (2019) Barriers to genome editing with CRISPR in bacteria. J Ind Microbiol Biotechnol 46:1327–1341. https://doi.org/10.1007/s10295-019-02195-1

Vogel P, Stafforst T (2019) Critical review on engineering deaminases for site-directed RNA editing. Curr Opin Biotechnol 55:74–80. https://doi.org/10.1016/j.copbio.2018.08.006

Wang X et al (2018) Efficient base editing in methylated regions with a human APOBEC3A-Cas9 fusion. Nat Biotechnol 36(10):946–949. https://doi.org/10.1038/nbt.4198

Watanabe T, Ochiai H, Sakuma T, Horch HW, Hamaguchi N, Nakamura T et al (2012) Non-transgenic genome modifications in a hemimetabolous insect using zinc-finger and TAL effector nucleases. Nat Commun 3:1–8. https://doi.org/10.1038/ncomms2020

Weber E, Gruetzner R, Werner S, Engler C, Marillonnet S (2011) Assembly of designer TAL effectors by Golden Gate cloning. PLoS One 6:e19722. https://doi.org/10.1371/journal.pone.0019722

Weber T, Wefers B, Wurst W, Sander S, Rajewsky K, Kühn R (2015) Increasing the efficiency of homology-directed repair for CRISPR-Cas9-induced precise gene editing in mammalian cells. Nat Biotechnol 33:543–548. https://doi.org/10.1038/nbt.3198

Weinstock DM, Jasin M (2006) Alternative pathways for the repair of RAG-induced DNA breaks. Mol Cell Biol 26:131–139. https://doi.org/10.1128/MCB.26.1.131-139.2006

West SC (1996) The RuvABC proteins and Holliday junction processing in *Escherichia coli*. J Bacteriol 178:1237–1241. https://doi.org/10.1128/jb.178.5.1237-1241.1996

Whitby MC, Lloyd RG (1998) Targeting Holliday junctions by the RecG branch migration protein of *Escherichia coli*. J Biol Chem 273:19729–19739. https://doi.org/10.1074/jbc.273.31.19729

Wood AJ, Lo T-W, Zeitler B, Pickle CS, Ralston EJ, Lee AH et al (2011) Targeted genome editing across species using ZFNs and TALENs. Science 333:307–307. https://doi.org/10.1126/science.1207773

Wright DA, Townsend JA, Winfrey RJ Jr, Irwin PA, Rajagopal J, Lonosky PM et al (2005) High-frequency homologous recombination in plants mediated by zinc-finger nucleases. Plant J 44:693–705. https://doi.org/10.1111/j.1365-313X.2005.02551.x

Wright DA et al (2014) TALEN-mediated genome editing: prospects and perspectives. Biochem J 462(1):15–24. https://doi.org/10.1042/BJ20140295

Wu J et al (2020) Engineering herbicide-resistant oilseed rape by CRISPR/Cas9-mediated cytosine base-editing. Plant Biotechnol J 18(9):1857–1859. https://doi.org/10.1038/s41598-021-87669-y

Wyman C, Kanaar R (2006) DNA double-strand break repair: all's well that ends well. Annu Rev Genet 40:363–383. https://doi.org/10.1146/annurev.genet.40.110405.090451

Xie K, Minkenberg B, Yang Y (2015) Boosting CRISPR/Cas9 multiplex editing capability with the endogenous tRNA-processing system. Proc Natl Acad Sci 112:3570–3575. https://doi.org/10.1073/pnas.1420294112

Xu G-L, Bestor TH (1997) Cytosine methylation targetted to pre-determined sequences. Nat Genet 17:376–378. https://doi.org/10.1038/ng1297-376

Xu R et al (2020) Development of plant prime-editing systems for precise genome editing. Plant Commun 1(3):100043. https://doi.org/10.1016/j.xplc.2020.100043

Yang L, Guell M, Byrne S, Yang JL, De Los Angeles A, Mali P et al (2013) Optimization of scarless human stem cell genome editing. Nucleic Acids Res 41:9049–9061. https://doi.org/10.1093/nar/gkt555

Yin K et al (2017) Progress and prospects in plant genome editing. Nat Plant 3(8):1–6. https://doi.org/10.1038/nplants.2017.107

Young JJ, Cherone JM, Doyon Y, Ankoudinova I, Faraji FM, Lee AH et al (2011) Efficient targeted gene disruption in the soma and germ line of the frog Xenopus tropicalis using engineered zinc-finger nucleases. Proc Natl Acad Sci 108:7052–7057. https://doi.org/10.1073/pnas.1102030108

Yu C, Liu Y, Ma T, Liu K, Xu S, Zhang Y et al (2015) Small molecules enhance CRISPR genome editing in pluripotent stem cells. Cell Stem Cell 16:142–147. https://doi.org/10.1016/j.stem.2015.01.003

Zaman QU et al (2019) Genome editing opens a new era of genetic improvement in polyploid crops. Crop J 7(2):141–150. https://doi.org/10.1016/j.cj.2018.07.004

Zhan X et al (2020) Genome editing for plant research and crop improvement. J Integr Plant Biol 63(I1):3–33. https://doi.org/10.1111/jipb.13063

Zhang L, Spratt SK, Liu Q, Johnstone B, Qi H, Raschke EE et al (2000) Synthetic zinc finger transcription factor action at an endogenous chromosomal site: activation of the human erythropoietin gene. J Biol Chem 275:33850–33860. https://doi.org/10.1074/jbc.M005341200

Zhang F, Maeder ML, Unger-Wallace E, Hoshaw JP, Reyon D, Christian M et al (2010) High frequency targeted mutagenesis in *Arabidopsis thaliana* using zinc finger nucleases. Proc Natl Acad Sci 107:12028–12033. https://doi.org/10.1073/pnas.0914991107

Zhang F, Wen Y, Guo X (2014) CRISPR/Cas9 for genome editing: progress, implications and challenges. Hum Mol Genet 23:R40–R46. https://doi.org/10.1093/hmg/ddu125

Zhang Y et al (2019) The emerging and uncultivated potential of CRISPR technology in plant science. Nat Plant 5(8):778–794. https://doi.org/10.1038/s41477-019-0461-5

Zhang Y et al (2020) A CRISPR way for accelerating improvement of food crops. Nat Food 1(4):200–205

Zhao L et al (2007) The restriction fold turns to the dark side: a bacterial homing endonuclease with a PD-(D/E)-XK motif. EMBO J 26(9):2432–2442. https://doi.org/10.1038/sj.emboj.7601672

Zupan J, Muth TR, Draper O, Zambryski P (2000) The transfer of DNA from *Agrobacterium tumefaciens* into plants: a feast of fundamental insights. Plant J 23:11–28. https://doi.org/10.1046/j.1365-313x.2000.00808.x

Part II
Developments in the Field of CRISPR/Cas Technology

Recent Advances and Application of CRISPR Base Editors for Improvement of Various Traits in Crops

P. Sushree Shyamli, Sandhya Suranjika, Seema Pradhan, and Ajay Parida

Abstract The rapidly increasing demands for quality food and cash crops have highlighted the need to develop crop varieties with trait-specific changes at molecular level to increase yield as well as quality of important crops. This in turn has led to the development of techniques to induce beneficial variations in the genomic structure of plants. The success of such method relies on their flexibility and feasibility, which allows customization using available resources. CRISPR-Cas systems have emerged as front-runners for such activities as they meet these requirements. Traditionally, these systems have been used to produce gene knockouts for functional characterization of important genes. However, they have now evolved to facilitate the introduction of single-base changes to introduce stable genetic changes in the plant species. These CRISPR base-editing technologies have contributed to a breakthrough in the field of plant biology and have the potential to revolutionize strategic crop improvement programs. In this chapter, we provide an account of the advances made in the development of these base editors and their application in crop improvement and hope that it offers some clarity regarding the possibilities of CRISPR base editors in securing food security in the future.

Keywords CRISPR · Base editors · CBE · ABE · Crop improvement

1 Introduction

An exponential growth in the world's population has created an imminent crisis leading to decreasing food productivity and posing the critical challenge to provide food for every individual. As per FAO (2017), the human population is estimated to reach about 9.1 billion by 2050. In order to meet the growing demand and to feed

P. S. Shyamli
Regional Centre for Biotechnology, Faridabad, New Delhi, Haryana, India

S. Suranjika
School of Biotechnology, KIIT University, Bhubaneswar, Odisha, India

S. Pradhan · A. Parida (✉)
Institute of Life Sciences, NALCO Square, Chandrasekharpur, Bhubaneswar, Odisha, India

© The Author(s), under exclusive license to Springer Nature Switzerland AG 2022
S. H. Wani, G. Hensel (eds.), *Genome Editing*,
https://doi.org/10.1007/978-3-031-08072-2_5

the world's ever-increasing population, the production of food needs to increase by a minimum of 60%. In addition to this, nearly one billion people in the world are projected to suffer from malnourishment. At the same time, our agricultural systems are deteriorating due to several reasons such as reduced agricultural land availability, increasing uncertainties of climate change, and increasing biotic and abiotic stresses (Peng et al. 2018). With the global population expected to surpass the projected statistics, modern-day agriculture is already facing massive challenges. Crops with higher yields and better quality with fewer inputs have become a necessity (Shiferaw et al. 2013).

The state of agricultural productivity has seen some improvements in the developed countries, somewhat meeting their food necessities, which also resulted in increased stress on food production affecting the consumption patterns (Voss-Fels et al. 2019). Therefore, it is necessary to develop scientific technologies for crop improvement and increase production while taking into account the growing needs of large number of small and marginal farmers who cannot afford expensive means for agriculture. To this end, conventional breeding and in some instances molecular breeding has been widely used in crop improvement. The progress has been slow as it takes several years to screen phenotypes and genotypes of viable crops and develop them into commercial varieties. Despite many innovations in plant breeding (Varshney et al. 2006), the present annual yield in major crops needs to be doubled to fulfil the rising demand for crop production (Li et al. 2018a, b, c, d, e).

Over the last decade, techniques such as modern speed breeding, high-throughput phenotyping (HTP), and genomic selection methods have accelerated the process of plant breeding (Hopkins 2003). To develop crops with desirable characteristics, molecular methods and genetic engineering techniques are being put in use (Araus et al. 2018; Majid et al. 2017; Watson et al. 2018; Zhang et al. 2014). Techniques such as high-throughput molecular markers, cis-genesis, intra-genesis, rapid gene isolation, large-scale sequencing, polyploidy breeding, genomics, and mutation breeding for the advancement of commercially important crop species' breeding are included (Mujjassim et al. 2019; Murovec et al. 2017; Muth et al. 2008). However, conventional breeding techniques often fall short when it comes to genome enhancement for developing new plant varieties. This led to the switch from conventional plant breeding to molecular marker-assisted breeding for the selection of superior lines/varieties (Dreher et al. 2002). The success of a breeding strategy largely depends on the choice of plant species. For example, diploid crops with shorter reproductive cycles prove to be far better for genetic improvement compared to polyploid crops (Abreu and Souza 2010; Kandemir and Saygili 2015; Lammerts Van Bueren et al. 2011). The advancements in major crops' yield have reached a plateau and even declined in many cases in most parts of the world despite substantial success during the last century (Acreche et al. 2008; Sadras and Lawson 2011).

One of the key goals of molecular biology is gene manipulation. It includes recognizing and manipulating genes by using the available high-quality genome assemblies. Next-generation sequencing (NGS) platforms have now made it possible for researchers to construct vast gene libraries and work out entire genomes

(Wendler et al. 2014). The finding of molecular markers and novel regulatory sequences have been made easier because of the advancements in these platforms (Metzker 2010). Molecular biology has facilitated the study of important aspects of plant development such as the detection of diverse cytoplasmic male sterility sources in hybrid breeding. Cloning of some fertility restorer genes in rice, maize, and sorghum has also been done (Dwivedi et al. 2008). The structure of plant germplasm has been determined using bulked segregant analysis (Zou et al. 2016), association mapping, genome resequencing (Bolger et al. 2014; Edwards and Batley 2010), and fine gene mapping in most studies. Germplasm structure analysis have enabled the identification of single nucleotide polymorphisms (SNPs), single base-pair polymorphisms based on single sequence repeats (SSRs), and unique biomarkers related to quantitative trait loci (QTL) for germplasm augmentation, genome manipulation, and generating gene libraries (Rashmin et al. 2015). Conventional mutagenesis, attempted in many species, requires large-scale screening, and therefore, it is time-consuming and can produce non-desirable mutations in many instances (McCallum et al. 2000). A direct approach for tracking mutations is marker-assisted screening (MAS). It helps to improve backcrossing efficiency (Nadeem et al. 2018) and for determining the homogeneity of the progeny phenotypes. In general, genome cleavage techniques work by making blunt ends, double-stranded breaks (DSBs), or overhangs of target nucleotide fragment, by site-directed incorporation/substitution of genes, homologous recombination, or knockout mutations (Lloyd et al. 2005). The DSBs, formed as a result of the activity of sequence-specific nucleases (SSNs), are repaired by the non-homologous end joining (NHEJ) mechanism, directing DNA substitutions at target sites (Symington and Gautier 2011). Genome editing is largely possible due to three main SSN systems. One is zinc finger nucleases (ZFNs), which forms the basis for DNA manipulation; the second system involves transcription activator-like effector nucleases (TALENs); and more recently, clustered regularly interspaced short palindromic repeats/associated protein 9 (CRISPR-Cas9) system is being used (Bolger et al. 2014; Lloyd et al. 2005; Wood et al. 2011). ZFNs are not cost-effective as they involve non-specific nucleotide recognition and designing and transformation of the constructs are not easy in plants (Pater et al. 2013; Schneider et al. 2016). Since most restriction enzymes are derived from bacteria, TALENs were also isolated from the prokaryotic plant pathogen *Xanthomonas* (Li et al. 2011). However, a lot of time and accuracy to edit the target sequence is needed as TALENs comprise huge and repetitive constructs (Char et al. 2015). Another promising nuclease (CRISPR-Cas9) was found in a bacterial immune system (Mahfouz et al. 2014) soon after the finding of TALENs. Due to its high effectiveness and accuracy in inducing site-directed breaks in double-stranded DNA, this system has been extensively used in recent plant genome-editing studies and has started replacing the TALEN and ZFN systems (Zhang et al. 2017). In recent times, a CRISPR-associated endonuclease, Cpf1 from *Prevotella* and *Francisella1*, has emerged as an alternative tool for precise genome editing. It has higher specificity, potency, and enormous possibilities for comprehensive applications (Li et al. 2018d; Zaidi et al. 2017). To generate point mutations, CRISPR-nCas9 (Cas9 nickase) or dCas9 (deactivated Cas9) combined with cytidine deaminase has proved to be a powerful tool for base editing.

In this chapter, we have summarized the methods used for crop improvement mainly focusing on the amelioration of different CRISPR-based base-editing platforms. We discuss the CRISPR base editors and their efficiencies in both DNA and RNA toward the transformation and advancement of desired phenotypes in plants. This chapter highlights the potential applications of base editing technology in crop improvement using specific case studies, their impediments, and the future implications of this novel emerging technology.

2 The Discovery of CRISPR (Clustered Regularly Interspaced Short Palindromic Repeats)-Based Genome-Editing System, Its Components, and Mode of Action

The timeline of how CRISPR-based editing was discovered in bacteria is very well known. Its usefulness in genome editing was first demonstrated in mammalian cells by Jinek et al. 2012. However, it was only after Doudna and Charpentier (2014) developed and fine-tuned the genome-editing technology that the method found its true potential and wide acceptability. According to Makarova et al. 2011, CRISPR and their associated proteins (Cas) are encoded by most archaea and many bacteria as an adaptive immunity feature to protect them from invading pathogens like viruses. The modular nature of the CRISPR-Cas system was initially interpreted to function in DNA repair, since the domains of Cas proteins exhibited properties similar to nucleases, polymerases, a helicase, and RNA binding proteins (Jansen et al. 2002; Makarova et al. 2002). This misconception was dispelled by the perceptive observation that some of the unique spacers in the CRISPR were nearly identical to the fragments of the genetic material of the invading virus or plasmid, thereby leading to the hypothesis that CRISPR-Cas systems may be involved in defence (Bolotin et al. 2005; Mojica et al. 2005; Pourcel et al. 2005). Later on, a lot of work was done to explore the homology-dependent cleavage mechanism and thus, the technology of CRISPR-Cas9 mediated cleavage became a promising genome-editing tool (Mojica et al. 2005; Liu et al. 2017). Plant and animal genome-editing studies have seen advancement due to the use of CRISPR in recent times. Nearly 5000 articles have been published between 2010 and 2018, describing the applications and implications of CRISPR. Compared to ZFNs/TALENs, CRISPR-Cas9 methodology have been more widely used to edit plant genomes in the last half decade, including model species such as *Arabidopsis*, rice, tobacco and a number of other crop species, reflective of its ease of use (Zhang et al. 2020).

Like any adaptive system, the CRISPR-based editing systems are comprised of two parts; one highly conserved and the other variable between organisms. CRISPR genome editing is more simple, unlike ZFNs and TALENs, and involves designing a guide RNA (gRNA) of about 20 nucleotides complementary to the DNA stretch within the target gene (Mishra et al. 2020). A typical CRISPR-Cas system consists of a series of inter-related mechanisms that work in a highly coordinated manner to modify the genome. Such coordination is also seen in the functioning of each component of the CRISPR-Cas system responsible for the execution of these mechanisms. The CRISPR-Cas loci are comprised of: (i) a CRISPR array which is made up of up to several hundred short (25–35 bp), direct, often palindromic repeats separated by unique spacers (30–40 bp) and (ii) adjacent *cas* genes and accessory genes ordered into operon(s) (Yin et al. 2017). The CRISPR-Cas9 gene knockout system has four important features: (a) synthetic guide RNA (about 18–20 nucleotides) binding to target DNA, (b) Cas9 cleavage at 3–4 nucleotides after the adjacent proto spacer motif (PAM) (generally, 50 NGG identifies the PAM sequence) (Jinek et al. 2012), (c) selection of a suitable binary vector and sgRNA cloning, and (d) transforming the construct in explants via *Agrobacterium* or micro-projectile gene bombardment. *Agrobacterium*-mediated transformation is favored in most cases given its effectiveness and secure delivery (Wang et al. 2018a).

3   Classification of CRISPR/Cas Systems

Robust classification of any system depends on the presence of "universal markers", which, in essence, are motifs that are present in all members of the system, providing them with unique features that distinguish them as a group. However, such markers are difficult to find in CRISPR-Cas systems due to the fast evolution of CRISPR-Cas loci, which also lends them their adaptability, therefore making them very efficient tools of defence. However, this also makes it very difficult to develop a comprehensive classification system for their categorization. The earliest attempts at classification were based on sequence similarities in the loci and conserved Cas proteins such as Cas1 (Makarova et al. 2011, 2015). This formed the basis for the classification of the CRISPR-Cas loci into two broad classes, class 1 and class2, based on the structure of their effector modules involved in crRNA processing and interference. Class 1 systems have effector modules comprised of multiple Cas proteins while the class 2 systems have a single effector module comprised of multiple domains (Makarova et al. 2015). The classification has since evolved to include a number of sub-categories or types and comprises six types divided into two broad classes: class 1 with types I, III, and IV and class 2 with types II, type V, and type VI (Murugan et al. 2017; Koonin and Makarova 2019). The system of classification is still evolving and has resulted in addition of many subtypes as well as restructuring to include more subtle characteristics. At present, classification accounts for 2 classes, 6 types, and 33 subtypes, which have been beautifully explained by Makarova et al. 2020. In their review, Makarova et al. (2020) set forth three significant developments that have led to this increase in the subtypes. Firstly, a number of RNA targeting type VI and subtypes of type V CRISPR-Cas systems have been discovered after the focus on class 2 systems due to their widespread applications in molecular biology. Secondly, several derived CRISPR-Cas systems were discovered with functions distinct from the previous systems and include variants of type I and type V. The third major development was the identification of gene families involved in roles other than adaptive immunity and associated, in particular, with type III variants. These efforts have made it possible to outline a strategy for classification and nomenclature of CRISPR-Cas systems.

4   An Overview of CRISPR-Cas9 in Crop Improvement

Since its discovery and successful customization, the CRISPR-Cas technology has been applied to improve various aspects of crop improvement such as disease resistance, abiotic stress tolerance, and genetic enhancement. For example, bacterial blight is among the most destructive afflictions of cultivated rice and is caused by *Xanthomonas oryzae pv. Oryzae*. The expression of *OsSWEET13*, which is a disease susceptibility gene, is necessary for causing this infection (Zhou et al. 2015). CRISPR-Cas9 technology was used to develop two knockout mutants of *OsSWEET13* targeting its promoter. This led to an improved resistance to bacterial blight disease in indica rice, IR24 (Zhou et al. 2015). In one such study, to examine the role of plant annexins in protection from environmental stresses and plant development, *OsAnn3* CRISPR knockouts were built to study the role played by the rice annexin gene (*OsAnn3*) under cold stress (Shen et al. 2017). It was interesting to note that the survival of T1 mutant lines decreased compared to wild-type plants under cold treatment, which reiterates the observation that several vital traits such as yield and abiotic stress tolerance are controlled by two or more genes. Several studies have attempted to map these quantitative regions (quantitative trait loci or QTL) that control agronomically important traits. Introgression into elite lines for developing better performing varieties can be possible with the identification of such QTL regions. However, introgression of closely linked QTLs is tedious, and introducing non-target regions into the elite line may cause deleterious effects. Hence, CRISPR-Cas9 system can prove to be a powerful tool to initiate and study rare mutations in crop plants.

Many important agronomic traits are determined by point mutations or a few base changes in a gene (Doebley et al. 2006; Ma et al. 2015). To correct the point mutations in the target gene CRISPR-Cas9-mediated gene replacements via homology-directed repair (HDR) has been reported as a feasible approach and has the potential for accelerating crop improvement (Li et al. 2018a, b, c, d, e; Sun et al. 2016; Wang et al. 2017a). In plants, however, low efficiency of template DNA delivery and occasional occurrence of HDR have always been a roadblock in achieving successful transformants (Ran et al. 2017). Moreover, CRISPR-Cas9 system is appropriate for gene knockout or knock-in, but cannot convert one base into another. For stable, unambiguous, and precise genome editing in crops, the need for different approaches was emphasized upon due to such limitations. "Base editing" has emerged as an alternative tool to HDR-mediated replacement which enables precise editing of plant genome by converting one single base to another in a programmed manner, without requiring a donor template or interruption of a gene (Komor et al. 2016). Generally, DNA base editors consist of a fusion of a catalytically inactive nuclease and a catalytically active base-modification enzyme that acts only on single-stranded DNAs (ssDNAs). Base editing is limited to four transition mutations ($C \rightarrow T$, $G \rightarrow A$, $A \rightarrow G$, and $T \rightarrow C$). It is an articulate technology for producing elite trait variations in crop plants, engineering novel traits in agriculturally important crops, and a key to food security (Eid et al. 2018). The base-editing approach has been efficiently standardized and ingrained in several crops including wheat, maize, rice, and tomato (Lu and Zhu 2017; Tang et al. 2019; Zong et al. 2017). So far, two types of DNA editors, the cytosine and adenine base editors (CBEs and ABEs) have emerged as tools for efficient genome modification in eukaryotic genomes (Hua et al. 2018; Liu et al. 2018; Qin et al. 2020; Zong et al. 2017).

5   Base Editors: Structure and Mechanisms

It has been inferred from extensive genome-wide organization studies that SNPs are an important sect of genetic variations that confer many important agronomic traits to plants (Li et al. 2017b, 2018a; Lu and Zhu 2017; Veillet et al. 2019). However, incorporating these favorable single nucleotide changes into commercial cultivars is a challenging endeavor and may take several years. The conception of CRISPR-Cas systems provided a faster, more accurate means to incorporate these variations. However, in many laboratories, this system remains unfeasible due to occurrences of off-target indel formation. This necessitated further improvement of this system to make it more precise and conducive to customizations. Thus, base editors coupled with CRISPR-Cas system are becoming a popular technology for converting one target nucleotide into another without DSBs and DRT (DNA damage repair/tolerance) (Komor et al. 2016; Li et al. 2018e; Nishida et al. 2016).

6   The Development and Evolution of Base Editors

Base editors are engineered ribonucleoprotein complexes that act as tools for single base substitution in cells and organisms. Single base substitutions are now a preferred method of plant genome modification since it can avoid off-target indels. These modifications are possible through base editors (Gaudelli et al. 2017; Komor et al. 2016). The DNA target module is either a catalytically inactive Cas9 nuclease (dCas9) or a Cas9 nickase guided by a sgRNA molecule. An "R-shaped loop" was created by the binding of dCas9-sgRNA to the target DNA, where a stretch of approximately 5–8 nucleotides of the target DNA gets unpaired. For cytosine modification, this small single-stranded domain acts as an editing or catalytic window for dCas9-tethered deaminase. The frequency of indels is limited by the use of base editors as they are capable of making single-base changes or substitutions without creating a DSB in the DNA. There are two types of DNA base editors: cytosine base editors (CBEs) and adenine base editors (ABEs).

6.1 Cytosine Base Editors (CBE)

The simplicity involved in the manufacture of the first-generation cytosine base editor (CBE1) could be misleading. It involved fusing cytidine deaminase, a rat apo-lipoprotein B mRNA-editing enzyme (rAPOBEC1), to the amino terminus of dCas9 using a 16-amino-acid linker (Komor et al. 2016). This CBE1 is capable of converting the cytidine (C) residue in the non-target strand to uracil (U), which is then recognized as Thymine (T) by cell replication machinery, resulting in a C·G to T·A transition (Komor et al. 2016). However, like any engineered system, the first generation of CBEs presented scope for improvement. The drawback was that base-excision repair usually reverses the conversion of C·G to T·A, and uracil DNA N-glycosylase (UNG) generally removes U in the C·G to U·A pair (Kunz et al. 2009). A second-generation cytosine base editor (CBE2) was developed to counter this problem. The activity of UNG was diminished by the inclusion of a uracil DNA glycosylase inhibitor (UGI) into CBE1 (Di Noia and Neuberger 2002). The system was further improved by the development of third-generation CBEs (CBE3) by replacing the dCas9 in CBE2 with Cas9 nickase (nCas9) containing the D10A mutation (Bharat et al. 2020). The D10A mutation causes the modified nCas9 to nick the gRNA-complementary DNA strand or the target strand at a site that flanks the U/G mismatch. The mismatch repair system of the cell recognizes this nick as it is a preferred substrate and then excises the newly synthesized DNA strands according to the existence of nicks (Kunkel and Erie 2015). The U/G mismatch and the flanking nick generated by CBE3 are thus, recognized in the target strand and G of U/G mismatch is excised. Consequently, the remaining U-containing non-target strand is used as a template for DNA re-synthesis to install a U/A pair, which would then be converted to a T/A pair after DNA replication or repair. This led to a considerable improvement in the efficiency of CBE3 over CBE2 (Komor et al. 2016). However, the inherent ability of cells to reject any type of change in the genome makes incorporating even single base modifications very difficult. Although CBE3 provides a go-around for the mismatch repair system, researchers were now faced with the base excision repair machinery, which is responsible for removing damaged bases with DNA glycosylases (Carter and Parsons 2016). This lead to low C-to-T editing efficiency as the U in the editing intermediate (U/G pair) could still be excised by UDG. Therefore, using a number of different methods, the fourth-generation cytosine base editor (CBE4) was engineered. In one strategy, Komor et al. (2017) fused another copy of UGI into CBE3 to keep the U in place leading to higher efficiency. Another strategy involves streamlining of codon to enhance the expression of the base editors, which increased base editing frequency by up to nine-fold (Koblan et al. 2018; Zafra et al. 2018). Yet another barrier to efficient base editing is the process of DNA methylation which suppresses the cytosine deamination catalyzed by mouse APOBEC1 (mA1) (Nabel et al. 2012). This led to the screening of efficient base editors that could induce editing in highly methylated regions of the genome. As a result, the human hA3A-derived CBE (hA3A-BE3) was identified to have the highest efficiency (Wang et al. 2018b). Another approach to optimize the CBE was to decrease indel formation at the time of base editing. This has been achieved by:

(a) creating a mutation in rAPOBEC1 to narrow the catalytic window (5th–sixth nucleotide) compared to CBE3 and therefore improve the accuracy of editing (Kim et al. 2017c),
(b) to generate an activation-induced cytidine deaminase (AID) base editor, combining Lamprey cytidine deaminase (pmCDA1), which is similar to rAPOBEC1 in structure and function, with activation-induced cytidine deaminase (AID), nCas9 (D10A), and UGI showed a low frequency of indel mutations and high on-target activity (Nishida et al. 2016),
(c) fusing Gam, a DNA-binding protein from bacteriophage Mu, with CBE3 to reduce indel formation during base editing as Gam can form a complex with ends of DSBs and protect them from degradation (di Fagagna et al. 2003).

Unlike SpCas9, which requires G/C rich PAM sequences (Kim et al. 2017a, b, c; Li et al. 2018c, 2019a, b, c; Tang et al. 2017; Wang et al. 2017a, b; Xu et al. 2017) Cpf1, a class 2, type V CRISPR-Cas system, is capable of recognizing the thymidine-rich TTTN PAM. Therefore, complementing the more popular SpCas9 system (NGG PAM) with Cpf1 could enable editing of AT-rich regions such as 5′ and 3′UTRs and promoter regions (Zetsche et al. 2015; Zong et al. 2017). Moreover, the PAM sequences could also be altered to improve accuracy and specificity (Hu et al. 2018; Nishimasu et al. 2018; Hua et al. 2019a).

6.2 Adenine Base Editor (ABE)

Another group of base editors facilitate the conversion of A·T to G·C in genomic DNA and are known as Adenine base editors (ABE). Generally, ABEs are single polypeptide chains comprised of: a) a wild type *E.coli* tRNA-specific adenosine deaminase (TadA) monomer, b) a synthetic *E. coli* TadA monomer (TadA*) which catalyzes deoxyadenosine deamination, and c) a Cas9 (D10A) nickase (Gaudelli et al. 2017). These components were combined to develop ABE7.10 which can effectively convert A•T-to-G•C while minimizing the levels of unwanted by-products, such as indels, in many organisms including plants (Kang et al. 2018; Ryu et al. 2018). In native conditions, the *E. coli* TadA homodimerises to deaminate adenosine located in a transfer RNA (tRNA) anticodon loop generating inosine (I) (Losey et al. 2006). Thus, ABE functions by binding to a target DNA sequence in a guide RNA-dependent manner, creating a small bubble of single-stranded DNA where the TadA-TadA* acts to deaminate an A located in this bubble to generate

I. Inosine (I) is recognized as G, following DNA repair or replication, and finally the original A·T base pair is replaced with a G·C base pair at the target site (Li et al. 2019a; Wolf et al. 2002). ABEs have emerged as a more popular choice for base editing as numerous studies have reported that the occurrence of RNA-dependent off-target DNA mutations were much lower in ABE7.10 as compared to CBEs (Jin et al. 2019; Zuo et al. 2019).

As in the case of CBEs, the efficiency of ABEs has also been improved using codon optimization and nuclear localization sequence optimization (Koblan et al. 2018; Zafra et al. 2018). Rees et al. (2019) were able to greatly reduce low-level cellular RNA editing from ABEs by introducing the E59A or E59Q mutation into TadA and the V106W mutation in TadA* without heavily compromising on-target DNA editing. Recently, ABE-P1S, a simplified base editor containing ecTadA*7.10-nSpCas9 (D10A) is known to show much higher editing efficiency in rice than the widely used ecTadA-ecTadA*7.10-nSpCas9 (D10A) fusion. The ecTadA*7.10-nCas9 fusion may ameliorate the editing efficiency of other ABEs containing SaCas9 or the SaKKH-Cas9 variant (Hua et al. 2020). Exploiting more efficient ABEs will facilitate the application of adenine base editing in crop improvement.

7   RNA Base Editors

Despite their widespread use, the Cas9 in DNA base editors presents the drawback of requiring a protospacer adjacent motif (PAM) at the editing site (Kim et al. 2017a, b, c) which led researchers to look out for alternatives that could produce a similar effect post-transcriptionally. The process of RNA editing has been widely studied in humans and is mediated by Adenosine deaminases acting on RNA (ADAR) which convert Adenosine (A) to Inosine (I) (Gilbert et al. 2016; Zhao et al. 2016). This is read as guanosine by the splicing and translation apparatuses, thereby leading to a modified amino acid sequence (Nishikura 2016). Cox et al. (2017) came up with the idea of editing RNA transcripts to alter their coding potential in a programmable manner and named it "RNA Editing for Programmable A to I Replacement" (REPAIR). A type VI CRISPR-associated RNA-guided RNase, Cas13 has RNA binding abilities and a previous study also reported that the Cas13a enzyme from *Leptotrichia wadei* (LwaCas13a) could cleave RNA and could be expressed in mammalian and plant cells for targeted knockdown of transcripts (Abudayyeh et al. 2017). The E488Q-mutated ADAR2 deaminase domain (ADAR$_{DD}$) was fused to catalytically inactive PspCas13b (dCas13) to generate the first generation of the REPAIR system. REPAIR is capable of editing full-length transcripts containing disease-relevant mutations in a mammalian cell and has no strict sequence limitations (Cox et al. 2017). The same group also improved the system and developed REPAIRv1 and REPAIRv2, which demonstrated higher precision than other RNA-editing platforms (Stafforst and Schneider 2012; Cox et al. 2017).

In addition to the REPAIR system, Abudayyeh et al. (2019) were also responsible for the development of an efficient RNA-editing system for precise C-to-U conversion called RNA Editing for Specific C to U Exchange (RESCUE). To accomplish this, they mutated the adenine deaminase domain of ADAR2, through directed evolution, to turn it into a cytidine deaminase and fused it to catalytically inactive Cas13 (dCas13).

8   Application of Base Editors in Crop Improvement

Transgenic breeding, where genes from other species/ organisms are introduced into crop plants of interest, has been one of the more adopted and used methods of crop improvement. Despite being standardized and widely applied, this method has given only a few transgenic crops to be utilized at field conditions (Raman 2017). The reason for such low output is the strict regulation of these genetically modified crops since this method integrates foreign DNA at random positions in the plant genome, sometimes leading to deleterious effects. However, crop improvement does not necessarily mean integrating long pieces of DNA into the plant genome. Single nucleotide changes in the genome are responsible for several agronomically important traits. Hence, base editing can be used as a powerful tool to introduce/ correct point mutations and accelerate crop improvement. Since base editors, especially RNA base editors, could generate DNA-free editing, these plants could be potentially free of the GMO tag and the regulations that are imposed with it (Voytas and Gao 2014). This is one of the major incentives for the popularization of base editors. To edit specific genes conferred by single nucleotide polymorphisms, cytosine and adenine base editors have been successfully used in a wide range of major crops and model plants (Hua et al. 2018; Li et al. 2018a, b, c, d, e; Lu and Zhu 2017; Ren et al. 2018).

8.1   CBE-Mediated Base Editing in Plants

The effective customization of CBEs using codon optimization, linker addition, tandem UGIs, etc., has been applied to develop stable and improved plants. The method has proved to be effective and even led to desired changes that could be carried into the next generation of offspring. To assess the C·G to T·A base-editing activity of xCas9-3.7 in plants, in one study, SpCas9 was replaced with xCas9-3.7 in CBE3, giving rise to xBE3. xBE3 has the ability to edit sites with NGN, GAA, and GAT PAMs. However, CBE and ABE containing the xCas9 variant failed to edit most target sites in plants. Cas9-NG variants, Cas9-NGv1 and Cas9-NG in rice protoplasts and stable lines, showed reduced activity at canonical NGG PAM sites but showed noticeable mutation frequency at non-canonical NG PAM sites (Hua et al. 2019a; Ren et al. 2019; Zhong et al. 2019). Zeng et al. (2020) were able to develop Plant high-efficiency CBEs (PhieCBEs) with expanded target ranges by fusing genetically evolved cytidine deaminases with two Cas9n-NG variants. In yet another commendable effort, Jin et al. (2020) were able to design CBEs with more precision in their genome-wide activity by designing their deaminase domains to have lower affinities for ssDNA. All these efforts reiterate the versatility of base editors. Some of the examples of successful gene editing are listed in Table 1.

Table 1 Application of CBEs in crop improvement

Base editor used	Plant	Target	Effect	References
CBE	Rice	*OsPDS, OsSBEIIb*	High amylose rice	Li et al. (2017b)
CBE3	Rice	*NRT1.1B* and *SLR1*	Improved nitrogen use efficiency	Lu and Zhu (2017) and Hu et al. (2015)
CBE	Tomato	Acetolactate synthase (ALS) gene	Marker-free and herbicide-resistant	Shimatani et al. (2017)
nCas9-cytidine deaminase fusion	Rice, wheat, and corn	*OsCDC48, OsSPL14, TaLOX2, ZmCENH3* genes	Efficient and site-specific C to T base editing in rice, wheat, and maize, with a deamination window covering 7 bases of the protospacer and produces virtually no indel mutations.	Zong et al. (2017)
Codon-optimized cytidine deaminase-Cas9n-UGI (CBE3)	Arabidopsis	Acetolactate synthase (ALS) gene	Herbicide resistance	Chen et al. (2017) and Dong et al. (2020)
CBE	Rice	Pi-d2andOsFLS2	Blast resistance	Ren et al. (2018)
CBE	Watermelon	Acetolactate synthase (ALS) gene	Herbicide resistance	Tian et al. (2018)
CBE	tomato and potato	Acetolactate synthase (ALS) gene	Herbicide resistance	Veillet et al. (2019)
CBE	Cotton	*GhCLA* and *GhPEBP* genes	Visible phenotype (editing of *GhCLA* can generate an albino phenotype and *GhPEBP* participates in the multiplex-branch developmental process	Qin et al. (2020)
A3A-PBE	Oilseed rape	*ALS* gene	Herbicide tolerance	Cheng et al. (2021) and Wu et al. (2020)

8.2 ABE-Mediated Base Editing in Plants

When it comes to minimizing off-target mutations, ABEs have proved to be superior to CBEs. In one study, researchers observed that there was a higher rate of C > T mutation in BE3 and HF1-BE3 plants relative to controls which could be attributed to APOBEC1 and/or UGI. On the other hand, ABE is derived by fusing a nCas9 protein with an engineered RNA adenosine deaminase, and therefore, it is possible that the engineered RNA adenosine deaminase does not show excessive DNA base editing. This minimizes the generation of genome-wide A > G SNVs outside the sgRNA targeting windows (Jin et al. 2019) making ABEs a preferred choice for genome editing in plants. In recent years, a number of changes have been introduced into the backbone of ABEs to cater to the specific needs of plant biologists. For example, to efficiently introduce A·T to G·C conversion, fluorescence-tracking adenine base editor (rBE14) using the nCas9 (D10A)-guided TadA: Tad A7.10 heterodimer has been developed in *OsMPK6, OsSERK2*, and *OsWRKY45* at respective frequencies of 16.7%, 32.1%, and 62.3% in rice (Yan et al. 2018). A·T to G·C conversion was enabled at frequencies up to 7.5% in protoplasts and 59.1% in regenerated rice and wheat crops, where they could produce high precision substitutions at the targeted loci with low indels. This was done using a new plant ABE based on an evolved tRNA adenosine deaminase fused to the nCas9 (Li et al. 2018a). A plant ABE (designated ABE-P1 for adenine base editor, plant version) was developed by Hua et al. (2018) by fusing recombinant *E. coli* TadA (ecTadA*7.10) protein to the N terminus of the nCas9 (D10A). The editing efficiency of ABE-P1 was evaluated in *OsSPL1* and *OsSLR14* gene loci of rice, and the editing efficiencies were 26.0% and 12.5%, respectively. ABE-P1 also enabled multiplex base editing with high efficiency (Hua et al. 2018). A similar approach was used to synthesize four plant-compatible ABE binary vectors (pcABEs) (Kang et al. 2018). The target regions for pcABE7.10 were sequenced and showed that the efficiency A·T to G·C editing activity was up to 4.1% and 8.8% in *Arabidopsis* and *Brassica napus* (rapeseed) respectively. To broaden their targetable sites, new ABEs were developed using engineered *SpCas9* variants. This was emulated in the editing efficiencies of two target genes, *OsSPL17* and *OsSPL14* in rice. Through such results, the applications of ABE in plants further expanded as it exhibited that ABEs containing SpCas9-NG worked efficiently in rice with broadened PAM compatibility (Hua et al. 2019a, b). Moreover, it has been shown that adenine and cytosine base editing could be simultaneously performed in rice (Hua et al. 2019a). Some of the other attempts at crop improvement and study of plant developmental processes using ABEs are listed in Table 2.

Table 2 Application of ABEs in crop improvement and functional characterization

Base editor used	Plant	Target gene	Effect	References
Enhanced ABE system	Rice	Carboxyl transferase domain of OsACC	Herbicide resistance (although accompanied by severe growth retardation and sterility)	Liu et al. (2020)
ABE	Rice	*OsWSL5,OsZEBRA3(Z3)*	Visible phenotype of white stripe leaf and light green/dark green leaf pattern, respectively.	Molla et al. (2020)
ABE	Rice	*OsTubA2* gene	Produced rice lines resistant to otherwise lethal levels of trifluralin and pendimethalin without fitness penalty	Liu et al. (2021)

8.3 Best of Both Worlds: Combining CBEs and ABEs for More Efficient Crop Gene Editing

Together, ABEs and CBEs enable the introduction of all four transitions (C·G to T·A or A·T to G·C) at target loci in the genome in a user-defined way, greatly expanding the capability of base editing. Therefore, it was logical to combine the editing capabilities of CBEs and ABEs to develop a highly efficient base editor that could facilitate directed mutagenesis in the plant genomes. This has led to engineering of saturated targeted endogenous mutagenesis editors (STEMEs) that combined dual base editors with a single sgRNA to achieve C:G > T:A and A:T > G:C substitutions by fusing cytidine deaminase with adenosine deaminase nCas9 (D10A) and uracil DNA glycosylase inhibitor (UGI) (Li et al. 2020). A similar approach was used to develop the base-editing-mediated gene evolution (BEMGE) method, employing both Cas9n-based cytosine and adenine base editors as well as a single-guide RNA (sgRNA) library tiling the full-length coding region, for developing novel rice germplasms with mutations in any endogenous gene (Kuang et al. 2020). These novel methods of base editing have proved to be useful in targeting any endogenous gene in any crop related to a trait of interest (e.g., plant height, leaf angle, seed size, seed protein content) in plants to identify novel alleles of varying strength. Thus, BEMGE is a powerful tool that can be used in breeding programs to identify functional genetic variants with precision and develop crops with specific traits in the near future.

9 RNA Editing in Plants

Despite their usefulness being established in some animal systems, RNA-editing methods (REPAIR and RESCUE) have not yet been utilized in plant systems effectively. However, there have been attempts at editing the mRNA in plants using base editors to disrupt gene activity post transcriptionally. This method of editing relies on the knowledge that although the internal sequences of introns exhibit great variations, the donor and acceptor sites, i.e., the GT and AG dinucleotides at the 5′ and 3′ ends of an intron, are highly conserved across plants and animals. Introducing point mutations at either the donor site GT or acceptor site AG of an intron would cause mRNA mis-splicing, leading to gene disruption (Brown 1996). Inspired by those natural mRNA mis-splicing mutants, Li et al. (2019c) applied this principle and modified CBEs to mutate these sites in *OsGL1*-1 and *OsNAL1* genes in rice, thereby inducing mRNA mis-splicing and target gene inactivation.

10   Limitations and Prospects

The CRISPR base editing technology has been used extensively not only in animal systems but also in several plants, since it permits accurate and effective conversion of single-base at targeted genomic sites. Limited PAM sites, high off-target activity, and wide editing window are some of the limiting factors for the pervasive use of base editors (Shimatani et al. 2017; Tian et al. 2018; Zong et al. 2017). A lot of efforts have been put in the past 2 years to upgrade the specificity of base editors and minimize their limitations. Further research efforts are required for improving the existing ABE and CBE tools, developing novel base editors, and applying REPAIR and RESCUE systems in plants. The CBE system can also engineer base substitutions simultaneously at multiple loci (Shimatani et al. 2018) and is being used widely for crop improvement. It uses multiplex base editors that allow editing of crop plants with various desired SNPs at multiple gene loci. Despite the low efficiency of editing, the RNP method provides a potential path to avoid regulatory obstacles and to make DNA-free base editing useful in advertising crops with improved agronomic traits (Rees et al. 2017).

10.1   Off-Target Activity of Base Editors

The application of base editing technique is limited by off-target editing. This occurs when additional cytosines are edited at a site proximal to the target base in the same DNA molecule. The genome-wide off-target activity of CBE has been demonstrated in rice and mouse cells (Jin et al. 2019; Zuo et al. 2019). However, in some studies it was shown that ABEs displayed unexpected cytosine deamination activity in human cells in addition to off-target single-nucleotide conversions at the DNA or RNA level in the presence or absence of sgRNAs (Kim et al. 2019). In a recent study, it was observed that CBEs BE3 and high-fidelity BE3 (HF1-BE3) induce unexpected and unpredictable genome-wide off-target mutations in rice crop. These mutations were usually the C to T type of single nucleotide variants (SNVs) (Jin et al. 2019). It was suggested that to minimize the off-target mutations, it is necessary to optimize the level of cytidine deaminase and UGI (Zuo et al. 2019). These CBE limitations could be resolved by:

1. Using different cytosine deaminases and Cas-nuclease fusions to generate more efficient CBEs.
2. Developing DNA base editors that have reduced undesirable RNA-editing activity.
3. Reducing sgRNA-independent deamination activity by the generation of cytosine deaminase (Zhou et al. 2013).
4. Regulating the expression level of cytosine deaminase-UGI protein complex (Grünewald et al. 2019b).

Gene therapy in animal cells has become a major concern due to induced off-target edits by CBE and ABE base editors at the DNA and RNA level (Grünewald et al. 2019a). However, these effects might not create any hurdle in plants if the off-target effects did not affect major agronomic traits and the desired base substitutions are acquired.

10.2 Generation of Indels

As compared to genome editing, DNA base editing yields a lower rate of indel formation; however, indels are still present. It has been conjectured (Kim et al. 2017b) that nCas9 (D10A) cleavage of single-stranded DNA is responsible for the production of these indels. Usually, the frequency of introduced indels is lower for ABEs than for CBEs (Gaudelli et al. 2017). Komor et al. indicated that the number of indels can be greatly reduced by fusing SunTag to the N-terminus of the Cas9 (D10A) nickase (Jiang et al. 2018) or by fusing bacteriophage Mu-derived Gam (Mu-GAM) protein to BE4 (Komor et al. 2017).

10.3 Need for a More Relaxed PAM Requirement

ABE and CBE base editors without any gene disruption can mediate efficient base editing. However, the presence of Protospacer Adjacent Motif (PAM) sequences and the editing window at the target locus limit the applications of DNA base editors. The PAM sequences determine the editing frequencies within the genome as most editors are based on the CRISPR-Cas9 system. Only the NGG PAM site was recognized by the first generation of base editors. It was engineered based on SpCas9 (Komor et al. 2016). To further expand the technology, new types of Cas protein and engineered Cas9 variants with altered PAM sequences have been introduced into base editors (Hu et al. 2018; Nishimasu et al. 2018). For example, a T-rich PAM sequence is recognized by LbCpf1-BE0, a Cpf1-based CBE that catalyzes C·G to T·A conversion in a narrow window (eighth-13th nucleotide of the protospacer) (Li et al. 2017a). The use of modified SpCas9 to create a modified adenine and cytosine base editors considerably broadened the range of targets (Hua et al. 2019b). xCas9 and SpCas9-NG base editors in plants helped in achieving C·G to T·A and A·T to G·C conversions at target sites with NG PAMs (Hua et al. 2019a, b; Ren et al. 2019; Zhong et al. 2019). Another drawback of DNA base editors are large editing windows. To overcome this limitation, the linker sequences were modified and the non-essential sequences were removed by developing high-precision base editors with narrowed editing windows that narrowed down from ~5 to as few as 1–2 nucleotides with engineered YEE-BE3 base editors. These YEE-BE3 base editors contain mutated cytidine deaminase domains that enable discrimination of neighboring C nucleotides, which would otherwise be edited with similar efficiency in HEK293T cells (Kim et al. 2017a, b, c). The specificity of base editing has increased significantly by the development of improved base editors and is expected to be used for precise single-base substitutions in plants.

10.4 Limited Application of RNA Base-Editing Technology in Plants

RESCUE and REPAIR systems have been used to edit RNA transcripts containing disease-relevant mutations in mammalian cells. However, the use of RNA base editing in crop improvement awaits exploration. The use of CRISPR base-editing technology in RNA transcripts in plants will greatly expand by these new systems. RNA base editing will be significant for plant biological research, such as in functional analysis of genes involved in a complex metabolic pathway. However, it is difficult to distinguish phenotypic variation caused by base changes in RNA transcripts from the position effect of a base editor transgene, if such effect is present.

10.5 Future Perspectives of the Emerging Technology

Since their invention, base editors are being used for basic research purposes and are being continually optimized for use in medicine and agriculture. For molecular breeding of crops, improved base editors will prove to be valuable tools. Newly engineered variants need to be adapted to improve the existing CBE and ABE base editors to expand the scope of base editing and increase the efficiency of editing in an extensive variety of crops. The plant ABE system has been successfully adapted in a wide range of crops (Li et al. 2018a, b, c, d, e; Yan et al. 2018). However, engineered Cas9 variants recognizing different PAM sequences or Cpf1 can be used for improving and extending the plant ABE system (Hu et al. 2018; Kim et al. 2017a, b, c). Moreover, to correct point mutations, the sgRNAs could be ligated with different aptamers MS2, PP7, COM, and boxB to facilitate simultaneous base conversions

(C-T and A-G) (Ma et al. 2016; Zalatan et al. 2015). Precise conversion of nucleotides is possible through protein delivery of base editors (Rees et al. 2017). Thus, highly accurate plant base editors should be generated and delivered through RNP delivery to establish DNA free strategy. In several crops, base editors have already been used in correcting point mutations related to agricultural traits. They were used to edit genes in rice, wheat, and corn with C: G to T: A editing efficiencies of 12–44%. However, RNA editing in plants is yet to be explored. At present, the REPAIR system enables A to I conversion in RNA editing. In future, C to U conversions may also be possible by fusion of dCas13 with other catalytic RNA-editing domains such as APOBEC (Cox et al. 2017). Catalytically inactive dCas9 or dCpf1 could enable conversion of A to I on DNA substrates either through formation of DNA–RNA hetero-duplex targets (Zheng et al. 2017) or mutagenesis of ADAR domain (Cox et al. 2017). The use of REPAIR system in plants is unexplored, although it is being utilized to correct disease-relevant mutations in human. Stable expression of CRISPR base editors is required for use of RNA editing, and therefore, it may not be highly beneficial in crop bioengineering. However, it could be useful for functional gene analysis. There are vast applications of this system that remains unexplored and researchers may find it interesting to unearth the many applications, which can be used for crop plants in the near future.

11 Conclusion

Genome editing has become a powerful tool since the development of the first CRISPR-Cas editing system less than a decade ago. Base editing is a novel platform within the genome designing tool kits, competent in modifying crops precisely and enhancing the quality of crops in the future. Base editors are proving to be an effective genome-editing approach, as instead of causing DSBs in DNA, it enables nucleotide substitutions. The process uses CRISPR components such as, guide RNAs, a mutant Cas9 and deaminase enzymes, to mutate single bases without altering the nearby sequence and without a DSB or donor template. For base editing in plants as well as in animals, cytidine and adenine deaminase-based base editors are continuously being upgraded and used. Base editing seems to be the method of choice for fixing point mutations as it removes the unwanted insertion and/or deletions caused by gene editing. However, due to their innate properties like PAM sequence dependency, base editors cannot act upon all bases in a genome. Moreover, the unexpectedly high number of off-targets introduced by CBEs is a major issue. Adopting the Cas9 variants and narrowing down the catalytic window to improve the existing CBE and ABE base editors can enhance DNA specificity and lower the off-target activity. This emerging base-editing technology is still in its infancy and needs optimization as highly precise base editors can have many applications and scope in model plants and crops for precision breeding.

References

Abreu GB, Souza J (2010) Strategies to improve mass selection in maize Selection of maize progenies for tryptophan content and grain yield View project Re-indução de longevidade em sementes osmocondionadas de tomate. View project

Abudayyeh OO, Gootenberg JS, Franklin B, Koob J, Kellner MJ, Ladha A, Joung J, Kirchgatterer P, Cox DB, Zhang FA (2019) Cytosine deaminase for programmable single-base RNA editing. Science. 365(6451):382–386. https://doi.org/10.1126/science.aax7063

Abudayyeh OO, Gootenberg JS, Essletzbichler P, Han S, Joung J, Belanto JJ, Verdine V, Cox DBT, Kellner MJ, Regev A, Lander ES, Voytas DF, Ting AY, Zhang F (2017) RNA targeting with CRISPR-Cas13. Nature 550(7675):280–284. https://doi.org/10.1038/nature24049

Acreche MM, Briceño-Félix G, Sánchez JAM, Slafer GA (2008) Physiological bases of genetic gains in Mediterranean bread wheat yield in Spain. Eur J Agron 28(3):162–170. https://doi.org/10.1016/j.eja.2007.07.001

Araus JL, Kefauver SC, Zaman-Allah M, Olsen MS, Cairns JE (2018) Translating high-throughput phenotyping into genetic gain. Trends Plant Sci 3(5):451–466. Elsevier Ltd. https://doi.org/10.1016/j.tplants.2018.02.001

Bharat SS, Li S, Li J, Yan L, Xia L (2020) Base editing in plants: current status and challenges. Crop J 8(3):384–395. Crop Science Society of China/Institute of Crop Sciences. https://doi.org/10.1016/j.cj.2019.10.002

Bolger ME, Weisshaar B, Scholz U, Stein N, Usadel B, Mayer KFX (2014) Plant genome sequencing – applications for crop improvement. Curr Opin Biotechnol 26:31–37. Elsevier Current Trends. https://doi.org/10.1016/j.copbio.2013.08.019

Bolotin A, Quinquis B, Sorokin A, Ehrlich SD (2005) Clustered regularly interspaced short palindrome repeats (CRISPRs) have spacers of extrachromosomal origin. Microbiology 151(8):2551–2561. https://doi.org/10.1099/mic.0.28048-0

Brown JWS (1996) Arabidopsis intron mutations and pre-mRNA splicing. Plant J 10(5):771–780. https://doi.org/10.1046/j.1365-313X.1996.10050771.x

Carter RJ, Parsons JL (2016) Base excision repair, a pathway regulated by posttranslational modifications. Mol Cell Biol 36(10):1426–1437. https://doi.org/10.1128/mcb.00030-16

Char SN, Unger-Wallace E, Frame B, Briggs SA, Main M, Spalding MH, Vollbrecht E, Wang K, Yang B (2015) Heritable site-specific mutagenesis using TALENs in maize. Plant Biotechnol J 13(7):1002–1010. https://doi.org/10.1111/pbi.12344

Chen Y, Wang Z, Ni H, Xu Y, Chen Q, Jiang L (2017) CRISPR/Cas9-mediated base-editing system efficiently generates gain-of-function mutations in Arabidopsis. Sci China Life Sci 60(5):520–523. https://doi.org/10.1007/s11427-017-9021-5

Cheng H, Hao M, Ding B, Mei D, Wang W, Wang H, Zhou R, Liu J, Li C, Hu Q (2021) Base editing with high efficiency in allotetraploid oilseed rape by A3A-PBE system. Plant Biotechnol J 19(1):87–97. https://doi.org/10.1111/pbi.13444

Cox DBT, Gootenberg JS, Abudayyeh OO, Franklin B, Kellner MJ, Joung J, Zhang F (2017) RNA editing with CRISPR-Cas13. Science 358(6366):1019–1027. https://doi.org/10.1126/science.aaq0180

di Fagagna F, d'Adda, Weller GR, Doherty AJ, Jackson SP (2003) The Gam protein of bacteriophage Mu is an orthologue of eukaryotic Ku. EMBO Rep 4(1):47–52. https://doi.org/10.1038/sj.embor.embor709

Di Noia J, Neuberger MS (2002) Altering the pathway of immunoglobulin hypermutation by inhibiting uracil-DNA glycosylase. Nature 419(6902):43–48. Nature Publishing Group. https://doi.org/10.1038/nature00981

Doebley JF, Gaut BS, Smith BD (2006) The molecular genetics of crop domestication. Cell 127(7):1309–1321. Cell Press. https://doi.org/10.1016/j.cell.2006.12.006

Dong H, Wang D, Bai Z, Yuan Y, Yang W, Zhang Y, Ni H, Jiang L (2020) Generation of imidazolinone herbicide resistant trait in Arabidopsis. PLoS One 15(5):e0233503. https://doi.org/10.1371/journal.pone.0233503

Doudna JA, Charpentier E (2014) The new frontier of genome engineering with CRISPR-Cas9. Science 346(6213):1077

Dreher K, Morris M, Khairallah M (2002) Cost-effective Compared with Conventional Plant Breeding Methods?. Economic and social issues in agricultural biotechnology 1:203

Dwivedi S, Perotti E, Ortiz R (2008) Towards molecular breeding of reproductive traits in cereal crops. Plant Biotechnol J 6(6):529–559. https://doi.org/10.1111/j.1467-7652.2008.00343.x

Edwards D, Batley J (2010) Plant genome sequencing: applications for crop improvement. Plant Biotechnol J 8(1):2–9. John Wiley & Sons, Ltd. https://doi.org/10.1111/j.1467-7652.2009.00459.x

Eid A, Alshareef S, Mahfouz MM (2018) CRISPR base editors: genome editing without double-stranded breaks. Biochem J 475(11):1955–1964. Portland Press Ltd. https://doi.org/10.1042/BCJ20170793

FAO—Food and Agriculture Organization of the United Nations (2017). How to feed the world in 2050 (FAO, Rome). http://www.fao.org/fileadmin/templates/wsfs/docs/expert_paper/How_to_Feed_the_World_in_2050.pdf.

Gaudelli NM, Komor AC, Rees HA, Packer MS, Badran AH, Bryson DI, Liu DR (2017) Programmable base editing of T to G C in genomic DNA without DNA cleavage. Nature 551(7681):464–471. https://doi.org/10.1038/nature24644

Gilbert WV, Bell TA, Schaening C (2016) Messenger RNA modifications: form, distribution, and function. Science 352(6292):1408–1412. https://doi.org/10.1126/science.aad8711

Grünewald J, Zhou R, Garcia SP, Iyer S, Lareau CA, Aryee MJ, Joung JK (2019a) Transcriptome-wide off-target RNA editing induced by CRISPR-guided DNA base editors. Nature 569(7756):433–437. https://doi.org/10.1038/s41586-019-1161-z

Grünewald J, Zhou R, Iyer S, Lareau CA, Garcia SP, Aryee MJ, Joung JK (2019b) CRISPR DNA base editors with reduced RNA off-target and self-editing activities. Nat Biotechnol 37(9):1041–1048. https://doi.org/10.1038/s41587-019-0236-6

Hopkins A (2003) A vision for the future of genomics research. Nature 422:835–847

Hu B, Wang W, Ou S, Tang J, Li H, Che R, Zhang Z, Chai X, Wang H, Wang Y, Liang C (2015) Variation in NRT1. 1B contributes to nitrate-use divergence between rice subspecies. Nat Genet 47(7):834–838. https://doi.org/10.1038/ng.3337

Hu JH, Miller SM, Geurts MH, Tang W, Chen L, Sun N, Zeina CM, Gao X, Rees HA, Lin Z, Liu DR (2018) Evolved Cas9 variants with broad PAM compatibility and high DNA specificity. Nature 556(7699):57–63. https://doi.org/10.1038/nature26155

Hua K, Tao X, Yuan F, Wang D, Zhu JK (2018) Precise A·T to G·C Base editing in the Rice genome. Mol Plant 11(4):627–630. Cell Press. https://doi.org/10.1016/j.molp.2018.02.007

Hua K, Tao X, Han P, Wang R, Zhu JK (2019a) Genome engineering in Rice using Cas9 variants that recognize NG PAM sequences. Mol Plant 12(7):1003–1014. https://doi.org/10.1016/j.molp.2019.03.009

Hua K, Tao X, Zhu J-K (2019b) Expanding the base editing scope in rice by using Cas9 variants. Plant Biotechnol J 17(2):499–504. https://doi.org/10.1111/pbi.12993

Hua K, Tao X, Liang W, Zhang Z, Gou R, Zhu J (2020) Simplified adenine base editors improve adenine base editing efficiency in rice. Plant Biotechnol J 18(3):770–778. https://doi.org/10.1111/pbi.13244

Jansen R, van Embden J. DA, Gaastra W, Schouls LM (2002) Identification of genes that are associated with DNA repeats in prokaryotes. Mol Microbiol 43(6). https://doi.org/10.1046/j.1365-2958.2002.02839.x

Jiang W, Feng S, Huang S, Yu W, Li G, Yang G, Liu Y, Zhang Y, Zhang L, Hou Y, Chen J, Chen J, Huang X (2018) BE-PLUS: a new base editing tool with broadened editing window and enhanced fidelity. Cell Res 28(8):855–861. https://doi.org/10.1038/s41422-018-0052-4

Jin S, Zong Y, Gao Q, Zhu Z, Wang Y, Qin P, Liang C, Wang D, Qiu JL, Zhang F, Gao C (2019) Cytosine, but not adenine, base editors induce genome-wide off-target mutations in rice. Science 364(6437):292–295. https://doi.org/10.1126/science.aaw7166

Jin S, Fei H, Zhu Z, Luo Y, Liu J, Gao S, Zhang F, Chen YH, Wang Y, Gao C (2020) Rationally designed APOBEC3B Cytosine Base editors with improved specificity. Mol Cell 79(5):728–740. e6. https://doi.org/10.1016/j.molcel.2020.07.005

Jinek M, Chylinski K, Fonfara I, Hauer M, Doudna JA, Charpentier E (2012) A programmable dual-RNA-guided DNA endonuclease in adaptive bacterial immunity. Science 337(6096):816–821. https://doi.org/10.1126/science.1225829

Kandemir N, Saygili İ (2015) Apomixis: new horizons in plant breeding. Turk J Agric For 39(4):549–556. https://doi.org/10.3906/tar-1409-74

Kang BC, Yun JY, Kim ST, Shin YJ, Ryu J, Choi M, Woo JW, Kim JS (2018) Precision genome engineering through adenine base editing in plants. Nat Plants 4(7):427–431. https://doi.org/10.1038/s41477-018-0178-x

Kim H, Kim ST, Ryu J, Kang BC, Kim JS, Kim SG (2017a) CRISPR/Cpf1-mediated DNA-free plant genome editing. Nat Commun 8(1):1–7. https://doi.org/10.1038/ncomms14406

Kim K, Ryu SM, Kim ST, Baek G, Kim D, Lim K, Chung E, Kim S, Kim JS (2017b) Highly efficient RNA-guided base editing in mouse embryos. Nat Biotechnol 35(5):435–437. https://doi.org/10.1038/nbt.3816

Kim YB, Komor AC, Levy JM, Packer MS, Zhao KT, Liu DR (2017c) Increasing the genome-targeting scope and precision of base editing with engineered Cas9-cytidine deaminase fusions. Nat Biotechnol 35(4):371–376. https://doi.org/10.1038/nbt.3803

Kim HS, Jeong YK, Hur JK, Kim JS, Bae S (2019) Adenine base editors catalyze cytosine conversions in human cells. Nat Biotechnol 37(10):1145–1148. https://doi.org/10.1038/s41587-019-0254-4

Koblan LW, Doman JL, Wilson C, Levy JM, Tay T, Newby GA, Maianti JP, Raguram A, Liu DR (2018) Improving cytidine and adenine base editors by expression optimization and ancestral reconstruction. Nat Biotechnol 36(9):843–848. https://doi.org/10.1038/nbt.4172

Komor AC, Kim YB, Packer MS, Zuris JA, Liu DR (2016) Programmable editing of a target base in genomic DNA without double-stranded DNA cleavage. Nature 533(7603):420–424. https://doi.org/10.1038/nature17946

Komor AC, Zhao KT, Packer MS, Gaudelli NM, Waterbury AL, Koblan LW, Kim YB, Badran AH, Liu DR (2017) Improved base excision repair inhibition and bacteriophage Mu Gam protein yields C:G-to-T:A base editors with higher efficiency and product purity. Sci Adv 3(8):eaao4774. https://doi.org/10.1126/sciadv.aao4774

Koonin EV, Makarova KS (2019) Origins and evolution of CRISPR-Cas systems. Philos Trans R Soc B Biol Sci 374(1772) Royal Society Publishing. https://doi.org/10.1098/rstb.2018.0087

Kuang Y, Li S, Ren B, Yan F, Spetz C, Li X, Zhou X, Zhou H (2020) Base-editing-mediated artificial evolution of OsALS1 in planta to develop novel herbicide-tolerant rice germplasms. Mol Plant 13(4):565–572. https://doi.org/10.1016/j.molp.2020.01.010

Kunkel TA, Erie DA (2015) Eukaryotic mismatch repair in relation to DNA replication. Annu Rev Genet 49(1):291–313. https://doi.org/10.1146/annurev-genet-112414-054722

Kunz C, Focke F, Saito Y, Schuermann D, Lettieri T, Selfridge J, Schär P (2009) Base excision by thymine DNA glycosylase mediates DNA-directed cytotoxicity of 5-fluorouracil. PLoS Biol 7(4):e1000091. https://doi.org/10.1371/journal.pbio.1000091

Lammerts Van Bueren ET, Jones SS, Tamm L, Murphy KM, Myers JR, Leifert C, Messmer MM (2011) The need to breed crop varieties suitable for organic farming, using wheat, tomato and broccoli as examples: a review. NJAS – Wageningen J Life Sci 58(3–4):193–205. https://doi.org/10.1016/j.njas.2010.04.001

Li T, Huang S, Jiang WZ, Wright D, Spalding MH, Weeks DP, Yang B (2011) TAL nucleases (TALNs): hybrid proteins composed of TAL effectors and FokI DNA-cleavage domain. Nucleic Acids Res 39(1):359–372. https://doi.org/10.1093/nar/gkq704

Li J, Sun Y, Du J, Zhao Y, Xia L (2017a) Generation of targeted point mutations in Rice by a modified CRISPR/Cas9 system. Mol Plant 10:526–529. https://doi.org/10.1016/j.molp.2016.12.001

Li J, Sun Y, Du J, Zhao Y, Xia L (2017b) Generation of targeted point mutations in Rice by a modified CRISPR/Cas9 system. Mol Plant 10(3):526–529. Cell Press. https://doi.org/10.1016/j.molp.2016.12.001

Li C, Zong Y, Wang Y, Jin S, Zhang D, Song Q, Zhang R, Gao C (2018a) Expanded base editing in rice and wheat using a Cas9-adenosine deaminase fusion. Genome Biol 19(1):1–9. https://doi.org/10.1186/s13059-018-1443-z

Li H, Rasheed A, Hickey LT, He Z (2018b) Fast-forwarding genetic gain. Trends Plant Sci 23(3):184–186. Elsevier Ltd. https://doi.org/10.1016/j.tplants.2018.01.007

Li S, Li J, Zhang J, Du W, Fu J, Sutar S, Zhao Y, Xia L (2018c) Synthesis-dependent repair of Cpf1-induced double strand DNA breaks enables targeted gene replacement in rice. J Exp Bot 69(20):4715–4721. https://doi.org/10.1093/jxb/ery245

Li S, Zhang X, Wang W, Guo X, Wu Z, Du W, Zhao Y, Xia L (2018d) Expanding the scope of CRISPR/Cpf1-mediated genome editing in Rice. Mol Plant 11(7):995–998. Cell Press. https://doi.org/10.1016/j.molp.2018.03.009

Li X, Wang Y, Liu Y, Yang B, Wang X, Wei J, Lu Z, Zhang Y, Wu J, Huang X, Yang L, Chen J (2018e) Base editing with a Cpf1-cytidine deaminase fusion. Nat Biotechnol 36(4):324–327. https://doi.org/10.1038/nbt.4102

Li G, Liu X, Huang S, Zeng Y, Yang G, Lu Z, Zhang Y, Ma X, Wang L, Huang X, Liu J (2019a) Efficient generation of pathogenic A-to-G mutations in human Tripronuclear embryos via ABE-Mediated Base editing. Mol Ther Nucl Acids 17:289–296. https://doi.org/10.1016/j.omtn.2019.05.021

Li S, Li J, He Y, Xu M, Zhang J, Du W, Zhao Y, Xia L (2019b) Precise gene replacement in rice by RNA transcript-templated homologous recombination. Nat Biotechnol 37(4):445–450. https://doi.org/10.1038/s41587-019-0065-7

Li Z, Xiong X, Wang F, Liang J, Li JF (2019c) Gene disruption through base editing-induced messenger RNA missplicing in plants. New Phytol 222(2):1139–1148. https://doi.org/10.1111/nph.15647

Li C, Zhang R, Meng X, Chen S, Zong Y, Lu C, Qiu JL, Chen YH, Li J, Gao C (2020) Targeted, random mutagenesis of plant genes with dual cytosine and adenine base editors. Nat Biotechnol 38(7):875–882. https://doi.org/10.1038/s41587-019-0393-7

Liu X, Wu S, Xu J, Sui C, Wei J (2017) Application of CRISPR/Cas9 in plant biology. Acta Pharm Sin B 7(3):292–302. Chinese Academy of Medical Sciences. https://doi.org/10.1016/j.apsb.2017.01.002

Liu Z, Lu Z, Yang G, Huang S, Li G, Feng S, Liu Y, Li J, Yu W, Zhang Y, Chen J, Sun Q, Huang X (2018) Efficient generation of mouse models of human diseases via ABE- and BE-mediated base editing. Nat Commun 9(1):1–8. https://doi.org/10.1038/s41467-018-04768-7

Liu X, Qin R, Li J, Liao S, Shan T, Xu R, Wu D, Wei P (2020) A CRISPR-Cas9-mediated domain-specific base-editing screen enables functional assessment of ACCase variants in rice. Plant Biotechnol J 18(9):1845–1847. https://doi.org/10.1111/pbi.13348

Liu L, Kuang Y, Yan F, Li S, Ren B, Gosavi G, Spetz C, Li X, Wang X, Zhou X, Zhou H (2021) Developing a novel artificial rice germplasm for dinitroaniline herbicide resistance by base editing of OsTubA2. Plant Biotechnol J 19(1):5–7. https://doi.org/10.1111/pbi.13430

Lloyd A, Plaisier CL, Carroll D, Drews GN (2005) Targeted mutagenesis using zinc-finger nucleases in Arabidopsis. Proc Natl Acad Sci U S A 102(6):2232–2237. https://doi.org/10.1073/pnas.0409339102

Losey HC, Ruthenburg AJ, Verdine GL (2006) Crystal structure of Staphylococcus aureus tRNA adenosine deaminase TadA in complex with RNA. Nat Struct Mol Biol 13(2):153–159. https://doi.org/10.1038/nsmb1047

Lu Y, Zhu JK (2017) Precise editing of a Target Base in the Rice genome using a modified CRISPR/Cas9 system. Mol Plant 10(3):523–525. Cell Press. https://doi.org/10.1016/j.molp.2016.11.013

Ma X, Zhang Q, Zhu Q, Liu W, Chen Y, Qiu R, Wang B, Yang Z, Li H, Lin Y, Xie Y, Shen R, Chen S, Wang Z, Chen Y, Guo J, Chen L, Zhao X, Dong Z, Liu YG (2015) A robust CRISPR/Cas9 system for convenient, high-efficiency multiplex genome editing in monocot and dicot plants. Mol Plant 8(8):1274–1284. https://doi.org/10.1016/j.molp.2015.04.007

Ma Y, Zhang J, Yin W, Zhang Z, Song Y, Chang X (2016) Targeted AID-mediated mutagenesis (TAM) enables efficient genomic diversification in mammalian cells. Nat Methods 13(12):1029–1035. https://doi.org/10.1038/nmeth.4027

Mahfouz MM, Piatek A, Stewart CN (2014) Genome engineering via TALENs and CRISPR/Cas9 systems: challenges and perspectives. Plant Biotechnol J 12(8):1006–1014. https://doi.org/10.1111/pbi.12256

Majid A, Parray GA, Wani SH, Kordostami M, Sofi NR, Waza AS, Shikari AB, Gulzar S (2017) Genome editing and its necessity in agriculture. Int J Curr Microbiol App Sci 6(11):5435–5443. https://doi.org/10.20546/ijcmas.2017.611.520

Makarova KS, Aravind L, Grishin NV, Rogozin IB, Koonin EV. A. (2002) A DNA repair system specific for thermophilic Archaea and bacteria predicted by genomic context analysis. Nucleic Acids Res 30:482–496. https://doi.org/10.1093/nar/30.2.482

Makarova KS, Haft DH, Barrangou R, Brouns SJJ, Charpentier E, Horvath P, Moineau S, Mojica FJM, Wolf YI, Yakunin AF, Van Der Oost J, Koonin EV (2011) Evolution and classification of the CRISPR-Cas systems. Nat Rev Microbiol 9(6):467–477. Nature Publishing Group. https://doi.org/10.1038/nrmicro2577

Makarova KS, Wolf YI, Alkhnbashi OS, Costa F, Shah SA, Saunders SJ, Barrangou R, Brouns SJJ, Charpentier E, Haft DH, Horvath P, Moineau S, Mojica FJM, Terns RM, Terns MP, White MF, Yakunin AF, Garrett RA, Van Der Oost J et al (2015) An updated evolutionary classification of CRISPR-Cas systems. Nat Rev Microbiol 13(11):722–736. https://doi.org/10.1038/nrmicro3569

Makarova KS, Wolf YI, Iranzo J, Shmakov SA, Alkhnbashi OS, Brouns SJJ, Charpentier E, Cheng D, Haft DH, Horvath P, Moineau S, Mojica FJM, Scott D, Shah SA, Siksnys V, Terns MP, Venclovas Č, White MF, Yakunin AF et al (2020) Evolutionary classification of CRISPR–Cas systems: a burst of class 2 and derived variants. Nat Rev Microbiol 18(2):67–83. Nature Research. https://doi.org/10.1038/s41579-019-0299-x

McCallum CM, Comai L, Greene EA, Henikoff S (2000) Targeted screening for induced mutations. Nat Biotechnol 18(4):455–457. https://doi.org/10.1038/74542

Metzker ML (2010) Sequencing technologies the next generation. Nat Rev Genet 11(1):31–46. https://doi.org/10.1038/nrg2626

Mishra R, Joshi RK, Zhao K (2020) Base editing in crops: current advances, limitations and future implications. Plant Biotechnol J 18(1):20–31. https://doi.org/10.1111/pbi.13225

Mojica FJM, Díez-Villaseñor C, García-Martínez J, Soria E (2005) Intervening sequences of regularly spaced prokaryotic repeats derive from foreign genetic elements. J Mol Evol 60(2):174–182. https://doi.org/10.1007/s00239-004-0046-3

Molla KA, Shih J, Yang Y (2020) Single-nucleotide editing for zebra3 and wsl5 phenotypes in rice using CRISPR/Cas9-mediated adenine base editors. aBIOTECH 1(2):106–118. https://doi.org/10.1007/s42994-020-00018-x

Mujjassim NE, Mallik M, Rathod NK, Nitesh SD (2019) Cisgenesis and intragenesis a new tool for conventional plant breeding: A review. J Pharmacogn Phytochem 8:2485-2489

Murovec J, Pirc Ž, Yang B (2017) New variants of CRISPR RNA-guided genome editing enzymes. Plant Biotechnol J 15(8):917–926. https://doi.org/10.1111/pbi.12736

Murugan K, Babu K, Sundaresan R, Rajan R, Sashital DG (2017) The revolution continues: newly discovered systems expand the CRISPR-Cas toolkit. Mol Cell 68(1):15–25. Cell Press. https://doi.org/10.1016/j.molcel.2017.09.007

Muth J, Hartje S, Twyman RM, Hofferbert H-R, Tacke E, Prfer D (2008) Precision breeding for novel starch variants in potato. Plant Biotechnol J 6(6):576–584. https://doi.org/10.1111/j.1467-7652.2008.00340.x

Nabel CS, Jia H, Ye Y, Shen L, Goldschmidt HL, Stivers JT, Zhang Y, Kohli RM (2012) AID/APOBEC deaminases disfavor modified cytosines implicated in DNA demethylation. Nat Chem Biol 8(9):751–758. https://doi.org/10.1038/nchembio.1042

Nadeem MA, Nawaz MA, Shahid MQ, Doğan Y, Comertpay G, Yıldız M, Hatipoğlu R, Ahmad F, Alsaleh A, Labhane N, Özkan H, Chung G, Baloch FS (2018) DNA molecular markers in plant breeding: current status and recent advancements in genomic selection and genome editing. Biotechnol Biotechnol Equip 32(2):261–285. https://doi.org/10.1080/13102818.2017.1400401

Nishida K, Arazoe T, Yachie N, Banno S, Kakimoto M, Tabata M, Mochizuki M, Miyabe A, Araki M, Hara KY, Shimatani Z, Kondo A (2016) Targeted nucleotide editing using hybrid prokaryotic and vertebrate adaptive immune systems. Science 353(6305). https://doi.org/10.1126/science.aaf8729

Nishikura K (2016) A-to-I editing of coding and non-coding RNAs by ADARs. Nat Rev Mol Cell Biol 17(2):83–96. Nature Publishing Group. https://doi.org/10.1038/nrm.2015.4

Nishimasu H, Shi X, Ishiguro S, Gao L, Hirano S, Okazaki S, Noda T, Abudayyeh OO, Gootenberg JS, Mori H, Oura S, Holmes B, Tanaka M, Seki M, Hirano H, Aburatani H, Ishitani R, Ikawa M, Yachie N et al (2018) Engineered CRISPR-Cas9 nuclease with expanded targeting space. Science 361(6408):1259–1262. https://doi.org/10.1126/science.aas9129

Pater S, Pinas JE, Hooykaas PJJ, Zaal BJ (2013) <scp>ZFN</scp> −mediated gene targeting of the Arabidopsis *protoporphyrinogen oxidase* gene through *Agrobacterium* -mediated floral dip transformation. Plant Biotechnol J 11(4):510–515. https://doi.org/10.1111/pbi.12040

Peng J, Botha-Oberholster A-M, Jaganathan D, Venkataraman G, Ramasamy K, Sellamuthu G, Jayabalan S (2018) CRISPR for crop improvement: an update review. Front Plant Sci 9. https://doi.org/10.3389/fpls.2018.00985

Pourcel C, Salvignol G, Vergnaud G. (2005) CRISPR elements in Yersinia pestis acquire new repeats by preferential uptake of bacteriophage DNA, and provide additional tools for evolutionary studies. Microbiology. 151(3):653–663. https://doi.org/10.1099/mic.0.27437-0

Qin L, Li J, Wang Q, Xu Z, Sun L, Alariqi M, Manghwar H, Wang G, Li B, Ding X, Rui H, Huang H, Lu T, Lindsey K, Daniell H, Zhang X, Jin S (2020) High-efficient and precise base editing of C•G to T•A in the allotetraploid cotton (*Gossypium hirsutum*) genome using a modified <scp>CRISPR</scp> /Cas9 system. Plant Biotechnol J 18(1):45–56. https://doi.org/10.1111/pbi.13168

Raman R (2017) The impact of genetically modified (GM) crops in modern agriculture: a review. GM Crops Food 8(4):195–208. https://doi.org/10.1080/21645698.2017.1413522

Ran Y, Liang Z, Gao C (2017) Current and future editing reagent delivery systems for plant genome editing. Sci China Life Sci 60(5):490–505. Science in China Press. https://doi.org/10.1007/s11427-017-9022-1

Rashmin D, Dhingani RM, Umrania VV, Tomar RS, Parakhia MV, Golakiya BA (2015) Introduction to QTL mapping in plants. www.annalsofplantsciences.com

Rees HA, Komor AC, Yeh WH, Caetano-Lopes J, Warman M, Edge ASB, Liu DR (2017) Improving the DNA specificity and applicability of base editing through protein engineering and protein delivery. Nat Commun 8(1):1–10. https://doi.org/10.1038/ncomms15790

Rees HA, Wilson C, Doman JL, Liu DR (2019) Analysis and minimization of cellular RNA editing by DNA adenine base editors. Sci Adv 5(5):eaax5717. https://doi.org/10.1126/sciadv.aax5717

Ren B, Yan F, Kuang Y, Li N, Zhang D, Zhou X, Lin H, Zhou H (2018) Improved base editor for efficiently inducing genetic variations in rice with CRISPR/Cas9-guided hyperactive hAID mutant. Mol Plant 11(4):623–626. https://doi.org/10.1016/j.molp.2018.01.005

Ren B, Liu L, Li S, Kuang Y, Wang J, Zhang D, Zhou X, Lin H, Zhou H (2019) Cas9-NG greatly expands the targeting scope of the genome-editing toolkit by recognizing NG and other atypical PAMs in Rice. Mol Plant 12(7):1015–1026. https://doi.org/10.1016/j.molp.2019.03.010

Ryu SM, Koo T, Kim K, Lim K, Baek G, Kim ST, Kim HS, Kim DE, Lee H, Chung E, Kim JS (2018) Adenine base editing in mouse embryos and an adult mouse model of Duchenne muscular dystrophy. Nat Biotechnol 36(6):536–539. https://doi.org/10.1038/nbt.4148

Sadras VO, Lawson C (2011) Genetic gain in yield and associated changes in phenotype, trait plasticity and competitive ability of South Australian wheat varieties released between 1958 and 2007. Crop Pasture Sci 62(7):533. https://doi.org/10.1071/CP11060

Schneider K, Schiermeyer A, Dolls A, Koch N, Herwartz D, Kirchhoff J, Fischer R, Russell SM, Cao Z, Corbin DR, Sastry-Dent L, Ainley WM, Webb SR, Schinkel H, Schillberg S (2016) Targeted gene exchange in plant cells mediated by a zinc finger nuclease double cut. Plant Biotechnol J 14(4):1151–1160. https://doi.org/10.1111/pbi.12483

Shen L, Hua Y, Fu Y, Li J, Liu Q, Jiao X, Xin G, Wang J, Wang X, Yan C, Wang K (2017) Rapid generation of genetic diversity by multiplex CRISPR/Cas9 genome editing in rice. Sci China Life Sci 60(5):506–515. https://doi.org/10.1007/s11427-017-9008-8

Shiferaw B, Smale M, Braun HJ, Duveiller E, Reynolds M, Muricho G (2013) Crops that feed the world 10. Past successes and future challenges to the role played by wheat in global food security. Food Sec 5(3):291–317. https://doi.org/10.1007/s12571-013-0263-y

Shimatani Z, Kashojiya S, Takayama M, Terada R, Arazoe T, Ishii H, Teramura H, Yamamoto T, Komatsu H, Miura K, Ezura H, Nishida K, Ariizumi T, Kondo A (2017) Targeted base editing in rice and tomato using a CRISPR-Cas9 cytidine deaminase fusion. Nat Biotechnol 35(5):441–443. https://doi.org/10.1038/nbt.3833

Shimatani Z, Fujikura U, Ishii H, Matsui Y, Suzuki M, Ueke Y, Taoka K-i, Terada R, Nishida K, Kondo A (2018) Inheritance of co-edited genes by CRISPR-based targeted nucleotide substitutions in rice. Plant Physiol Biochem 131:78–83. https://doi.org/10.1016/j.plaphy.2018.04.028

Stafforst T, Schneider MF (2012) An RNA–deaminase conjugate selectively repairs point mutations. Angew Chem Int Ed 51(44):11166–11169. https://doi.org/10.1002/anie.201206489

Sun Y, Zhang X, Wu C, He Y, Ma Y, Hou H, Guo X, Du W, Zhao Y, Xia L (2016) Engineering herbicide-resistant Rice plants through CRISPR/Cas9-mediated homologous recombination of Acetolactate synthase. Mol Plant 9(4):628–631. Cell Press. https://doi.org/10.1016/j.molp.2016.01.001

Symington LS, Gautier J (2011) Double-Strand break end resection and repair pathway choice. Annu Rev Genet 45(1):247–271. https://doi.org/10.1146/annurev-genet-110410-132435

Tang X, Lowder LG, Zhang T, Malzahn AA, Zheng X, Voytas DF, Zhong Z, Chen Y, Ren Q, Li Q, Kirkland ER, Zhang Y, Qi Y (2017) A CRISPR-Cpf1 system for efficient genome editing and transcriptional repression in plants. Nat Plants 3(3):1–5. https://doi.org/10.1038/nplants.2017.18

Tang X, Ren Q, Yang L, Bao Y, Zhong Z, He Y, Liu S, Qi C, Liu B, Wang Y, Sretenovic S, Zhang Y, Zheng X, Zhang T, Qi Y, Zhang Y (2019) Single transcript unit <scp>CRISPR</scp> 2.0 systems for robust Cas9 and Cas12a mediated plant genome editing. Plant Biotechnol J 17(7):1431–1445. https://doi.org/10.1111/pbi.13068

Tian S, Jiang L, Cui X, Zhang J, Guo S, Li M, Zhang H, Ren Y, Gong G, Zong M, Liu F, Chen Q, Xu Y (2018) Engineering herbicide-resistant watermelon variety through CRISPR/Cas9-mediated base-editing. Plant Cell Rep 37(9):1353–1356. https://doi.org/10.1007/s00299-018-2299-0

Varshney RK, Hoisington DA, Tyagi AK (2006) Advances in cereal genomics and applications in crop breeding. Trends Biotechnol 24(11):490–499. Elsevier Ltd. https://doi.org/10.1016/j.tibtech.2006.08.006

Veillet F, Perrot L, Chauvin L, Kermarrec M-P, Guyon-Debast A, Chauvin J-E, Nogué F, Mazier M (2019) Transgene-free genome editing in tomato and potato plants using agrobacterium-mediated delivery of a CRISPR/Cas9 Cytidine Base editor. Int J Mol Sci 20(2):402. https://doi.org/10.3390/ijms20020402

Voss-Fels KP, Stahl A, Hickey LT (2019) Q&A: modern crop breeding for future food security. BMC Biol 17(1):18. BioMed Central Ltd. https://doi.org/10.1186/s12915-019-0638-4

Voytas DF, Gao C (2014) Precision genome engineering and agriculture: opportunities and regulatory challenges. PLoS Biol 12(6):e1001877. https://doi.org/10.1371/journal.pbio.1001877

Wang M, Lu Y, Botella JR, Mao Y, Hua K, Zhu J, kang. (2017a) Gene targeting by homology-directed repair in Rice using a Geminivirus-based CRISPR/Cas9 system. Mol Plant 10(7):1007–1010. Cell Press. https://doi.org/10.1016/j.molp.2017.03.002

Wang M, Mao Y, Lu Y, Tao X, Zhu J (2017b) Multiplex gene editing in Rice using the CRISPR-Cpf1 system. Mol Plant 10:1011–1013. https://doi.org/10.1016/j.molp.2017.03.001

Wang X, Tu M, Wang D, Liu J, Li Y, Li Z, Wang Y, Wang X (2018a) CRISPR/Cas9-mediated efficient targeted mutagenesis in grape in the first generation. Plant Biotechnol J 16(4):844–855. https://doi.org/10.1111/pbi.12832

Wang X, Li J, Wang Y, Yang B, Wei J, Wu J, Wang R, Huang X, Chen J, Yang L (2018b) Efficient base editing in methylated regions with a human apobec3a-cas9 fusion. Nat Biotechnol 36(10):946. https://doi.org/10.1038/nbt.4198

Watson A, Ghosh S, Williams MJ, Cuddy WS, Simmonds J, Rey MD, Asyraf Md Hatta M, Hinchliffe A, Steed A, Reynolds D, Adamski NM, Breakspear A, Korolev A, Rayner T, Dixon LE, Riaz A, Martin W, Ryan M, Edwards D et al (2018) Speed breeding is a powerful tool to accelerate crop research and breeding. Nat Plants 4(1):23–29. https://doi.org/10.1038/s41477-017-0083-8

Wendler N, Mascher M, Nöh C, Himmelbach A, Scholz U, Ruge-Wehling B, Stein N (2014) Unlocking the secondary gene-pool of barley with next-generation sequencing. Plant Biotechnol J 12(8):1122–1131. https://doi.org/10.1111/pbi.12219

Wolf J, Gerber AP, Keller W (2002) TadA, an essential tRNA-specific adenosine deaminase from Escherichia coli. EMBO J 21(14):3841–3851. https://doi.org/10.1093/emboj/cdf362

Wood AJ, Lo TW, Zeitler B, Pickle CS, Ralston EJ, Lee AH, Amora R, Miller JC, Leung E, Meng X, Zhang L, Rebar EJ, Gregory PD, Urnov FD, Meyer BJ (2011) Targeted genome editing across species using ZFNs and TALENs. Science 333(6040):307. American Association for the Advancement of Science. https://doi.org/10.1126/science.1207773

Wu J, Chen C, Xian G, Liu D, Lin L, Yin S, Sun Q, Fang Y, Zhang H, Wang Y (2020) Engineering herbicide-resistant oilseed rape by CRISPR/Cas9-mediated cytosine base-editing. Plant Biotechnol J 18(9):1857–1859. https://doi.org/10.1111/pbi.13368

Xu R, Qin R, Li H, Li D, Li L, Wei P, Yang J (2017) Generation of targeted mutant rice using a CRISPR-Cpf1 system. Plant Biotechnol J 15(6):713–717. https://doi.org/10.1111/pbi.12669

Yan F, Kuang Y, Ren B, Wang J, Zhang D, Lin H, Yang B, Zhou X, Zhou H (2018) Highly efficient A·T to G·C base editing by Cas9n-Guided tRNA adenosine deaminase in rice. Mol Plant 11:631–634. https://doi.org/10.1016/j.molp.2018.02.008

Yin K, Gao C, Qiu JL (2017) Progress and prospects in plant genome editing. Nat Plants 3(8):1–6. Palgrave Macmillan Ltd. https://doi.org/10.1038/nplants.2017.107

Zafra MP, Schatoff EM, Katti A, Foronda M, Breinig M, Schweitzer AY, Simon A, Han T, Goswami S, Montgomery E, Thibado J, Kastenhuber ER, Sánchez-Rivera FJ, Shi J, Vakoc CR, Lowe SW, Tschaharganeh DF, Dow LE (2018) Optimized base editors enable efficient editing in cells, organoids and mice. Nat Biotechnol 36(9):888–896. https://doi.org/10.1038/nbt.4194

Zaidi SSEA, Mahfouz MM, Mansoor S (2017) CRISPR-Cpf1: a new tool for plant genome editing. Trends Plant Sci 22(7):550–553. Elsevier Ltd. https://doi.org/10.1016/j.tplants.2017.05.001

Zalatan JG, Lee ME, Almeida R, Gilbert LA, Whitehead EH, La Russa M, Tsai JC, Weissman JS, Dueber JE, Qi LS, Lim WA (2015) Engineering complex synthetic transcriptional programs with CRISPR RNA scaffolds. Cell 160(1–2):339–350. https://doi.org/10.1016/j.cell.2014.11.052

Zeng D, Liu T, Tan J, Zhang Y, Zheng Z, Wang B, Zhou D, Xie X, Guo M, Liu YG, Zhu Q (2020) PhieCBEs: plant high-efficiency Cytidine Base editors with expanded target range. Mol Plant 13(12):1666–1669. https://doi.org/10.1016/j.molp.2020.11.001

Zetsche B, Gootenberg JS, Abudayyeh OO, Slaymaker IM, Makarova KS, Essletzbichler P, Volz SE, Joung J, Van Der Oost J, Regev A, Koonin EV (2015) Cpf1 is a single RNA-guided endonuclease of a class 2 CRISPR-Cas system. Cell. 163(3):759–771. https://doi.org/10.1016/j.cell.2015.09.038

Zhang F, Wen Y, Guo X (2014) CRISPR/Cas9 for genome editing: Progress, implications and challenges. Hum Mol Genet 23(R1):R40–R46. https://doi.org/10.1093/hmg/ddu125

Zhang Y, Xie X, Liu YG, Zhang Y, Xie X, Liu YG, Ma X (2017) CRISPR/Cas9-based genome editing in plants. Prog Mol Biol Transl Sci 149:133–150. Elsevier B.V. https://doi.org/10.1016/bs.pmbts.2017.03.008

Zhang D, Zhang Z, Unver T, Zhang B (2020) CRISPR/Cas: a powerful tool for gene function study and crop improvement. J Adv Res 29:207–221

Zhao BS, Roundtree IA, He C (2016) Post-transcriptional gene regulation by mRNA modifications. Nat Rev Mol Cell Biol 18(1):31–42. Nature Publishing Group. https://doi.org/10.1038/nrm.2016.132

Zheng Y, Lorenzo C, Beal PA (2017) DNA editing in DNA/RNA hybrids by adenosine deaminases that act on RNA. Nucleic Acids Res 45(6):3369–3377. https://doi.org/10.1093/nar/gkx050

Zhong Z, Sretenovic S, Ren Q, Yang L, Bao Y, Qi C, Yuan M, He Y, Liu S, Liu X, Wang J, Huang L, Wang Y, Baby D, Wang D, Zhang T, Qi Y, Zhang Y (2019) Improving plant genome editing with high-Fidelity xCas9 and non-canonical PAM-targeting Cas9-NG. Mol Plant 12(7):1027–1036. https://doi.org/10.1016/j.molp.2019.03.011

Zhou W, Karcher D, Bock R (2013) Importance of adenosine-to-inosine editing adjacent to the anticodon in an Arabidopsis alanine tRNA under environmental stress. Nucleic Acids Res 41(5):3362–3372. https://doi.org/10.1093/nar/gkt013

Zhou J, Peng Z, Long J, Sosso D, Liu B, Eom JS, Huang S, Liu S, Vera Cruz C, Frommer WB, White FF, Yang B (2015) Gene targeting by the TAL effector PthXo2 reveals cryptic resistance gene for bacterial blight of rice. Plant J 82(4):632–643. https://doi.org/10.1111/tpj.12838

Zong Y, Wang Y, Li C, Zhang R, Chen K, Ran Y, Qiu JL, Wang D, Gao C (2017) Precise base editing in rice, wheat and maize with a Cas9-cytidine deaminase fusion. Nat Biotechnol 35(5):438–440. https://doi.org/10.1038/nbt.3811

Zou C, Wang P, Xu Y (2016) Bulked sample analysis in genetics, genomics and crop improvement. Plant Biotechnol J 14(10):1941–1955. Blackwell Publishing Ltd. https://doi.org/10.1111/pbi.12559

Zuo E, Sun Y, Wei W, Yuan T, Ying W, Sun H, Yuan L, Steinmetz LM, Li Y, Yang H (2019) Cytosine base editor generates substantial off-target single-nucleotide variants in mouse embryos. Science 364(6437):289–292. https://doi.org/10.1126/science.aav9973

New Cas Endonuclease Variants Broadening the Scope of the CRISPR Toolbox

Goetz Hensel

Abstract The biotechnological usage of the bacterial-derived CRISPR/Cas system allowed specific and straightforward manipulation of plant genomes for the first time. Different Cas proteins have other binding prerequisites associated with the mechanistic interaction between Cas proteins and genomic DNA. The gene-specific protospacer-associated motif (PAM) for the common SpCas9 is NGG. Although this is a common motif frequently found in plant genic sequences, it still limits a flexible targeting of any genomic position. Further analysis of related and unrelated species identified similar proteins with different PAM requirements. In addition, the in vitro evolution of canonical Cas proteins expanded the toolbox further. This chapter introduces the natural origin of Cas proteins, summarises PAM requirements and Cas engineering, and highlights the new functionalities through fusion with exogenous protein domains, base and prime editing.

Keywords SpCas9 · Base editors · Prime editing · Cas evolution

1 Introduction

The biotechnological use of sequence-specific endonucleases (SSNs) has revolutionised the directed study of the inherited traits of organisms. Systems such as clustered regularly interspaced short palindromic repeats (CRISPR)/Cas9 are widely used to (1) study gene functions in different organisms or (2) make application-oriented modifications, such as improving agronomically essential plant traits. In terms of the effort required to create the functional units, they are the most convenient gene modification tools currently available. In contrast to protein-mediated binding of zinc finger nucleases (ZFN) or transcription activator-like effector nucleases (TALEN), Cas proteins are RNA-directed endonucleases; these

G. Hensel (✉)
Centre for Plant Genome Engineering, Institute of Plant Biochemistry, Heinrich-Heine-University, Düsseldorf, Germany

Centre of Region Haná for Biotechnological and Agricultural Research, Czech Advanced Technology and Research Institute, Palacký University Olomouc, Olomouc, Czech Republic
e-mail: goetz.hensel@hhu.de

© The Author(s), under exclusive license to Springer Nature Switzerland AG 2022
S. H. Wani, G. Hensel (eds.), *Genome Editing*,
https://doi.org/10.1007/978-3-031-08072-2_6

133

cleave DNA in a targeted manner, most commonly used for site-directed mutagenesis or genome/gene editing. The following chapter will point out the different Cas variants, their origin, similarities and differences, and examples of their use.

2 The Natural Origin of Cas Proteins

In their natural environment, cas proteins are part of bacterial defence systems (Barrangou et al. 2007), organised in clusters of *cas* genes (Haft et al. 2005). For biotechnological use, only two components of these bacterial defence systems are used. For navigation in the target genome or site-specificity, a guide RNA (gRNA) is used. This can vary depending on the cas protein used. For binding to the target DNA, SpCas9 requires an approximately 130 nucleotide long hybrid RNA consisting of a 42 nucleotide long CRISPR RNA (crRNA) and an 89 bp long trans-activating crRNA (tracrRNA) (Jinek et al. 2012). In contrast, AsCas12a only requires a 42 nucleotide long tracrRNA (Zhong et al. 2019), which, for example, significantly reduces the cost of synthetic production.

The second component is the double-strand break-inducing Cas protein. Whole-genome sequencing has identified more than 45 Cas families, further subdivided based on functional elements (Haft et al. 2005). At least 4 *Cas* genes (*cas1–4*) have been associated with other subtypes, with *cas1* considered a universal marker for CRISPR systems (Haft et al. 2005). Barrangou and co-workers (Barrangou et al. 2007) have experimentally demonstrated that new spacers are integrated into the CRISPR array after bacteriophage infection, leading to resistance to this bacteriophage.

2.1 Cas Proteins Have Different PAM Requirements

Cas proteins require a sequence called protospacer adjacent motif (PAM) for their binding to double-stranded DNA, which varies depending on the origin of the Cas protein (Fig. 1A). For the most commonly used *Streptococcus pyogenes* Cas9 protein (SpCas9), this sequence is NGG, where the N is variable and represents all four nucleotides (Jinek et al. 2012). In addition, the PAM is located downstream from the gRNA recognition sequence. The limitation in selecting the target sequence due to the presence of this PAM is overcome by using Cas proteins from other organisms such as FbCas12a (TTTV, (Zetsche et al. 2015)) or a targeted modification of the Cas9 protein (xCas (Hu et al. 2018a)), or SpCas9-NG (Nishimasu et al. 2018). This means that almost any genomic sequence is now accessible.

Cas endonucleases differ not only in terms of their PAM but also the type of double-strand break induced. While SpCas9 produces smooth cut ends, sticky ends are obtained when using LbCas12a. Another difference is the localisation of the PAM concerning the gRNA recognition sequence. While for SpCas9, xCas and

A

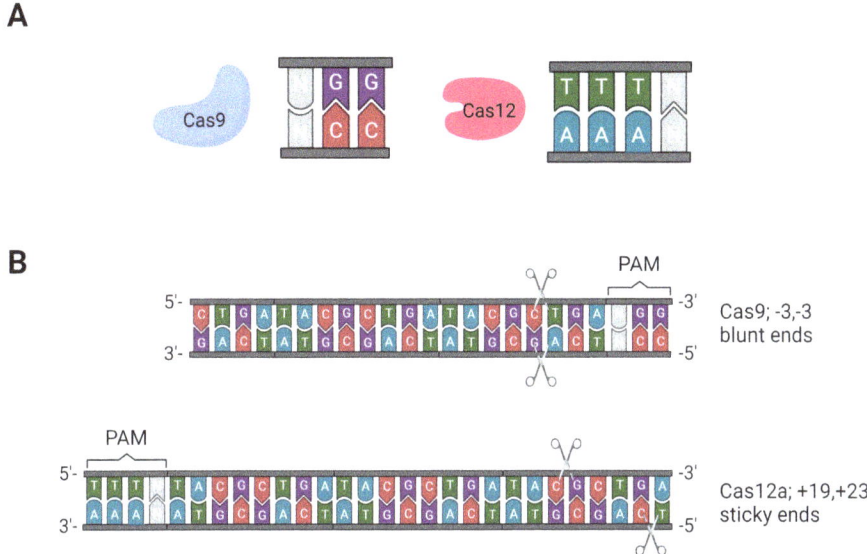

B

Fig. 1 Illustration of PAM requirements and cutting patterns of Cas proteins. (**a**) Enzymes as SpCas9 and Cas proteins of other species need an NGG for which the N stands for any other of the four nucleotides (left) or TTTN for FbCas12a (right). (**b**) The nuclease domains of SpCas9 induce a double-strand break three nucleotides upstream of the PAM, while FbCas12a induce sticky ends while cutting 19/23 nucleotides downstream of the PAM

SpCas9-NG, the PAM is always located downstream, for FbCas12a and AsCas12a, the PAM is located upstream of the gRNA (Fig. 1B).

2.2 Protein Engineering of Cas Proteins

By elucidating the structure of Cas proteins, various functional domains could be described. For example, SpCas9 contains nuclease (NUC) and recognition (REC) domains (Jiang and Doudna 2017). These have been modified using directed evolution methods (for review see (Liu et al. 2020)). By mutating specific amino acids in the previously named nuclease domains, SpCas9 could be converted into a single-strand break-targeting nuclease (Cas-nickase), designated nCas, or into a catalytically inactivated nuclease, designated dCas (Mali et al. 2013; Qi et al. 2013).

To increase target specificity, changes were made in the REC domain of the Cas9 protein. Changes to amino acids D1135V, R1335Q and T1337R resulted in a variant called SpCas9-VQR (Kleinstiver et al. 2015). SpCas9-VQR had higher editing efficiency than the SpCas9 wild-type protein in rice, for example (Hu et al. 2018b). To further reduce off-targets, further modifications were suggested. SpCas9-HF1 is a quadruple mutant altering all Cas9-mediated DNA contact sites to the target strand's phosphate backbone (Kleinstiver et al. 2016). The authors showed by *EGFP* disruption assay that the enzyme retains on-target while harbouring reduced off-target activity.

As previously mentioned, PAM requirements limit the use of CRISPR technology. Hu and colleagues (Hu et al. 2018a) used phage-assisted continuous evolution (PACE) to rapidly generate Cas9 variants that accept an expanded range of PAM sequences. They developed an xCas9 3.7 enzyme with comparable on-target activity to SpCas9 at NGG PAM sites and substantially higher efficiencies at NG, GAA and GAT PAM sites. The GUIDE-seq (Tsai et al. 2015) assay could not confirm the expected higher off-target activity.

In a similar approach, Nishimazu and colleagues (Nishimasu et al. 2018) developed the SpCas9-NG protein. It has additional amino acid substitutions in predicted protein interaction sites with the sugar-phosphate backbone of the PAM duplex and the interaction with the ribose moiety of the second G. Modifying the enzyme in the following amino acids (R1335V/L1111R/D1135V/G1218R/E1219F/ A1322R/ T1337R) lead to the VRVRFRR variant designated as SpCas9-NG. While having lower activities at NGG PAM sites, it still expands the applicability of the CRISPR/Cas toolbox (Molla et al. 2020a).

2.3   New Cas Functionality Through Fusion with Exogenous Protein Domains

Using only the target sequence specificity of the gRNA/Cas protein complex, the fusion of the dCas protein with the domain of transcriptional regulators such as activators or repressors can enable the regulation of gene expression (Pan et al. 2021a). This offers advantages over conventional overexpression of target genes and is demonstrated for the first time in human cells (Maeder et al. 2013; Perez-Pinera et al. 2013). Initial success for gene activation in plants was achieved by fusing dCas9 with VP64 (Piatek et al. 2015; Lowder et al. 2015). VP64 is a tetrameric repeat derived from the VP16 protein of the herpes simplex virus (Sadowski et al. 1988). Further improvements here have since led to a system called CRISPR-Act3.0, which has elucidated a significant increase in effective target gene activation, causing simultaneous activation of many enzyme-coding genes in the β-carotene biosynthetic pathway and the proanthocyanidin biosynthetic pathway in rice, as well as multigene activation in Arabidopsis (Pan et al. 2021b).

Gene inactivation (CRISPR interference, CRISPRi) involves another application of CRISPR/Cas technology. Here, binding of the dCas protein to the promoter of a target gene near the transcription start site (TSS) prevents the function of the RNA polymerase or the binding of transcription factors. In plants, there are few examples of the use of CRISPR-mediated gene inactivation (Piatek et al. 2015; Lowder et al. 2015). In analogy to gene activation, chimeric Arabidopsis's SUPERMAN Repression Domain X (SRDX) domains are coupled to the dCas (Hiratsu et al. 2003). Results from Arabidopsis (Piatek et al. 2015; Lowder et al. 2015) and rice

(Tang et al. 2017; Ming et al. 2020) achieved a 40% reduction compared to the wild-type control.

The latest development is epigenetic regulation using CRISPR-based programmable epigenome editor protein, termed CRISPRoff-V1, composed of ZNF 10KRAB, Dnmt3A (D3A) and Dnmt3L (D3L) protein domains fused to catalytically inactive *S. pyogenes* dCas9 (Nuñez et al. 2021). In this process, gene regulation is blocked by methylation. This change can also be transferred to the progeny.

Fusion with reporter genes can also be used for live-cell imaging of specific genomic regions (CRISPR imaging, (Khosravi et al. 2020)). This technique allows overcoming the limitations associated with the long-used fluorescence in situ hybridisation (FISH) technology. FISH requires cell fixation and a DNA denaturation step that can lead to altered chromatin structure, preventing the visualisation of dynamic processes (Kozubek et al. 2000; Boettiger et al. 2016). For CRISPR imaging, the target-specific property of CRISPR/Cas technology is used to visualise specific chromosomal regions using fluorescent reporter genes. Dreissig and colleagues (Dreissig et al. 2017) observed dynamic telomere movements in *Nicotiana benthamiana*. A corresponding review article provides deeper insight into this (Singh and Jain 2021).

A summary of CRISPR approaches in rice, which also refers to the use of possible Cas variants, was recently published (Ganie et al. 2021).

2.4   Base Editors and Prime Editing

Another variant of the Cas proteins was the fusion with so-called base editors. These are fusion proteins that allow base exchange in the vicinity of the single-strand break by combining catalytically compromised Cas proteins with single-strand deaminases. There are currently three types, adenine, cytosine and cytosine transversion base editors (Molla et al. 2020b). Adenine base editors (ABEs) catalyse the exchange of A-T base pairings into G-C base pairings. In contrast, cytosine base editors (CBEs) allow conversion of C-G into T-A base pairings. After binding of the RNA-protein complex to the target sequence, the DNA double-strand is bent apart (Jiang and Doudna 2017), making an R-shaped single strand accessible to the deaminase. In CBEs, the cytosine is converted to uracil, which is read as thymine by polymerases (Komor et al. 2016).

Similarly, in ABEs, adenosines located within the R-loop are converted into inosines using laboratory-adapted TadA* deoxyadenosine deaminases (Gaudelli et al. 2017). These are then read as guanines by cellular polymerase. In this way, several identical nucleotides can also be edited within the R-loop. However, mismatches generated in this way are mutagenic, which has led to the development of corresponding repair mechanisms in many organisms (Wood 1996). Uracil is removed from the genomic DNA faster than inosine (Gaudelli et al. 2017). Therefore, improved CBEs employ uracil glycosylase inhibitor proteins (UGIs), which extend the half-life of uracil and thus support base exchange (Komor et al. 2017; Wang et al. 2017). Similar experiments with corresponding inhibitors for ABEs have not achieved an increase in base editing, suggesting that this process is slower in mammalian cells (Lau et al. 2000).

Several deaminases are most commonly used for CBEs are APOBEC (Komor et al. 2016) and CDA (Nishimasu et al. 2018). APOBEC stands for apolipoprotein B mRNA-editing enzyme, catalytic polypeptide-like. There are several family members of APOBEC enzymes that have a preference for TC motifs (APOBEC1) or CC motifs (ABPOBEC A3G) (Liu et al. 2019). Base editors, such as CDA (cytosine deaminase), have a larger 'processing window' and higher activity than APOBEC enzymes (Nishida et al. 2016). Different classes of deaminases used in generating base editors could be found in an earlier review article (Molla and Yang 2019).

Initial experiments here have used SpCas9 as the endonuclease (Kim et al. 2017b). In subsequent years, other combinations of CBEs with xCas9 (Hu et al. 2018a; Zhong et al. 2019; Tan et al. 2020), SaCas9 (Kim et al. 2017a) or LbCas12a (Li et al. 2018) have been successfully tested. Something similar is also true for ABEs. A good summary was published by Anzalone and colleagues (Anzalone et al. 2020).

Limitations of the base editors are the restriction in terms of the 'editing window' and that not all theoretically possible base substitutions can be made. An extension of base editors is the so-called prime editing (Anzalone et al. 2019). This system allows an exchange of all possible base pairings and thus further improves targeted gene editing. Furthermore, short DNA sequences can also be inserted at defined positions. Prime editors are fusion proteins between a Cas9 nickase and a reverse transcriptase. Here, the synthetic target sequence is brought along as an extension of the gRNA, mainly at the 3′ end. The gRNA modified in this way is called pegRNA. After binding to the target sequence, a single-strand break is generated with Cas9 nickase. The released 3′ end is used as a template for reverse transcription with the pegRNA extension. To facilitate DNA repair, the pegRNA also contains a region of homology to the target site. Although many of these systems are being developed in human cell cultures, there are also applications in plants. A summary of this was recently published by Molla and colleagues (Molla et al. 2021).

References

Anzalone AV, Randolph PB, Davis JR, Sousa AA, Koblan LW, Levy JM et al (2019) Search-and-replace genome editing without double-strand breaks or donor DNA. Nature 576(7785):149–157. https://doi.org/10.1038/s41586-019-1711-4

Anzalone AV, Koblan LW, Liu DR (2020) Genome editing with CRISPR-Cas nucleases, base editors, transposases and prime editors. Nat Biotechnol 38(7):824–844. https://doi.org/10.1038/s41587-020-0561-9

Barrangou R, Fremaux C, Deveau H, Richards M, Boyaval P, Moineau S et al (2007) CRISPR provides acquired resistance against viruses in prokaryotes. Science 315(5819):1709–1712. https://doi.org/10.1126/science.1138140

Boettiger AN, Bintu B, Moffitt JR, Wang S, Beliveau BJ, Fudenberg G et al (2016) Super-resolution imaging reveals distinct chromatin folding for different epigenetic states. Nature 529(7586):418–422. https://doi.org/10.1038/nature16496

Dreissig S, Schiml S, Schindele P, Weiss O, Rutten T, Schubert V et al (2017) Live-cell CRISPR imaging in plants reveals dynamic telomere movements. Plant J Cell Mol Biol 91(4):565–573. https://doi.org/10.1111/tpj.13601

Ganie SA, Wani SH, Henry R, Hensel G (2021) Improving rice salt tolerance by precision breeding in a new era. Curr Opinion Plant Biol 60:101996. https://doi.org/10.1016/j.pbi.2020.101996

Gaudelli NM, Komor AC, Rees HA, Packer MS, Badran AH, Bryson DI, Liu DR (2017) Programmable base editing of a•T to G•C in genomic DNA without DNA cleavage. Nature 551(7681):464–471. https://doi.org/10.1038/nature24644

Haft DH, Selengut J, Mongodin EF, Nelson KE (2005) A guild of 45 CRISPR-associated (Cas) protein families and multiple CRISPR/Cas subtypes exist in prokaryotic genomes. PLoS Comput Biol 1(6):e60. https://doi.org/10.1371/journal.pcbi.0010060

Hiratsu K, Matsui K, Koyama T, Ohme-Takagi M (2003) Dominant repression of target genes by chimeric repressors that include the EAR motif, a repression domain, in Arabidopsis. Plant J Cell Mol Biol 34(5):733–739. https://doi.org/10.1046/j.1365-313x.2003.01759.x

Hu JH, Miller SM, Geurts MH, Tang W, Chen L, Sun N et al (2018a) Evolved Cas9 variants with broad PAM compatibility and high DNA specificity. Nature 556(7699):57–63. https://doi.org/10.1038/nature26155

Hu X, Meng X, Liu Q, Li J, Wang K (2018b) Increasing the efficiency of CRISPR-Cas9-VQR precise genome editing in rice. Plant Biotechnol J 16(1):292–297. https://doi.org/10.1111/pbi.12771

Jiang F, Doudna JA (2017) CRISPR-Cas9 structures and mechanisms. Ann Rev Biophy 46:505–529. https://doi.org/10.1146/annurev-biophys-062215-010822

Jinek M, Chylinski K, Fonfara I, Hauer M, Doudna JA, Charpentier E (2012) A programmable dual-RNA-guided DNA endonuclease in adaptive bacterial immunity. Science 337(6096):816–821. https://doi.org/10.1126/science.1225829

Khosravi S, Ishii T, Dreissig S, Houben A (2020) Application and prospects of CRISPR/Cas9-based methods to trace defined genomic sequences in living and fixed plant cells. Chromosom Res Int J Mol Supramol Evol Aspect Chromosom Biol 28(1):7–17. https://doi.org/10.1007/s10577-019-09622-0

Kim S, Bae T, Hwang J, Kim J-S (2017a) Rescue of high-specificity Cas9 variants using sgRNAs with matched 5′ nucleotides. Genome Biol 18(1):218. https://doi.org/10.1186/s13059-017-1355-3

Kim YB, Komor AC, Levy JM, Packer MS, Zhao KT, Liu DR (2017b) Increasing the genome-targeting scope and precision of base editing with engineered Cas9-cytidine deaminase fusions. Nat Biotechnol 35(4):371–376. https://doi.org/10.1038/nbt.3803

Kleinstiver BP, Prew MS, Tsai SQ, Topkar VV, Nguyen NT, Zheng Z et al (2015) Engineered CRISPR-Cas9 nucleases with altered PAM specificities. Nature 523(7561):481–485. https://doi.org/10.1038/nature14592

Kleinstiver BP, Pattanayak V, Prew MS, Tsai SQ, Nguyen NT, Zheng Z, Joung JK (2016) High-fidelity CRISPR-Cas9 nucleases with no detectable genome-wide off-target effects. Nature 529(7587):490–495. https://doi.org/10.1038/nature16526

Komor AC, Kim YB, Packer MS, Zuris JA, Liu DR (2016) Programmable editing of a target base in genomic DNA without double-stranded DNA cleavage. Nature 533(7603):420–424. https://doi.org/10.1038/nature17946

Komor AC, Zhao KT, Packer MS, Gaudelli NM, Waterbury AL, Koblan LW et al (2017) Improved base excision repair inhibition and bacteriophage Mu Gam protein yields C:G-to-T:A base editors with higher efficiency and product purity. Sci Adv 3(8):eaao4774. https://doi.org/10.1126/sciadv.aao4774

Kozubek S, Lukásová E, Amrichová J, Kozubek M, Lisková A, Slotová J (2000) Influence of cell fixation on chromatin topography. Anal Biochem 282(1):29–38. https://doi.org/10.1006/abio.2000.4538

Lau AY, Wyatt MD, Glassner BJ, Samson LD, Ellenberger T (2000) Molecular basis for discriminating between normal and damaged bases by the human alkyladenine glycosylase, AAG. Proc Natl Acad Sci U S A 97(25):13573–13578. https://doi.org/10.1073/pnas.97.25.13573

Li X, Wang Y, Liu Y, Yang B, Wang X, Wei J et al (2018) Base editing with a Cpf1-cytidine deaminase fusion. Nat Biotechnol 36(4):324–327. https://doi.org/10.1038/nbt.4102

Liu Z, Chen S, Shan H, Chen M, Song Y, Lai L, Li Z (2019) Highly precise base editing with CC context-specificity using engineered human APOBEC3G-nCas9 fusions. bioRxiv

Liu R, Liang L, Freed EF, Gill RT (2020) Directed evolution of CRISPR/Cas Systems for Precise Gene Editing. Trends Biotechnol. https://doi.org/10.1016/j.tibtech.2020.07.005

Lowder LG, Zhang D, Baltes NJ, Paul JW, Tang X, Zheng X et al (2015) A CRISPR/Cas9 toolbox for multiplexed plant genome editing and transcriptional regulation. Plant Physiol 169(2):971–985. https://doi.org/10.1104/pp.15.00636

Maeder ML, Linder SJ, Cascio VM, Fu Y, Ho QH, Joung JK (2013) CRISPR RNA-guided activation of endogenous human genes. Nat Methods 10(10):977–979. https://doi.org/10.1038/nmeth.2598

Mali P, Yang L, Esvelt KM, Aach J, Guell M, DiCarlo JE et al (2013) RNA-guided human genome engineering via Cas9. Science 339(6121):823–826. https://doi.org/10.1126/science.1232033

Ming M, Ren Q, Pan C, He Y, Zhang Y, Liu S et al (2020) CRISPR-Cas12b enables efficient plant genome engineering. Nat Plants 6(3):202–208. https://doi.org/10.1038/s41477-020-0614-6

Molla KA, Yang Y (2019) CRISPR/Cas-Mediated Base editing: technical considerations and practical applications. Trends Biotechnol 37(10):1121–1142. https://doi.org/10.1016/j.tibtech.2019.03.008

Molla KA, Karmakar S, Islam MT (2020a) Wide horizons of CRISPR-cas-derived technologies for basic biology, agriculture, and medicine. In: Islam MT, Bhowmik PK, Molla KA (eds) CRISPR-cas methods. Springer, New York, pp 1–23

Molla KA, Qi Y, Karmakar S, Baig MJ (2020b) Base editing landscape extends to perform Transversion mutation. Trend Genet TIG 36(12):899–901. https://doi.org/10.1016/j.tig.2020.09.001

Molla KA, Sretenovic S, Bansal KC, Qi Y (2021) Precise plant genome editing using base editors and prime editors. Nat Plants 7(9):1166–1187. https://doi.org/10.1038/s41477-021-00991-1

Nishida K, Arazoe T, Yachie N, Banno S, Kakimoto M, Tabata M et al (2016) Targeted nucleotide editing using hybrid prokaryotic and vertebrate adaptive immune systems. Science 353(6305). https://doi.org/10.1126/science.aaf8729

Nishimasu H, Shi X, Ishiguro S, Gao L, Hirano S, Okazaki S et al (2018) Engineered CRISPR-Cas9 nuclease with expanded targeting space. Science 361(6408):1259–1262. https://doi.org/10.1126/science.aas9129

Nuñez JK, Chen J, Pommier GC, Cogan JZ, Replogle JM, Adriaens C et al (2021) Genome-wide programmable transcriptional memory by CRISPR-based epigenome editing. Cell 184(9):2503–2519. https://doi.org/10.1016/j.cell.2021.03.025

Pan C, Sretenovic S, Qi Y (2021a) CRISPR/dCas-mediated transcriptional and epigenetic regulation in plants. Curr Opin Plant Biol 60:101980. https://doi.org/10.1016/j.pbi.2020.101980

Pan C, Wu X, Markel K, Malzahn AA, Kundagrami N, Sretenovic S et al (2021b) CRISPR-Act3.0 for highly efficient multiplexed gene activation in plants. Nat Plants 7(7):942–953. https://doi.org/10.1038/s41477-021-00953-7

Perez-Pinera P, Kocak DD, Vockley CM, Adler AF, Kabadi AM, Polstein LR et al (2013) RNA-guided gene activation by CRISPR-Cas9-based transcription factors. Nat Methods 10(10):973–976. https://doi.org/10.1038/nmeth.2600

Piatek A, Ali Z, Baazim H, Li L, Abulfaraj A, Al-Shareef S et al (2015) RNA-guided transcriptional regulation in planta via synthetic dCas9-based transcription factors. Plant Biotechnol J 13(4):578–589. https://doi.org/10.1111/pbi.12284

Qi LS, Larson MH, Gilbert LA, Doudna JA, Weissman JS, Arkin AP, Lim WA (2013) Repurposing CRISPR as an RNA-guided platform for sequence-specific control of gene expression. Cell 152(5):1173–1183. https://doi.org/10.1016/j.cell.2013.02.022

Sadowski I, Ma J, Triezenberg S, Ptashne M (1988) GAL4-VP16 is an unusually potent transcriptional activator. Nature 335(6190):563–564. https://doi.org/10.1038/335563a0

Singh V, Jain M (2021) Recent advancements in CRISPR-Cas toolbox for imaging applications. Critic Rev Biotechnol:1–24. https://doi.org/10.1080/07388551.2021.1950608

Tan J, Zhang F, Karcher D, Bock R (2020) Expanding the genome-targeting scope and the site selectivity of high-precision base editors. Nat Commun 11(1):629. https://doi.org/10.1038/s41467-020-14465-z

Tang X, Lowder LG, Zhang T, Malzahn AA, Zheng X, Voytas DF et al (2017) A CRISPR-Cpf1 system for efficient genome editing and transcriptional repression in plants. Nat Plants 3:17018. https://doi.org/10.1038/nplants.2017.18

Tsai SQ, Zheng Z, Nguyen NT, Liebers M, Topkar VV, Thapar V et al (2015) GUIDE-seq enables genome-wide profiling of off-target cleavage by CRISPR-Cas nucleases. Nat Biotechnol 33(2):187–197. https://doi.org/10.1038/nbt.3117

Wang L, Xue W, Yan L, Li X, Wei J, Chen M et al (2017) Enhanced base editing by co-expression of free uracil DNA glycosylase inhibitor. Cell Res 27(10):1289–1292. https://doi.org/10.1038/cr.2017.111

Wood RD (1996) DNA repair in eukaryotes. Ann Rev Biochem 65:135–167. https://doi.org/10.1146/annurev.bi.65.070196.001031

Zetsche B, Gootenberg JS, Abudayyeh OO, Slaymaker IM, Makarova KS, Essletzbichler P et al (2015) Cpf1 is a single RNA-guided endonuclease of a class 2 CRISPR-Cas system. Cell 163(3):759–771. https://doi.org/10.1016/j.cell.2015.09.038

Zhong Z, Sretenovic S, Ren Q, Yang L, Bao Y, Qi C et al (2019) Improving plant genome editing with high-fidelity xCas9 and non-canonical PAM-targeting Cas9-NG. Mol Plant 12(7):1027–1036. https://doi.org/10.1016/j.molp.2019.03.011

Multiplexed Genome Editing in Plants Using CRISPR/Cas-Based Endonuclease Systems

Nagaveni Budhagatapalli and Goetz Hensel

Abstract Agronomic relevant plant traits are mainly complex and therefore hard to manipulate. Targeted mutagenesis using CRISPR/Cas technology allows a simple introduction of a mutation in the preferred genomic location. By following sophisticated cloning procedures such as Golden Gate cloning or by using polycistronic tRNA-gRNA arrays, targeting multiple genomic positions became feasible. Basic and applied research got; therefore, numerous options manipulating plant genomes. This multiplex genome editing does not only include setting mutations to inactivate genes or gene families. By using fusions of functional domains with catalytically inactive Cas enzymes, genes can be activated, repressed, or altered in any other way. The present chapter summarizes strategies and the basic principles of multiplexed genome editing. It introduces the two-component transcriptional unit multiplex system, multiplexing using a single Pol III promoter, using bidirectional promoters, and highlights applications using the respective cloning procedure.

Keywords Targeted mutagenesis · Polycistronic tRNA-gRNA array · Two-component transcriptional unit · Multitarget · Multigene family

The original version of the chapter has been revised. A correction to this chapter can be found at https://doi.org/10.1007/978-3-031-08072-2_18

N. Budhagatapalli
Centre for Plant Genome Engineering, Institute of Plant Biochemistry, Heinrich-Heine-University, Düsseldorf, Germany

G. Hensel (✉)
Centre for Plant Genome Engineering, Institute of Plant Biochemistry, Heinrich-Heine-University, Düsseldorf, Germany

Centre of Region Haná for Biotechnological and Agricultural Research, Czech Advanced Technology and Research Institute, Palacký University Olomouc, Olomouc, Czech Republic
e-mail: goetz.hensel@hhu.de

1 Introduction

The biotechnological use of sequence-specific endonucleases (SSNs) has revolutionized studies of inherited traits of organisms. Systems such as clustered regularly interspaced short palindromic repeats (CRISPR)/CRISPR-associated (Cas) proteins Cas9 and Cas12a are widely used to (1) study gene functions in different organisms or (2) make application-oriented modifications, such as improving agronomically essential plant traits. They are currently the most convenient gene modification tools regarding the effort required to create two functional units (gRNA and Cas). Unlike DNA-directed zinc finger nucleases (ZFNs) or transcription activator-like effector nucleases (TALENs), Cas proteins are RNA-directed endonucleases. Cas proteins cleave DNA in a targeted manner most commonly used for site-directed mutagenesis or genome/gene editing. Cas proteins require an adjacent protospacer motif (PAM) sequence for their binding to double-stranded DNA, which varies depending on the origin of the Cas protein. For the most commonly used *Streptococcus pyogenes* Cas9 protein (SpCas9), the PAM sequence is NGG, where the N is variable and represents all four nucleotides (Jinek et al. 2012). Cas endonucleases differ in terms of their PAM and the type of double-strand break-induced. While SpCas9 generates smooth cut ends, sticky ends are obtained when using LbCas12a. The limitation in selecting the target sequence due to the presence of this PAM is overcome by using Cas proteins from other organisms or engineering the Cas9 protein (xCas9 (Hu et al. 2018), SpCas9-NG (Nishimasu et al. 2018)). Thus, almost any genomic sequence is accessible. A plethora of Cas variants with alternative PAM sequences have been listed earlier (Molla et al. 2020c).

SpCas9-based plant gene editing has been extensively studied and applied in many plant species (for review, see chapter "Genome Engineering as a Tool for Enhancing Crop Traits: Lessons from CRISPR/Cas9" and Kumlehn et al. (2018)). Although SpCas9 and FnCas12a have nearly identical sizes (1368 vs 1307 amino acids), Cas9 has two nuclease domains (HNH and RuvC) while Cas12a has only RuvC domain. For binding to the target DNA, Cas9 requires an approximately 130 nucleotide long hybrid RNA consisting of a 42-nucleotide long CRISPR RNA (crRNA) and an 89 bp long trans-activating crRNA (tracrRNA) (Jinek et al. 2012). In contrast, Cas12a requires only a 42 nucleotide long tracrRNA (Zhong et al. 2018), which significantly reduces the cost of synthetic production.

In addition, Cas proteins could be modified by mutating specific amino acids in the previously named nuclease domains, converting them into a nuclease targeting double-stranded DNA but cleaves only one strand (Cas nickase), designated nCas or none (dead nuclease designated dCas; (Mali et al. 2013; Qi et al. 2013)). Fusion of the dCas protein with the domain of transcriptional regulators such as activators or repressors allows regulation of gene expression (Tang et al. 2017; Qi et al. 2016). Fusion with reporter genes can also mark specific genomic regions (CRISPR imaging, (Khosravi et al. 2020)). A recent development is an epigenetic regulation using CRISPR-based programmable epigenome editor protein, termed CRISPRoff-V1, composed of ZNF 10KRAB, Dnmt3A (D3A), and Dnmt3L (D3L) protein domains fused to catalytically inactive *S. pyogenes* dCas9 (Nuñez et al. 2021). In this process, gene regulation is blocked by methylation. This alteration can also be transmitted to the progeny.

The interaction of multiple genes often characterizes agronomically important traits. These can be different members of a gene family or numerous genes in a network (e.g. (Zuo and Li 2014)). As a result, simultaneous or sequential targeting of different sites in the same target genome is essential. This process, called multiplexing, is easier to achieve with CRISPR/Cas technology. The sequence-specific navigation elements do not consist of several polypeptides but a short RNA molecule about 20 nucleotides long. These are easier to clone or synthesize. Since functional endonuclease elements are often transferred as part of a bacterial T-DNA, the size of these elements plays a crucial role here.

In Cas9 technology, target specificity is determined by an 18–20 bp guide RNA (gRNA) that can theoretically be designed for any target (Baltes and Voytas 2015). A single gRNA can be used to target alleles or homeoalleles of a gene or gene family in a conserved region (Sánchez-León et al. 2018). Multiplexing allows multiple gRNAs to target different genes (Shen et al. 2017).

To create stable mutants, transfer requires functional units that contain sequences for the cas gene and one or more gRNAs and usually plant selection marker elements. Alternatively, these units can be pre-assembled as a ribonucleoprotein complex (RNP) in the test tube, which has advantages in regulatory hurdles in some parts of the world (chapter "Genome Editing: A Review of the Challenges and Approaches"). Modular systems are used for ease of assembly, such as MoClo based on Golden Gate technology (Weber et al. 2011) or vector set published by Hahn et al. (2020). All building blocks, such as promoters, coding sequences, and terminators, are in individual vectors assembled arbitrarily in different modules. Alternatively, methods based on homology-dependent cloning systems (e.g. Gibson, overlap-PCR) can be used.

Cas proteins were expressed in plants in the first generation of vectors by polymerase II and gRNAs by polymerase III promoters, such as U3 and U6 promoters. Because of limitations regarding using elements from dicots in monocots, these regulatory building blocks must be specific to monocot or dicot species. Pol III-transcribed small nucleolar RNAs (snoRNAs) such as U3 and U6 require a specific ribonucleotide, A and G, respectively, to initiate their transcription. While this reduces the number of potential Cas/gRNA target sites in the genome, small tricks can quickly convert gRNAs that begin with C or T. For this purpose, if an A or G is present at position −19, the first nucleotide can be omitted, or it can be replaced by an A or G. Both approaches have already been successfully applied (Le Cong et al. 2013).

This chapter summarises the approaches used for multiplexing in targeted mutagenesis, gene activation/repression, base editing, and de-novo domestication. The methodological differences will be exemplified by model and crop plant species, while a comprehensive but not complete overview will be given in a table.

Targeted Mutagenesis

Although genome editing is predominant in the literature, the most common applications are targeted mutagenesis of one or more genes or gene alleles. Therefore, it is not surprising that most results on simultaneous targeting using CRISPR/Cas technology also describe targeted knockout. Below are descriptions of the most commonly used variations in the structure and combination of functional units. Table 1 summarises several examples for the successful knockout in a diverse set of plants.

It should be noted that multiplexing refers to the simultaneous targeting of several gene variants in a diploid plant, or the simultaneous targeting of a gene in the different subgenomes of a polyploid plant, or also the knockout of unrelated genes. Table 1 shows examples of all the cases mentioned above. Only two examples are given here as examples. Celiac disease is a human disease in which patients are allergic to specific storage proteins of the wheat grain (Sollid et al. 2012). In bread wheat, α-gliadins are encoded by approximately 100 genes, and pseudogenes (Ozuna et al. 2015) organized in tandem at the *Gli-2* loci of chromosomes 6A, 6B and 6D. In an attempt to target a conserved region within these loci using two gRNAs, scientists have shown wheat lines in which 35 of 45 different *α-gliadin* sequences were mutated (Sánchez-León et al. 2018). Since wheat is challenging to transform, Budhagatapalli and colleagues (Budhagatapalli et al. 2020) chose a different route. They took advantage of the fact that maize is easier to transform, and wheat can be pollinated with maize pollen. After fertilization, the chromosomal DNA of maize is not passed on, and the haploid embryo can then undergo genome duplication either spontaneously or using chemicals. It has been shown that both bread wheat and durum wheat can be mutated almost genotype-independently using this method (Budhagatapalli et al. 2020). Since wide hybridization has also been described for other plant species, this is another way to modify genes efficiently.

Multiplex Gene Activation/Repression

Due to its targeting accuracy and ease of cloning, CRISPR/Cas technology can be used not only for targeted induction of double-strand breaks. By inactivating RuvC (Cas12a) or both nuclease domains (HNH and RuvC, Cas9), one can deliver a dCas fusion protein to the desired target in the subject. The possibilities for such fusions are numerous and, as described in the introduction, range from activating or repressing genes, setting epigenetic modifications, or visualizing genomic regions by fusion with a reporter gene. Therefore, it stands to reason that such approaches would also be interested in simultaneously targeting multiple targets. To date, there are only examples in the literature from Arabidopsis and rice describing such results. In both cases, multiplexing was achieved by multiple gRNA arrays within a T-DNA (Lowder et al. 2015; Xiong et al. 2021). Thus, a fusion of the deactivated Cas9 with the VP16 or VP64 domain measured a two- to sevenfold increase in the transcriptional activity of the transcription factor *AtPAP1* (Lowder et al. 2015). The same authors also showed by fusing the SRDX domain to the deactivated Cas9 that the Arabidopsis RNA processing factor AtCSTF63 could be reduced in activity to 60% of the wild-type level. The latest development in this field has been CRISPR–Act3.0,

Table 1 Examples of multiplexing using CRISPR/Cas technology

Plant species	Speciality	Target gene(s), (No. of gRNAs)	Promoter::Cas variant	Transfer method	Multiplexing strategy	Mutation efficiency (mutants/plants analyzed) [%]	Explants analyzed	Reference
1. Targeted mutagenesis								
1.1 *Actinidia* Lindl. (kiwifruit)								
Constitutive								
	TCTU	*PDS* gRNA1 + gRNA2 (2) *PDS* gRNA3 + gRNA4 (2)	*35S::SpCas9*	*Agrobacterium tumefaciens*	Multiple gRNA arrays on one T-DNA	7.5 (9/120) 8.3 (1/12) 6.8 (19/280) 7.1 (2/28)	T_0 calli	Wang et al. (2018b)
	TCTU	*PDS* gRNA1 + gRNA2 (2) *PDS* gRNA3 + gRNA4 (2)	*35S::SpCas9*	*Agrobacterium tumefaciens*	tRNA-sgRNA (PTG)	90.8 (109/120) 83.3 (10/12) 73.8 (192/260) 65.4 (17/26)	T_0 calli	
1.2 *Arabidopsis thaliana*								
Root cup-specific								
	TCTU	*GFP* + *SMB* + *EXI1* + *GL1* + *ARF7* + *ARF19* (6)	*SMB::Cas9*	Floral dip	Multiple gRNA arrays on one T-DNA	55.0 (52/95)	T_1 plants	Bollier et al. (2021)
		SMB-1 *SMB-2* (2)				62.0 (13/21) 75.0 (9/12)	T_1 plants	Decaestecker et al. (2019)
Egg cell-specific								
	TCTU	*GLCAT* 14A,B,C (3) *GLCAT* 14A,B,C (4)	*EC1.1e1.2::zCas9*	Floral dip	Multiple gRNA arrays on one T-DNA	n.a.	n.a.	Zhang et al. (2020a)
Stomatal-lineage-specific								
								(continued)

Table 1 (continued)

Plant species	Speciality	Target gene(s), (No. of gRNAs)	Promoter::Cas variant	Transfer method	Multiplexing strategy	Mutation efficiency (mutants/plants analyzed) [%]	Explants analyzed	Reference
		GFP + PDS3 (2)	TMM::Cas9	Floral dip	Multiple gRNA arrays on one T-DNA	90.0 (18/20) ~80.0	T_1 plants T_2 protoplasts	Decaestecker et al. (2019)
			FAMA::Cas9	Floral dip		0 (0/21) 30.0–74.0	T_1 plants T_2 protoplasts	
		YDA (2)	TMM::Cas9	Floral dip		100.0 (40/40)	T_1 plants	
	Lateral-root-specific	GFP (1)	GATA23::Cas9	Floral dip	Cas9/multiple gRNA arrays	87.0 (20/23)	T_1 plants	Decaestecker et al. (2019)
	Constitutive	YDA (2)	PcUbi::Cas9	Floral dip	Multiple gRNA arrays on one T-DNA	94.0 (33/35)	T_1 plants	Decaestecker et al. (2019)
		ARF7 + ARF19 (2)	PcUbi::Cas9	Floral dip	Multiple gRNA arrays on one T-DNA	69.0 (18/26)	T_1 plants	
		PTOX (5)	OsUbi::Cas9	Floral dip	tRNA-sgRNA (PTG)	24.4 (30/123)	T_1 plants	Hui et al. (2019)
		PYL (6)	AtUbi1::hCas9	Floral dip	Multiple gRNA arrays on one T-DNA	13.0–93.0	T_1 plants	Zhang et al. (2016)
		TRY + CPC + ETC2 (6)	ZmUbi1::zCas9	Floral dip	Multiple gRNA arrays on one T-DNA	>90 (n.a.)	T_1 plants	Xing et al. (2014)

	CHLI1 + CHLI2 (2)	2x35S::zCas9	Floral dip	Multiple gRNA arrays on one T-DNA	67.0 (24/36)	T₁ plants	Xing et al. (2014)
	ERECTA + RPM1 + RPS4 + RPS6 + RPP4 + RPS2 + TMM + WER1 + RPP2a + FORKED1 + RPS5 + GL1 (24)	RPS5a::zCas9i	Floral dip	Multiple gRNA arrays on one T-DNA	One reported plant with mutations in all 12 targets	T₁ plants	Stuttmann et al. (2021)
	At05g55580 (3)		Floral dip		42.9 (6/14) 33.3 (3/9) 22.2 (2/9)	T₁ plants	Ma et al. (2015)

1.3 Brassica oleracea

Constitutive

TCTU	PDS (4) SERK3 (4) SERK3 + MS1 (8)	35S::Cas9	Agrobacterium tumefaciens	tRNA-sgRNA (PTG)	68.0 (17/25) 100.0 (29/29) 72.2 (13/18) 27.8 (5/18)	T₀ plants	Ma et al. (2019)

1.4 Cichorium intybus L. (chicory)

Constitutive

TCTU	PDS (3)	35S::pCas9	Agrobacterium tumefaciens	Multiple gRNA arrays on one T-DNA	31.3 (10/32)	Hairy roots	Bernard et al. (2019)

1.5 Citrus x sinensis

Constitutive

TCTU	PDS (3)	35S::csy4:Cas9	Agrobacterium tumefaciens	tRNA-sgRNA (PTG)	44.4 (4/9)	T₀ plants	Huang et al. (2020)

1.6 Cucumis melo (melon)

Constitutive

TCTU	PDS gRNA1 + gRNA2 (2)	35S::zCas9	Agrobacterium tumefaciens	Multiple gRNA arrays on one T-DNA	45.0 42.0	T₀ plants	Hooghvorst et al. (2019)

(continued)

Table 1 (continued)

Plant species	Speciality	Target gene(s), (No. of gRNAs)	Promoter::Cas variant	Transfer method	Multiplexing strategy	Mutation efficiency (mutants/plants analyzed) [%]	Explants analyzed	Reference
1.7 Glycine max (soybean)								
Constitutive								
	TCTU	*F3H1* + *F3H2* + *FNSII-1* (4)	*GmUbi3::dpCas9*	*Agrobacterium tumefaciens*	Multiple gRNA arrays on one T-DNA	55.6 (15/27) 48.1 (13/27) 81.5 (22/27)	T_0 plants	Zhang et al. (2020c)
1.8 Gossypium ssp. (cotton)								
Constitutive								
	TCTU	*MYB25* like A + *MYB25* like D (2)	*35S::zzCas9*	*Agrobacterium tumefaciens*	Multiple gRNA arrays on one T-DNA	100.0 (8/8)	T_0 plants	Li et al. (2017)
1.9 Hordeum vulgare (barley)								
Constitutive								
	TCTU	*CKX1* + *CKX3*	*ZmUbi1::Cas9i*	*Agrobacterium tumefaciens*	tRNA-sgRNA (PTG)	34.5 (49/142) 9.1 (13/142)	T_0 plants	Gasparis et al. (2018)
1.10 Musa cavendishii (banana)								
Constitutive								
		PDS (3)	*2x35S::Cas9*	*Agrobacterium tumefaciens*	tRNA-sgRNA (PTG)		T_0 plants	Ntui et al. (2020)
1.11 Nicotiana ssp.								
Constitutive								
Nicotiana tabacum		*PDS3* (1) + *AGAMOUS* (2)	*2x35S::Cas9*	*Agrobacterium tumefaciens*	Tandem array or tRNA		T_1 plants	Ellison et al. (2020)
Nicotiana attenuata		*EAH1, NEC5b, NEC3a, AOC, MYC2,* and *NEC1c* (2)	*35S::Cas9*	PEG	tRNA-sgRNA (PTG)	~15.0	Protoplasts	Oh et al. (2020)

Species		Target genes	Promoter::Cas9	Delivery	sgRNA	Efficiency	Generation	Reference
Nicotiana benthamiana		*FucT1 + 2* (3) *FucT1–4* (4) *XylT1 + 2* (3) *FucT1–4 +* *XylT1 + 2-*(7)	*35SPPPDK::Cas9p*	*Agrobacterium tumefaciens*	tRNA-sgRNA (PTG)	10.0–14.0 0–66.0 62.0–67.0 0 (for all 6 target genes)	T_0 plants	Jansing et al. (2019)

1.12 *Oryza glabarrima* (wild rice)

Constitutive

| | | *GN1A +*
GS3 +
GW2 (3) | *OsUbi::Cas9* | *Agrobacterium tumefaciens* | tRNA-sgRNA (PTG) | 3.9 (3/76)
25.0 (19/76)
0, but 21.1 triple mutants (16/76) | T_0 plants | Lacchini et al. (2020) |

1.13 *Oryza sativa* (rice)

Constitutive

	TCTU	*SR32 + SR33a + SR33 + SR40* (4)	*OsUbi::Cas9*	*Agrobacterium tumefaciens*	tRNA-sgRNA (PTG)	0 (none quadruple but 2 triple) (0/2)	T_0 plants	Butt et al. (2019)
	TCTU	*RS21a + RSZ21 + RSZ23* (3)	*OsUbi::Cas9*	*Agrobacterium tumefaciens*	tRNA-sgRNA (PTG)	100.0 (2/2)	T_0 plants	
	TCTU	*SC25 + SC32 + SC34* (3)	*OsUbi::Cas9*	*Agrobacterium tumefaciens*	tRNA-sgRNA (PTG)	100.0 (2/2)	T_0 plants	
	TCTU	*SCL30a + SCL30 + SCL57* (3)	*OsUbi::Cas9*	*Agrobacterium tumefaciens*	tRNA-sgRNA (PTG)	0 (0/2)	T_0 plants	
	TCTU	*RS2Z36 + RS2Z37 + RS2Z38 + RS2Z39* (4)	*OsUbi::Cas9*	*Agrobacterium tumefaciens*	tRNA-sgRNA (PTG)	50.0 (1/2)	T_0 plants	

(continued)

Table 1 (continued)

Plant species	Speciality	Target gene(s), (No. of gRNAs)	Promoter::Cas variant	Transfer method	Multiplexing strategy	Mutation efficiency (mutants/plants analyzed) [%]	Explants analyzed	Reference
		RS29 + RS33 (2)	OsUbi::Cas9	Agrobacterium tumefaciens	tRNA-sgRNA (PTG)	100.0 (2/2)	T_0 plants	Pathak et al. (2019)
		GUS PDS Chalk5 (2)	OsUbi::Cas9	Agrobacterium tumefaciens	tRNA-sgRNA (PTG)	1.8 (2/113) 6.2 (2/32) 7.5 (4/53)	Callus	
		PDS + DEP1 (2)	ZmUbi1::Cas9	Agrobacterium tumefaciens	Csy4 tRNA-sgRNA (PTG)	90.6 (29/32) 86.7–93.3 (26–28/30)	T_0 plants	Tang et al. (2019)
		PDS + DEP1 (3)	ZmUbi1::Cas9	Agrobacterium tumefaciens	Csy4 tRNA-sgRNA (PTG)	60.6–93.9 (20–31/33) 67.6–89.2 (25–33/37)	T_0 plants	
		PDS + DEP1 + YSA (6)	ZmUbi1::Cas9	Agrobacterium tumefaciens	tRNA-sgRNA (PTG)	60.5–97.4 (23–37/38)	T_0 plants	
		DEP1 + ROC5 (4)	ZmUbi1::LbCas12a	Agrobacterium tumefaciens	crRNA array	29.2–50 (7–12/24)	T_0 plants	
		RLK (4)	ZmUbi1::FnCas12a	Agrobacterium tumefaciens	crRNA array	43.8–75 (14–24/32)	T_0 plants	Wang et al. (2017)
		BEL (4)	ZmUbi1::LbCas12a	Agrobacterium tumefaciens	crRNA array	40.0–60.0 (12–18/30)	T_0 plants	
	TCTU (HH-HDV)	LEA1, LEA2 (8)	ZmUbi1::FnCas12a	Agrobacterium tumefaciens	crRNA array	0–70.8	T_0 plants	Wang et al. (2018a)
	SSTU (tRNA)	LEA1, LEA2 (8)	ZmUbi1::FnCas12a	Agrobacterium tumefaciens	crRNA array	0–70.8	T_0 plants	Wang et al. (2018a)
	TCTU (tRNA)	LEA Dehydrin (9)	ZmUbi1::LbCas12a	Agrobacterium tumefaciens	crRNA array	4.2–54.2	T_0 plants	

SSTU (tRNA)	LEA Dehydrin (9)	ZmUbi1::LbCas12a	Agrobacterium tumefaciens	crRNA array	4.2–70.8	T0 plants
TCTU	CYP81A (3)	ZmUbi1::Cas9	Agrobacterium tumefaciens	Multiple gRNA arrays on one T-DNA	70.8–79.2	T0 plants
SSTU	CYP81A (3)	ZmUbi1::Cas9	Agrobacterium tumefaciens	crRNA array	50–86.4	T0 plants
	EP3, GS3, GW2, BADH2, DEP1, Gn1a, LPA1, Hd1	35S::Cas9	Agrobacterium tumefaciens	Multiple gRNA arrays on one T-DNA	50.0 / 100.0 / 67.0 / 81.0 / 83.0 / 97.0 / 67.0 / 78.0	T0 plants	Shen et al. (2017)
	YSA + ROC5 (3)	35S::pcoCas9	Agrobacterium tumefaciens	Multiple gRNA arrays on one T-DNA	33.3–53.3	T0 plants	Lowder et al. (2015)
	FTL1 + FTL4 + FTL5 + FTL6 + FTL9 + FTL10 + FTL11 + FTL13 (8)	ZmUbi1::Cas9	Agrobacterium tumefaciens	Multiple gRNA arrays on one T-DNA	100.0 (3/3) / 100.0 (3/3) / 100.0 (3/3) / 100.0 (3/3) / 100.0 (3/3) / 100.0 (3/3) / 0 (0/3) / 100.0 (3/3)	T0 plants	Ma et al. (2015)
	FTL7 + FTL8 + FTL12 (3)	ZmUbi1::Cas9	Agrobacterium tumefaciens	Multiple gRNA arrays on one T-DNA	84.6 (22/26) / 91.7 (22/24) / 91.7 (22/24)	T0 plants	Ma et al. (2015)

(continued)

Table 1 (continued)

Plant species	Speciality	Target gene(s), (No. of gRNAs)	Promoter::Cas variant	Transfer method	Multiplexing strategy	Mutation efficiency (mutants/plants analyzed) [%]	Explants analyzed	Reference
		GSTU + MRP15 + AnP (3)	ZmUbi1::Cas9	Agrobacterium tumefaciens	Multiple gRNA arrays on one T-DNA	100 (5/5) 75 (3/4) 100 (4/4)	T₀ plants	Lemmon et al. (2018)
		Waxy + Waxy + Waxy	ZmUbi1::Cas9	Agrobacterium tumefaciens	Multiple gRNA arrays on one T-DNA	66.7 (2/3) 100 (2/2) 33.3 (1/3)	T₀ plants	

1.14 Physalis pruinosa (groundcherry)

Constitutive

| | | SP5G (2) CLV1 (2) | 35S::Cas9 | Agrobacterium tumefaciens | Multiple gRNA arrays on one T-DNA | Several | T₀ plants | |

1.15 Setaria viridis (green foxtail)

Constitutive

| | | Drm1a + Drm1b gRNA2 (3) | ZmUbi1:: Trex2:TaCas9 | Agrobacterium tumefaciens | tRNA-sgRNA (PTG) | 82.0 (9/11) 73.0 (8/11) | T₀ plants | Weiss et al. (2020) |
| | | Drm1a + Drm1b gRNA1 + Drm1b gRNA2 (3) | ZmUbi1:: Trex2:TaCas9 | Agrobacterium tumefaciens | tRNA-sgRNA (PTG) | 100.0 (11/11) 100.0 (11/11) 100.0 (11/11) | T₀ plants | |

1.16 Solanum lycopersicum (tomato)

Constitutive

| | TCTU | SP + O + FW2.2 + CycB + FAS + MULT (6) | 35S::ccsy4:AtCas9 | Agrobacterium tumefaciens | Multiple gRNA arrays on one T-DNA | 33.3 (3/10) quadruple mutants | T₀ plants | Zsögön et al. (2018) |
| | TCTU | CLV3 + FW2.2 + MULT + CycB (8) | 35S::ccsy4:AtCas9 | Agrobacterium tumefaciens | Multiple gRNA arrays on one T-DNA | 66.6 (2/3) quadruple mutants | T₀ plants | |

	Target	Cas	Delivery	gRNA	Efficiency (%)	Plants	Reference
TCTU	SGR1 + LCY-E + Blc + LCY-B1 + LCY-B2	ZmUbi1::Cas9	Agrobacterium tumefaciens	Multiple gRNA arrays on one T-DNA	6.9 4.5 0.9	T$_0$ plants	Li et al. (2018)

1.17 Saccharum spp. hybrid

Constitutive

	Target	Cas	Delivery	gRNA	Efficiency (%)	Plants	Reference
TCTU	MgCh gRNA1 + gRNA2	35S::Cas9	Biolistic	Two gRNA arrays on one T-DNA	14 (3/22)	T$_0$ plants	Eid et al. (2021)

1.18 Triticum aestivum (bread wheat)

Constitutive

	Target	Cas	Delivery	gRNA	Efficiency (%)	Plants	Reference
	GW2T6, GS3T11 and GSE5T10 (3)	ZmUbi1::LbCas12a	PEG	crRNA array	6.9 4.5 0.9	Protoplasts	Wang et al. (2021)
	GW2T6 + GS3T11 + SPL16T2 + GASR7T2 + AnIT11 + GSE5T10 + GW7T13 GW7T14 (8)	ZmUbi1::LbCas12a	PEG	crRNA array	9.9 4.5 0.9 1.6 2.1 0.6 1.0 0.6	Protoplasts	
TCTU	BRI1 gRNA1 + gRNA2	ZmUbi1::zCas9	Agrobacterium tumefaciens	One gRNA targeting three homeoalleles	3.6–10 (3/83, 2/44, 1/10) 50–66 (2/4, 2/3)	T$_0$ plants	Budhagatapalli et al. (2020)

1.19 Triticum durum (durum wheat)

Constitutive

	Target	Cas	Delivery	gRNA	Efficiency (%)	Plants	Reference
	α-Amylase/trypsin inhibitors (7)	ZmUbim3::wCas9	Particle bombardment	tRNA-sgRNA (PTG)	14.4 (14/97)	T$_1$ plants	Camerlengo et al. (2020)

(continued)

Table 1 (continued)

Plant species	Speciality	Target gene(s), (No. of gRNAs)	Promoter::Cas variant	Transfer method	Multiplexing strategy	Mutation efficiency (mutants/plants analyzed) [%]	Explants analyzed	Reference
1.20 *Zea mays* (maize)								
Constitutive								
	TCTU	*RPL* + *PPR* (2)	*ZmUbi1::zCas9*	*Agrobacterium tumefaciens*	Gly tRNA-sgRNA (PTG)	88.9 (16/18) 85.7 (12/14)	T₀ plants	Qi et al. (2016)
	TCTU	*lncRNAs* (4)	*ZmUbi1::zCas9*	*Agrobacterium tumefaciens*	Gly tRNA-sgRNA (PTG)	100 (11/11)	T₀ plants	
	TCTU	*O2* (3)	*ZmUbi1::zCas9*	*Agrobacterium tumefaciens*	Gly tRNA-sgRNA (PTG)	100.0 (3/3)	T₀ plants	Gong et al. (2021)
	TCTU	*O2* (3)	*ZmUbi1::zCas12a*	*Agrobacterium tumefaciens*	crRNA array	33.3 (1/3)	T₀ plants	
2. Gene activation								
2.1 *Arabidopsis thaliana*								
		PAP1 (3) *miR319* (3)	*AtUbi10::dCas9:VP64*	*Agrobacterium tumefaciens*	Multiple gRNA arrays on one T-DNA	2–seven-fold 3–7.5-fold	T₀ plants	Lowder et al. (2015)
		FIS2 (3)	*AtUbi10::dCas9:VP64*	*Agrobacterium tumefaciens*	Multiple gRNA arrays on one T-DNA	200–400-fold	T₀ plants	
2.2 *Oryza sativa*								
	TCTU	*GW7* and *ER1*	*AtUbi10::dCas9-TV*		Multiple gRNA arrays on one T-DNA	–		Xiong et al. (2021)
3. Gene repression								
Arabidopsis thaliana								

CSTF64 (3)	AtUbi10:dCas9:3X(SRDX)	Agrobacterium tumefaciens	Multiple gRNA arrays on one T-DNA	60% reduced	T0 plants	Lowder et al. (2015)
miR159A + miR159B (3)				50% reduced		
4. Base editing						
Oryza sativa						
ALS1 + GS1 gRNA 3 (2)	ZmUbi1::TadA:nCas9	Agrobacterium tumefaciens		93.7 (45/48) 75.0 (36/48)	T0 calli	Yan et al. (2021)
TubA2 + ACC gRNA 2 (2)		Agrobacterium tumefaciens		83.3 (35/42) 76.2 (32/42)	T0 calli	
ALS1 + GS1 gRNA 3 + TubA2 + ACC gRNA 2 (4)		Agrobacterium tumefaciens		89.6 (43/48) 56.3 (27/48) 89.6 (43/48) 83.3 (40/48)	T0 calli	
WSL5 ZEBRA3 (Z3)	OsUbi::TadA:nCas9	Agrobacterium tumefaciens	tRNA-sgRNA (PTG)	2.8 (4/142) 14.1 (20/142)	T0 plants	Molla et al. (2020b)
ALS NRT1.1B GRF4 Waxy	OsUbi::SaKKHn: PmCDA1:UGI	Agrobacterium tumefaciens	tRNA-sgRNA (PTG)	5.0–45.9 (1/20–17/37) 44.4 (8/18) 17.6 (3/17) 2.5–26.3 (1/40–5/19)	T0 plants	Zhang et al. (2020b)

EC1.1e1.2 At egg cell 1.1 with egg cell-specific enhancer AtEC2.1, EC1.2 – At egg-cell 1.2 (At2g21740), SMB – AtSOMBRERO (AT1G79580), zCas9 – maize codon-optimized Cas9, GLCAT – β-glucuronosyltransferases (GLCAT14A-At5g39990, GLCAT14B-At5g15050, and GLCAT14C-At2g37585), n.a. – not available, TMM – TOO MANY MOUTHS, FAMA – bHLH transcription factor (bHLH097), named after the goddess of rumour, PTG – Polycistronic tRNA–gRNA, PTOX – Plastid terminal oxidase, PYL – pyrabactin resistance 1-like, hCas9 – human-codon optimized *Cas9* gene, Cas9p – plant codon-optimized *Cas9*, 35SPPDK – hybrid promoter of CaMV 35S and fused to the maize C4PPDK basal promoter (Yoo et al. 2007), SSTU – simplified single transcriptional unit, TCTU – two-component transcriptional unit, HH – hammerhead ribozyme, HDV – hepatitis delta virus ribozyme, EP3 – ERECT PANICLE, GS3 – for grain length and weight, GW2 – major QTL associated with rice grain width and weight, BADH2 – betaine aldehyde dehydrogenase 2, DEP1 – dense and erect panicle 1, Gn1a – grain number, LPA1 – loose plant architecture 1, Hd1 – heading date 1, YSA – young seedling albino, ROC5 – Rice Outermost Cell-specific Gene5, dpCas9 – dicotyledon codon-optimized Cas9, SELF-PRUNING (SP), OVATE (O), FRUIT WEIGHT 2.2 (FW2.2), FASCIATED (FAS) or MULTIFLORA (MULT) and LYCOPENE BETA CYCLASE (CycB)

which allowed a robust activation of up to seven genes for metabolic engineering in rice (Pan et al. 2021). To test the system in dicot plants, the authors modified the expression of *AtFT1* and *AtTCL1*, genes involved in flowering time and trichome development. Using the CRISPR-Act3.0 system, early flowering Arabidopsis T3 mutant plants with reduced trichomes could be generated (Pan et al. 2021).

Multiplex Base Editing

After successfully establishing targeted double-strand breakage using CRISPR/Cas technology, various base editing systems have been introduced in recent years (Molla and Yang 2019). These are based on exchanging single or multiple bases in a specific section within the target region. These include adenine base editors, which cause an A-to-G exchange (Gaudelli et al. 2017). Furthermore, cytosine base editors exchange C-to-T (Ren et al. 2018). Recently a new set of base editors have been developed to install C-to-G editing (Molla et al. 2020a). It stands to reason that these systems are also of interest for multiplexing. The first paper on this has been published in rice (Yan et al. 2021). This work showed that when driving two targets simultaneously, between 75% and 94% of gene editing could be achieved on the single sequences and about 73% co-editing efficiency. If four targets are controlled simultaneously, this is still 56% (Yan et al. 2021). Another example for multiplex adenine base editing was provided by targeting *OsWSL5* and *OsZEBRA3* (*Z3*) in rice protoplasts and stable transgenic plants (Molla et al. 2020b). Through selfing and genetic segregation, transgene-free, base edited *wsl5* and *z3* mutants were obtained, displaying the expected variegated leaves phenotypes (Molla et al. 2020b). Examples for multiplexing using cytosine base editors were recently published. Zhang and colleagues (Zhang et al. 2020b) used the SaKKH Cas9 ortholog 15 target sites in *OsALS*, *OsNRT1.1B*, *OsGRF4*, and *OsWaxy*, were targeted. As a result, 2.5–45.9% of the target sites were modified (Zhang et al. 2020b).

De novo Domestication

Knowledge of more and more gene functions is leading to the ability to break new ground in breeding. The genetic diversity lost during the domestication process can now again be expanded by using CRISPR/Cas technology (Fernie and Yan 2019). This means that it is now possible to repeat domestication in fast motion. For example, it has already been shown that CRISPR/Cas-induced mutation of six genes can transform a wild form of tomato into cultivated tomato (Zsögön et al. 2018). Another research group has converted an African landrace into cultivated rice by mutating plant height via a knockout in *HTD1* and three gene loci relevant to grain size and yield (*GS3*, *GW2* and *GN1A*) (Lacchini et al. 2020). Another example has been given recently by modifying six agronomically important traits in allotetraploid rice *Oryza alta* (Yu et al. 2021). After establishing an efficient tissue culture and transformation system, CRISPR/Cas technology was used for de novo domestication of allotetraploid rice to develop into a new staple cereal to strengthen world food security (Yu et al. 2021). Thus, multiplexing here allows a significant improvement in our gene pool available for breeding. Therefore, traits of wild forms, such as drought tolerance or resistance to certain pests, can be transferred to cultivated forms without linkage drag.

2   Strategies for the Expression of Multiple gRNAs

In higher cultivated plants, multiplex genome editing can be achieved by combining the cas protein with multiple gRNAs as single or numerous transcription units in a T-DNA. The size of a gRNA expression unit varies from 250–500 bp, consisting of RNA Pol III promoter, gRNA sequence, and minimal termination signal (poly-A) or Pol III terminator. Due to plasmid size and transgene silencing by multiple promoters, U3 or U6, simultaneous expression of various gRNAs would be challenging (Ma et al. 2015). Therefore, several advanced strategies and cloning methods have been developed to combine multiple gRNAs into a single unit and to combine different cas and gRNA units.

In contrast to plants, highly efficient mutagenesis can be achieved in animals by co-injection of multiple gRNAs and/cas-coding mRNA into single cells. This is not applicable in higher plants, but transient expression of gRNA and cas, as genes in multiple constructs and ribonucleoprotein complex (RNP), could be achieved by transfection of protoplast cells. Regeneration of plants from protoplast cultures is complex, especially in monocotyledons. In tobacco, *PDS*-gene-edited plants regenerated from protoplasts have been reported (Lin et al. 2018).

Possible strategies (Hsieh-Feng and Yang 2020) for multiplexing in plants are shown in Fig. 1, and examples for different plant species are summarised in Table 1.

2.1   Two-Component Transcriptional Unit Multiplex System

Initial approaches for simultaneous targeting of multiple targets followed the need for the CRISPR/Cas system to consist of two components. One is the target-specific gRNA, and the other is the molecular scissors, the Cas protein. Both parts can be transferred into the cells as transcriptional units. In the form of pre-processed RNA, any number of units can be mixed. Alternatively, circular or linearized plasmid DNA can be used. The last option is to assemble Cas protein with synthetic or self-processed RNA, called ribonucleoprotein complex (RNP). RNA, plasmid DNA, or RNPs can then be biolistically transferred either PEG-mediated into protoplasts, by *Rhizobium radiobacter*, or into any plant cells after binding to tungsten or gold particles. Other transfer methods have been developed (liposomes, nanoparticles), but these cannot be discussed further here.

2.1.1   Cas9 and Multiple gRNA-TU Arrays

For simultaneous targeting of different targets, multiple gRNAs can be expressed as single gRNA cassettes, each transcribed by a separate RNA Pol III promoter. With the advancement of cloning methods, such as Golden Gate Cloning or Gibson Assembly, multiple gRNA expression units can now be assembled with individual

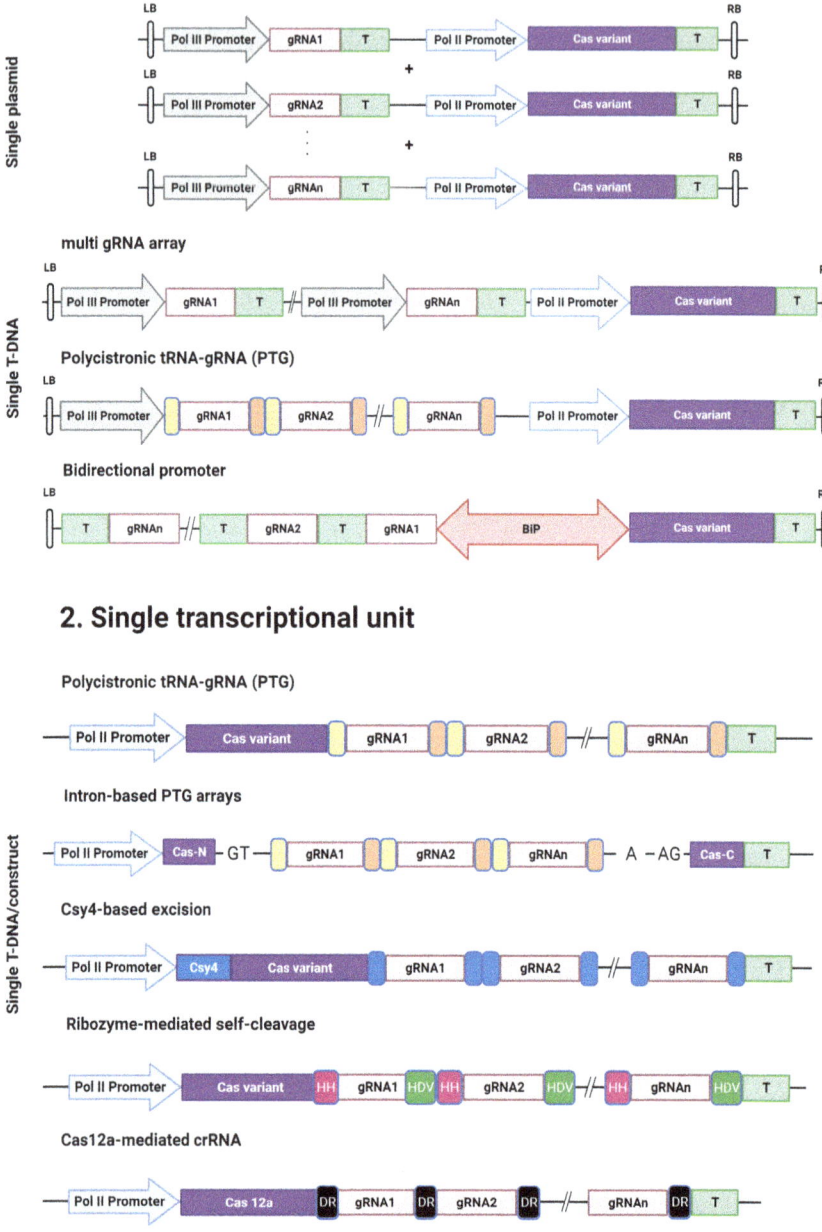

Fig. 1 Schematic overview of diverse options for multiplexed CRISPR/Cas units

gRNA modules (Lowder et al. 2015; Ma et al. 2015; Xing et al. 2014). In Arabidopsis and rice, up to eight gRNAs have been assembled for multiplex genome editing of members of a gene family (Ma et al. 2015). In tomato protoplasts, eight gRNAs have been reported to be expressed with individual Pol III promoters to simultaneously delete four genes, each 3 kb in size (Čermák et al. 2017). Recently, a paper was published in which up to 32 sgRNA transcription units (TUs) were assembled (Stuttmann et al. 2021). The authors generated an octuple mutant in *Nicotiana benthamiana* using a construct with ten sgRNA TUs.

Furthermore, using a 24 sgRNA TUs construct, they targeted twelve genes in *Arabidopsis thaliana*. They were able to detect a duodecuple mutant (12x) in the T2 generation, which showed the functionality of their approach (Stuttmann et al. 2021). However, one should remember that Pol III promoters have limited choice and availability, as they are not well characterized in higher plants. Another limitation of this method is that the insertion of larger T-DNA into plant cells and multiple gRNAs with repetitive promoter sequences is complicated and can lead to recombination and silencing.

2.1.2   Cas9 and Multiple gRNAs Using Single Pol III Promoter

The t-RNA processing system is a conserved and precise mechanism to produce diverse and most abundant snoRNA with tRNA from a single polycistronic gene in virtually all organisms. In polycistronic tRNA-gRNA (PTG), an array of multiple gRNAs is flanked by pre-tRNA with 5′-leader and 3′-trailer sequences that are recognized and cleaved by endogenous RNase P and RNase Z, respectively, at specific sites. The t-RNA gene sequences have box A and box B elements bound by transcription factor IIIC (TF IIIC), which recruits TF IIIB and RNA Pol III to the transcription start site at the Pol III promoter. Since t-RNA sequences have higher transcriptional activity, the engineering of the endogenous t-RNA system is thought to be suitable for processing multiple gRNAs. At the same time, this may overcome the limitation for the start nucleotide of gRNA for U3/U6 promoters. Therefore, Xie and colleagues (Xie et al. 2015) developed a system using a polycistronic tRNA-gRNA (PTG) gen to produce multiple target gRNAs simultaneously. After testing in protoplasts, they used the technique in stably transgenic rice plants. They demonstrated that multiple genomic loci were mutated simultaneously and that short chromosomal fragments could also be excised (Xie et al. 2015).

A similar study was conducted in maize (Qi et al. 2016). The tRNA processing system was established using the maize U6 promoter to mutate genes in single and multiplex approaches (Table 1). The authors noted that they demonstrated higher mutation efficiency in the multiplex approaches. In kiwi, Wang and colleagues compared the TCTU and PTG (Wang et al. 2018b). While using TCTU, only less than 10% of the generated calli had a detectable mutation at one or both target regions in the *PDS* gene. Calli developed with PTG had between 65 and over 90% of the target regions mutated. Additional examples are summarised in Table 1.

Since the Cas12a endonuclease, unlike Cas9, can also correctly process an array of multiple crRNAs in series, rice was used to investigate whether this leads to further simplification of the PTG system described previously. To this end, the authors performed a comparison with a TCTU and a so-called SSTU using FnCas12a and LbCas12a. The constructs used in this process contained a crRNA array of 8 and 9 targets, respectively. The mutation efficiencies obtained were comparable, making this system another simplification (Wang et al. 2018b).

2.1.3   Cas9 and gRNA Using Bidirectional Promoters

Other options include the use of bidirectional promoters (BiP), in which Cas and gRNA genes are expressed in opposite directions with a single Pol II promoter. BiPs described to date include the Mini 35S enhancer and the rice *BiP* promoter (*OsBiP1*). These BiP systems were tested and used for site-directed mutagenesis in Arabidopsis and rice. In direct comparison, *OsBiP1* showed higher and intense expression, resulting in 75.9–93.3% mutation efficiency in rice (Ren et al. 2019). Multiplexing with bidirectional promoters has not been described so far.

2.2   Single Transcriptional Unit

As previously described, the *cas* gene is expressed by a Pol II promoter, and the gRNA is expressed by a Pol III promoter. Since Pol III promoters are constitutively expressed, it is not possible to develop an inducible or tissue-specific CRISPR system using them. To overcome the limitations of Pol III promoters, single or multiple gRNAs can be expressed as a single long transcript driven by Pol II promoters. For proper processing, the gRNA functional units must then be flanked with interfaces for appropriate endogenous enzymes. In rice, it has been reported that the Pol II promoter *ZmUbi1* can be used to express a single crRNA ribozyme cassette and achieve mutation efficiency of up to 100% (Zhong et al. 2018; Tang et al. 2017).

In the intron-based single expression unit, gRNA(s) or crRNA(s) are inserted into introns. Such introns are then integrated into the open reading frame (ORF) of Cas proteins. After transcription and precise splicing of the intron-based single-expression unit, the Cas ORF is translated into a functional protein, and the gRNA matures independently. A flexible intron-based single expression system was introduced in rice, and multiplexing gRNAs was achieved by polycistronic tRNA-gRNA (PTG) and direct repeat (DR) and hammerhead (HH) hepatitis delta virus (HDV) ribozymes self-cleavage sites for Cas9 and Cas12a (Zhong et al. 2020). Details of the ribozymes are described in the following section.

2.2.1   Ribozymes-gRNA Ribozyme Array

Ribozymes are catalytically active RNA molecules that can, among other things, cut RNA. They are present in every cell and thus suitable to cut even long transcripts, such as gRNA arrays, to the correct length. Two of these ribozymes have been successfully used for multiplexing using CRISPR/Cas: one is the Hammerhead ribozyme and the Hepatitis Delta Virus (HDV) ribozyme. A synthetic gene, ribozyme-gRNA ribozymes (RGR), has successfully expressed multiple gRNAs driven by Pol II and III promoters in animals and plants (Lee et al. 2016; He et al. 2017). A gRNA is flanked by ribozyme-encoding sequences, HH at the 5′ end and HDV at the 3′ end; the RGR transcript is catalyzed by ribozymes that release the mature gRNA. Multiple RGR units connected by a linker of 12 bp achieved multiplexing genome editing in rice (He et al. 2017).

2.2.2   Csy4 gRNA Array

The CRISPR system Yersinia 4 (Csy4) is an endoribonuclease that processes RNA hairpins to release gRNAs. It was first identified in the bacterium *Pseudomonas aeruginosa*, which is involved in adaptive immunity. It is highly specific in the recognition, binding, and cleavage of a 20 bp sequence motif (Tsai et al. 2014). The use of Csy4 RNase-based multiplexing has been extensively reported in human cells and yeast (Tsai et al. 2014; Nissim et al. 2014). To express multiple gRNAs as a single RNA transcribed from a single promoter, each gRNA is flanked by Csy4 cleavage motifs as spacers (Čermák et al. 2017). Csy4 is fused to the cas protein in a single transcription unit (Tang et al. 2019).

3   Parameters Influencing Successful Multiplex Targeting

For successful simultaneous targeting of positions in a plant genome, some parameters should be considered. First, there is the question of the appropriate selection of regulatory elements. There is experimental evidence that specific promoters express differently in different plant species and tissues, which should be considered for the particular approach. It is equally important to have a sufficiently large number of molecules available. Processes in which multiple gRNAs compete for too small an amount of cas protein will be less efficient than those that take this aspect into account when selecting the expression system. The limitation of the construct's size, which mainly concerns viral vector-based expression systems, has been largely overcome. Also, potential problems associated with simple cloning and directed ordering within a construct have been overcome by cloning techniques such as Golden Gate or Gibson assembly. If one wants to cut fragments between 2 target sites, the size and position (+/−strand) between the interfaces can decide success. For example, if the distances are too small, the binding of the second gRNA/Cas complex may be blocked by the presence of the other. Although there has been a success in removing or reversing entire chromosomal segments in Arabidopsis (Schmidt et al. 2020), all the details that guarantee such a successful experiment are unknown. To this end, the cellular repair pathways (NHEJ, HDR, MMEJ, etc.) that are functional in the overhangs generated at the interfaces must also be considered.

Another issue is the secondary structure of the gRNA. Again, there is to note that successful binding to the cas protein is only guaranteed if the appropriate loops are formed. Online tools such as RNAfold (RNAfold web server (univie.ac.at)) can provide valuable help in this regard. Before going to the main experiment, the efficiency of guide RNAs can be checked in vitro (Budhagatapalli et al. 2016; Karmakar et al. 2021) to avoid potential failure. Furthermore, accessibility in a chromosomal context should also be considered. Thus, some genomic positions will be easier to access than others.

4   Conclusion

In summary, simultaneous targeting of multiple genomic targets has been successfully tested in some plants. This includes targeted mutation, tissue-specific mutation, and approaches targeting activation, repression of gene activity or even base editing. Limitations concerning the size of the constructs or the expression level could be overcome. At the same time, however, transferring the elements into the corresponding target cells remains. Without effective regeneration systems, only limited information about possible phenotypes is possible from widely used protoplast systems. The question of off-targets has been insufficiently investigated so far. Does the simultaneous targeting of multiple gene loci also lead to more off-targets? If so, it still might be less critical, as such unwanted mutations can be lost during the segregation of sexually propagating plants. Since selecting suitable gRNAs generally determines the number of off-targets, there should not be significant changes here. Also, few authors address position effects within the array. In any case, there is still a need for research here.

References

Baltes NJ, Voytas DF (2015) Enabling plant synthetic biology through genome engineering. Trends Biotechnol 33(2):120–131. https://doi.org/10.1016/j.tibtech.2014.11.008

Bernard G, Gagneul D, Santos HAD, Etienne A, Hilbert J-L, Rambaud C (2019) Efficient genome editing using CRISPR/Cas9 technology in chicory. Int J Mol Sci 20(5). https://doi.org/10.3390/ijms20051155

Bollier N, Buono RA, Jacobs TB, Nowack MK (2021) Efficient simultaneous mutagenesis of multiple genes in specific plant tissues by multiplex CRISPR. Plant Biotechnol J 19(4):651–653. https://doi.org/10.1111/pbi.13525

Budhagatapalli N, Schedel S, Gurushidze M, Pencs S, Hiekel S, Rutten T, Kusch S, Morbitzer R, Lahaye T, Panstruga R, Kumlehn J, Hensel G (2016) A simple test for the cleavage activity of customized endonucleases in plants. Plant Methods 12:18. https://doi.org/10.1186/s13007-016-0118-6

Budhagatapalli N, Halbach T, Hiekel S, Büchner H, Müller AE, Kumlehn J (2020) Site-directed mutagenesis in bread and durum wheat via pollination by cas9/guide RNA-transgenic maize used as haploidy inducer. Plant Biotechnol J 18(12):2376–2378. https://doi.org/10.1111/pbi.13415

Butt H, Piatek A, Li L, Reddy ASN, Mahfouz MM (2019) Multiplex CRISPR mutagenesis of the serine/arginine-rich (SR) gene family in Rice. Genes 10(8). https://doi.org/10.3390/genes10080596

Camerlengo F, Frittelli A, Sparks C, Doherty A, Martignago D, Larré C, Lupi R, Sestili F, Masci S (2020) CRISPR-Cas9 multiplex editing of the α-amylase/trypsin inhibitor genes to reduce allergen proteins in durum wheat. Front Sustain Food Syst 4. https://doi.org/10.3389/fsufs.2020.00104

Čermák T, Curtin SJ, Gil-Humanes J, Čegan R, Kono TJY, Konečná E, Belanto JJ, Starker CG, Mathre JW, Greenstein RL, Voytas DF (2017) A multipurpose toolkit to enable advanced genome engineering in plants. Plant Cell 29(6):1196–1217. https://doi.org/10.1105/tpc.16.00922

Decaestecker W, Buono RA, Pfeiffer ML, Vangheluwe N, Jourquin J, Karimi M, van Isterdael G, Beeckman T, Nowack MK, Jacobs TB (2019) CRISPR-TSKO: a technique for efficient mutagenesis in specific cell types, tissues, or organs in Arabidopsis. Plant Cell 31(12):2868–2887. https://doi.org/10.1105/tpc.19.00454

Eid A, Mohan C, Sanchez S, Wang D, Altpeter F (2021) Multiallelic, targeted mutagenesis of magnesium chelatase with CRISPR/Cas9 provides a rapidly Scorable phenotype in highly Polyploid sugarcane. Front Genome Edit 3. https://doi.org/10.3389/fgeed.2021.654996

Ellison EE, Nagalakshmi U, Gamo ME, Huang P-J, Dinesh-Kumar S, Voytas DF (2020) Multiplexed heritable gene editing using RNA viruses and mobile single guide RNAs. Nature plants 6(6):620–624. https://doi.org/10.1038/s41477-020-0670-y

Fernie AR, Yan J (2019) De novo domestication: an alternative route toward new crops for the future. Mol Plant 12(5):615–631. https://doi.org/10.1016/j.molp.2019.03.016

Gasparis S, Kała M, Przyborowski M, Łyżnik LA, Orczyk W, Nadolska-Orczyk A (2018) A simple and efficient CRISPR/Cas9 platform for induction of single and multiple, heritable mutations in barley (Hordeum vulgare L.). Plant Methods 14:111. https://doi.org/10.1186/s13007-018-0382-8

Gaudelli NM, Komor AC, Rees HA, Packer MS, Badran AH, Bryson DI, Liu DR (2017) Programmable base editing of A•T to G•C in genomic DNA without DNA cleavage. Nature 551(7681):464–471. https://doi.org/10.1038/nature24644

Gong C, Huang S, Song R, Qi W (2021) Comparative study between the CRISPR/Cpf1 (Cas12a) and CRISPR/Cas9 systems for multiplex gene editing in maize. Agriculture 11(5):429. https://doi.org/10.3390/agriculture11050429

Hahn F, Korolev A, Loures LS, Nekrasov V (2020) A modular cloning toolkit for genome editing in plants. BMC Plant Biol 20(1):179. https://doi.org/10.1186/s12870-020-02388-2

He Y, Zhang T, Yang N, Xu M, Yan L, Wang L, Wang R, Zhao Y (2017) Self-cleaving ribozymes enable the production of guide RNAs from unlimited choices of promoters for CRISPR/Cas9 mediated genome editing. J Genetics Genomics = Yi chuan xue bao 44(9):469–472. https://doi.org/10.1016/j.jgg.2017.08.003

Hooghvorst I, López-Cristoffanini C, Nogués S (2019) Efficient knockout of phytoene desaturase gene using CRISPR/Cas9 in melon. Sci Rep 9(1):17077. https://doi.org/10.1038/s41598-019-53710-4

Hsieh-Feng V, Yang Y (2020) Efficient expression of multiple guide RNAs for CRISPR/Cas genome editing. aBIOTECH 1(2):123–134. https://doi.org/10.1007/s42994-019-00014-w

Hu JH, Miller SM, Geurts MH, Tang W, Chen L, Sun N, Zeina CM, Gao X, Rees HA, Lin Z, Liu DR (2018) Evolved Cas9 variants with broad PAM compatibility and high DNA specificity. Nature 556(7699):57–63. https://doi.org/10.1038/nature26155

Huang X, Wang Y, Jin X, Wang N (2020) Development of multiplex genome editing toolkits for citrus with high efficacy in biallelic and homozygous mutations. Plant Mol Biol 104(3):297–307. https://doi.org/10.1007/s11103-020-01043-6

Hui L, Zhao M, He J, Hu Y, Huo Y, Hao H, Hao Y, Zhu W, Wang Y, Xu M, Aigen F (2019) A simple and reliable method for creating PCR-detectable mutants in Arabidopsis with the polycistronic tRNA–gRNA CRISPR/Cas9 system. Acta Physiol Plant 41(10). https://doi.org/10.1007/s11738-019-2961-3

Jansing J, Sack M, Augustine SM, Fischer R, Bortesi L (2019) CRISPR/Cas9-mediated knockout of six glycosyltransferase genes in Nicotiana benthamiana for the production of recombinant proteins lacking β-1,2-xylose and core α-1,3-fucose. Plant Biotechnol J 17(2):350–361. https://doi.org/10.1111/pbi.12981

Jinek M, Chylinski K, Fonfara I, Hauer M, Doudna JA, Charpentier E (2012) A programmable dual-RNA-guided DNA endonuclease in adaptive bacterial immunity. Science (New York, NY) 337(6096):816–821. https://doi.org/10.1126/science.1225829

Karmakar S, Behera D, Baig MJ, Molla KA (2021) In vitro Cas9 cleavage assay to check guide RNA efficiency. In: Tofazzal Islam M, Molla KA (eds) CRISPR-Cas methods. Springer US, New York, pp 23–39

Khosravi S, Ishii T, Dreissig S, Houben A (2020) Application and prospects of CRISPR/Cas9-based methods to trace defined genomic sequences in living and fixed plant cells. Chromosome Res 28(1):7–17. https://doi.org/10.1007/s10577-019-09622-0

Kumlehn J, Pietralla J, Hensel G, Pacher M, Puchta H (2018) The CRISPR/Cas revolution continues: from efficient gene editing for crop breeding to plant synthetic biology. J Integr Plant Biol 60(12):1127–1153. https://doi.org/10.1111/jipb.12734

Lacchini E, Kiegle E, Castellani M, Adam H, Jouannic S, Gregis V, Kater MM (2020) CRISPR-mediated accelerated domestication of African rice landraces. PLoS One 15(3):e0229782. https://doi.org/10.1371/journal.pone.0229782

Le Cong FAR, Cox D, Lin S, Barretto R, Habib N, Hsu PD, Wu X, Jiang W, Marraffini LA, Zhang F (2013) Multiplex genome engineering using CRISPR/Cas systems. Science (New York, NY) 339(6121):819–823. https://doi.org/10.1126/science.1231143

Lee CM, Cradick TJ, Bao G (2016) The Neisseria meningitidis CRISPR-Cas9 system enables specific genome editing in mammalian cells. Mol Ther 24(3):645–654. https://doi.org/10.1038/mt.2016.8

Lemmon ZH, Reem NT, Dalrymple J, Soyk S, Swartwood KE, Rodriguez-Leal D, van Eck J, Lippman ZB (2018) Rapid improvement of domestication traits in an orphan crop by genome editing. Nat Plants 4(10):766–770. https://doi.org/10.1038/s41477-018-0259-x

Li J, Norville JE, Aach J, McCormack M, Zhang D, Bush J, Church GM, Sheen J (2013) Multiplex and homologous recombination-mediated genome editing in Arabidopsis and Nicotiana benthamiana using guide RNA and Cas9. Nat Biotechnol 31: 688– 691

Li C, Unver T, Zhang B (2017) A high-efficiency CRISPR/Cas9 system for targeted mutagenesis in cotton (Gossypium hirsutum L.). Sci Rep 7:43902. https://doi.org/10.1038/srep43902

Li X, Wang Y, Chen S, Tian H, Daqi F, Zhu B, Luo Y, Zhu H (2018) Lycopene is enriched in tomato Fruit by CRISPR/Cas9-mediated multiplex genome editing. Front Plant Sci 9:559. https://doi.org/10.3389/fpls.2018.00559

Lin C-S, Hsu C-T, Yang L-H, Lee L-Y, Jin-Yuan F, Cheng Q-W, Fu-Hui W, Hsiao HC-W, Yesheng Zhang R, Zhang W-JC, Chen-Ting Y, Wang W, Liao L-J, Gelvin SB, Shih M-C (2018) Application of protoplast technology to CRISPR/Cas9 mutagenesis: from single-cell mutation detection to mutant plant regeneration. Plant Biotechnol J 16(7):1295–1310. https://doi.org/10.1111/pbi.12870

Lowder LG, Zhang D, Baltes NJ, Paul JW, Tang X, Zheng X, Voytas DF, Hsieh T-F, Zhang Y, Qi Y (2015) A CRISPR/Cas9 toolbox for multiplexed plant genome editing and transcriptional regulation. Plant Physiol 169(2):971–985. https://doi.org/10.1104/pp.15.00636

Ma X, Zhang Q, Zhu Q, Liu W, Chen Y, Qiu R, Wang B, Yang Z, Li H, Lin Y, Xie Y, Shen R, Chen S, Wang Z, Chen Y, Guo J, Chen L, Zhao X, Dong Z, Liu Y-G (2015) A robust CRISPR/Cas9 system for convenient, high-efficiency multiplex genome editing in monocot and dicot plants. Mol Plant 8(8):1274–1284. https://doi.org/10.1016/j.molp.2015.04.007

Ma C, Zhu C, Zheng M, Liu M, Zhang D, Liu B, Li Q, Si J, Ren X, Song H (2019) CRISPR/Cas9-mediated multiple gene editing in Brassica oleracea var. capitata using the endogenous tRNA-processing system. Hortic Res 6:20. https://doi.org/10.1038/s41438-018-0107-1

Mali P, Yang L, Esvelt KM, Aach J, Guell M, DiCarlo JE, Norville JE, Church GM (2013) RNA-guided human genome engineering via Cas9. Science (New York, NY) 339(6121):823–826. https://doi.org/10.1126/science.1232033

Molla KA, Yang Y (2019) CRISPR/Cas-Mediated Base editing: technical considerations and practical applications. Trends Biotechnol 37(10):1121–1142. https://doi.org/10.1016/j.tibtech.2019.03.008

Molla KA, Qi Y, Karmakar S, Baig MJ (2020a) Base editing landscape extends to per-form Transversion mutation. Trends Genet 36(12):899–901. https://doi.org/10.1016/j.tig.2020.09.001

Molla KA, Shih J, Yang Y (2020b) Single-nucleotide editing for zebra3 and wsl5 phenotypes in rice using CRISPR/Cas9-mediated adenine base editors. aBIOTECH 1(2):106–118. https://doi.org/10.1007/s42994-020-00018-x

Molla KA, Karmakar S, Tofazzal Islam M (2020c) Wide horizons of CRISPR-Cas-derived Technologies for Basic Biology, agriculture, and medicine. In: Tofazzal Islam M, Bhowmik PK, Molla KA (eds) CRISPR-Cas methods. Springer US, New York, pp 1–23

Nishimasu H, Shi X, Ishiguro S, Gao L, Hirano S, Okazaki S, Noda T, Abudayyeh OO, Gootenberg JS, Mori H, Oura S, Holmes B, Tanaka M, Seki M, Hirano H, Aburatani H, Ishitani R, Ikawa M, Yachie N, Zhang F, Nureki O (2018) Engineered CRISPR-Cas9 nuclease with expanded targeting space. Science (New York, NY) 361(6408):1259–1262. https://doi.org/10.1126/science.aas9129

Nissim L, Perli SD, Fridkin A, Perez-Pinera P, Timothy KL (2014) Multiplexed and programmable regulation of gene networks with an integrated RNA and CRISPR/Cas toolkit in human cells. Mol Cell 54(4):698–710. https://doi.org/10.1016/j.molcel.2014.04.022

Ntui VO, Tripathi JN, Tripathi L (2020) Robust CRISPR/Cas9 mediated genome editing tool for banana and plantain (Musa spp.). Curr Plant Biol 21:100128. https://doi.org/10.1016/j.cpb.2019.100128

Nuñez JK, Chen J, Pommier GC, Zachery Cogan J, Replogle JM, Adriaens C, Ramadoss GN, Shi Q, Hung KL, Samelson AJ, Pogson AN, Kim JYS, Chung A, Leonetti MD, Chang HY, Kampmann M, Bernstein BE, Hovestadt V, Gilbert LA, Weissman JS (2021) Genome-wide programmable transcriptional memory by CRISPR-based epigenome editing. Cell 184(9):2503–2519.e17. https://doi.org/10.1016/j.cell.2021.03.025

Oh Y, Lee B, Kim H, Kim S-G (2020) A multiplex guide RNA expression system and its efficacy for plant genome engineering. Plant Methods 16:37. https://doi.org/10.1186/s13007-020-00580-x

Ozuna CV, Iehisa JCM, Giménez MJ, Alvarez JB, Sousa C, Barro F (2015) Diversification of the celiac disease α-gliadin complex in wheat: a 33-mer peptide with six overlapping epitopes, evolved following polyploidization. Plant J Cell Mol Biol 82(5):794–805. https://doi.org/10.1111/tpj.12851

Pan C, Wu X, Markel K, Malzahn AA, Kundagrami N, Sretenovic S, Zhang Y, Cheng Y, Shih PM, Qi Y (2021) CRISPR-Act3.0 for highly efficient multiplexed gene activation in plants. Nat Plants 7(7):942–953. https://doi.org/10.1038/s41477-021-00953-7

Pathak B, Zhao S, Manoharan M, Srivastava V (2019) Dual-targeting by CRISPR/Cas9 leads to efficient point mutagenesis but only rare targeted deletions in the rice genome. 3 Biotech 9(4):158. https://doi.org/10.1007/s13205-019-1690-z

Qi LS, Larson MH, Gilbert LA, Doudna JA, Weissman JS, Arkin AP, Lim WA (2013) Repurposing CRISPR as an RNA-guided platform for sequence-specific control of gene expression. Cell 152(5):1173–1183. https://doi.org/10.1016/j.cell.2013.02.022

Qi W, Zhu T, Tian Z, Li C, Zhang W, Song R (2016) High-efficiency CRISPR/Cas9 multiplex gene editing using the glycine tRNA-processing system-based strategy in maize. BMC Biotechnol 16(1):58. https://doi.org/10.1186/s12896-016-0289-2

Ren B, Yan F, Kuang Y, Li N, Zhang D, Zhou X, Lin H, Zhou H (2018) Improved base editor for efficiently inducing genetic variations in Rice with CRISPR/Cas9-guided hyperactive hAID mutant. Mol Plant 11(4):623–626. https://doi.org/10.1016/j.molp.2018.01.005

Ren Q, Zhong Z, Wang Y, You Q, Li Q, Yuan M, He Y, Caiyan Qi X, Tang XZ, Zhang T, Qi Y, Zhang Y (2019) Bidirectional promoter-based CRISPR-Cas9 Systems for Plant Genome Editing. Front Plant Sci 10:1173. https://doi.org/10.3389/fpls.2019.01173

Sánchez-León S, Gil-Humanes J, Ozuna CV, Giménez MJ, Sousa C, Voytas DF, Barro F (2018) Low-gluten, nontransgenic wheat engineered with CRISPR/Cas9. Plant Biotechnol J 16(4):902–910. https://doi.org/10.1111/pbi.12837

Schmidt C, Fransz P, Rönspies M, Dreissig S, Fuchs J, Heckmann S, Houben A, Puchta H (2020) Changing local recombination patterns in Arabidopsis by CRISPR/Cas mediated chromosome engineering. Nat Commun 11(1):4418. https://doi.org/10.1038/s41467-020-18277-z

Shen L, Hua Y, Yaping F, Li J, Liu Q, Jiao X, Xin G, Wang J, Wang X, Yan C, Wang K (2017) Rapid generation of genetic diversity by multiplex CRISPR/Cas9 genome editing in rice. Sci China Life Sci 60(5):506–515. https://doi.org/10.1007/s11427-017-9008-8

Sollid LM, Qiao S-W, Anderson RP, Gianfrani C, Koning F (2012) Nomenclature and listing of celiac disease relevant gluten T-cell epitopes restricted by HLA-DQ molecules. Immunogenetics 64(6):455–460. https://doi.org/10.1007/s00251-012-0599-z

Stuttmann J, Barthel K, Martin P, Ordon J, Erickson JL, Herr R, Ferik F, Kretschmer C, Berner T, Keilwagen J, Marillonnet S, Bonas U (2021) Highly efficient multiplex editing: one-shot generation of 8× Nicotiana benthamiana and 12× Arabidopsis mutants. Plant J Cell Mol Biol 106(1):8–22. https://doi.org/10.1111/tpj.15197

Tang X, Lowder LG, Zhang T, Malzahn AA, Zheng X, Voytas DF, Zhong Z, Chen Y, Ren Q, Li Q, Kirkland ER, Zhang Y, Qi Y (2017) A CRISPR-Cpf1 system for efficient genome editing and transcriptional repression in plants. Nat Plants 3:17018. https://doi.org/10.1038/nplants.2017.18

Tang X, Ren Q, Yang L, Yu B, Zhong Z, He Y, Liu S, Qi C, Liu B, Wang Y, Sretenovic S, Zhang Y, Zheng X, Zhang T, Qi Y, Zhang Y (2019) Single transcript unit CRISPR 2.0 systems for robust Cas9 and Cas12a mediated plant genome editing. Plant Biotechnol J 17(7):1431–1445. https://doi.org/10.1111/pbi.13068

Tsai SQ, Wyvekens N, Khayter C, Foden JA, Thapar V, Reyon D, Goodwin MJ, Aryee MJ, Keith Joung J (2014) Dimeric CRISPR RNA-guided FokI nucleases for highly specific genome editing. Nat Biotechnol 32(6):569–576. https://doi.org/10.1038/nbt.2908

Wang M, Mao Y, Yuming L, Tao X, Zhu J-K (2017) Multiplex gene editing in Rice using the CRISPR-Cpf1 system. Mol Plant 10(7):1011–1013. https://doi.org/10.1016/j.molp.2017.03.001

Wang M, Mao Y, Yuming L, Wang Z, Tao X, Zhu J-K (2018a) Multiplex gene editing in rice with simplified CRISPR-Cpf1 and CRISPR-Cas9 systems. J Integr Plant Biol 60(8):626–631. https://doi.org/10.1111/jipb.12667

Wang Z, Wang S, Li D, Zhang Q, Li L, Zhong C, Liu Y, Huang H (2018b) Optimized paired-sgRNA/Cas9 cloning and expression cassette triggers high-efficiency multiplex genome editing in kiwifruit. Plant Biotechnol J 16(8):1424–1433. https://doi.org/10.1111/pbi.12884

Wang W, Tian B, Pan Q, Chen Y, He F, Bai G, Akhunova A, Trick HN, Akhunov E (2021) Expanding the range of editable targets in the wheat genome using the variants of the Cas12a and Cas9 nucleases. Plant Biotechnol J. https://doi.org/10.1111/pbi.13669

Weber E, Engler C, Gruetzner R, Werner S, Marillonnet S (2011) A modular cloning system for standardized assembly of multigene constructs. PLoS One 6(2):e16765. https://doi.org/10.1371/journal.pone.0016765

Weiss T, Wang C, Kang X, Zhao H, Gamo ME, Starker CG, Crisp PA, Zhou P, Springer NM, Voytas DF, Zhang F (2020) Optimization of multiplexed CRISPR/Cas9 system for highly efficient genome editing in Setaria viridis. Plant J Cell Mol Biol 104(3):828–838. https://doi.org/10.1111/tpj.14949

Xie K, Minkenberg B, Yang Y (2015) Boosting CRISPR/Cas9 multiplex editing capability with the endogenous tRNA-processing system. Proc Natl Acad Sci U S A 112(11):3570–3575. https://doi.org/10.1073/pnas.1420294112

Xing H-L, Dong L, Wang Z-P, Zhang H-Y, Han C-Y, Liu B, Wang X-C, Chen Q-J (2014) A CRISPR/Cas9 toolkit for multiplex genome editing in plants. BMC Plant Biol 14:327. https://doi.org/10.1186/s12870-014-0327-y

Xiong X, Liang J, Li Z, Gong B-Q, Li J-F (2021) Multiplex and optimization of dCas9-TV-mediated gene activation in plants. J Integr Plant Biol 63(4):634–645. https://doi.org/10.1111/jipb.13023

Yan D, Ren B, Liu L, Yan F, Li S, Wang G, Sun W, Zhou X, Zhou H (2021) High-efficiency and multiplex adenine base editing in plants using new TadA variants. Mol Plant 14(5):722–731. https://doi.org/10.1016/j.molp.2021.02.007

Yoo, SD., Cho, YH. & Sheen, J. Arabidopsis mesophyll protoplasts: aversatile cell system for transient gene expression analysis. Nat Protoc 2, 1565–1572 (2007). https://doi.org/10.1038/nprot.2007.199

Yu H, Lin T, Meng X, Huilong D, Zhang J, Liu G, Chen M, Jing Y, Kou L, Li X, Gao Q, Liang Y, Liu X, Fan Z, Liang Y, Cheng Z, Chen M, Tian Z, Wang Y, Chu C, Zuo J, Wan J, Qian Q, Han B, Zuccolo A, Wing RA, Gao C, Liang C, Li J (2021) A route to de novo domestication of wild allotetraploid rice. Cell 184(5):1156–1170.e14. https://doi.org/10.1016/j.cell.2021.01.013

Zhang J-P, Li X-L, Neises A, Chen W, Lin-Ping H, Ji G-Z, Jun-Yao Y, Jing X, Yuan W-P, Cheng T, Zhang X-B (2016) Different effects of sgRNA length on CRISPR-mediated gene knockout efficiency. Sci Rep 6:28566. https://doi.org/10.1038/srep28566

Zhang Y, Held MA, Showalter AM (2020a) Elucidating the roles of three β-glucuronosyltransferases (GLCATs) acting on arabinogalactan-proteins using a CRISPR-Cas9 multiplexing approach in Arabidopsis. BMC Plant Biol 20(1):221. https://doi.org/10.1186/s12870-020-02420-5

Zhang C, Wang F, Zhao S, Kang G, Jinling Song L, Li, and Jinxiao Yang. (2020b) Highly efficient CRISPR-SaKKH tools for plant multiplex cytosine base editing. Crop J 8(3):418–423. https://doi.org/10.1016/j.cj.2020.03.002

Zhang P, Hongyang D, Wang J, Yixiang P, Yang C, Yan R, Yang H, Cheng H, Deyue Y (2020c) Multiplex CRISPR/Cas9-mediated metabolic engineering increases soya bean isoflavone content and resistance to soya bean mosaic virus. Plant Biotechnol J 18(6):1384–1395. https://doi.org/10.1111/pbi.13302

Zhong Z, Zhang Y, Qi You X, Tang QR, Liu S, Yang L, Wang Y, Liu X, Liu B, Zhang T, Zheng X, Le Y, Zhang Y, Qi Y (2018) Plant genome editing using FnCpf1 and LbCpf1 nucleases at redefined and altered PAM sites. Mol Plant 11(7):999–1002. https://doi.org/10.1016/j.molp.2018.03.008

Zhong Z, Liu S, Liu X, Liu B, Tang X, Ren Q, Zhou J, Zheng X, Qi Y, Zhang Y (2020) Intron-based single transcript unit CRISPR systems for plant genome editing. Rice (New York, N.Y.) 13(1):8. https://doi.org/10.1186/s12284-020-0369-8

Zsögön A, Čermák T, Naves ER, Notini MM, Edel KH, Weinl S, Freschi L, Voytas DF, Kudla J, Peres LEP (2018) De novo domestication of wild tomato using genome editing. Nat Biotechnol. https://doi.org/10.1038/nbt.4272

Zuo J, Li J (2014) Molecular dissection of complex agronomic traits of rice: a team effort by Chinese scientists in recent years. Natl Sci Rev 1(2):253–276. https://doi.org/10.1093/nsr/nwt004

Transgene-Free Genome Editing in Plants

Thorben Sprink, Frank Hartung, and Janina Metje-Sprink

Abstract Genome editing has revolutionized genetics and breeding likewise. Especially in plant breeding, it opened new ways to address traits with never known specificity. In many cases genome editing tools are provided by classical transgenic methods, i.e., by *Agrobacterium*-based delivery, but it is also possible to perform genome editing without the use of a transgene by providing proteins, or nucleic acid protein complexes. These methods have the big advantage that transgene organisms can be avoided at any time; even transgenic intermediates are not needed. However, transgene-free methods are technically challenging, and editing rates are often lower compared to classical methods. Nevertheless, it offers great opportunities to produce plants without the need of any transgene, simplifying the regulatory processes in many jurisdictions around the globe. In this chapter, we present methods and delivery methods that can be used for transgene-free editing and present first promising examples.

Keywords Genome editing · Mito TALEs · Meganucleases · Plastid editing · Transgene-free methods

1 Introduction

For many years, scientists and breeders have tried to precisely edit the genome of plants, an aim that was a long time hard to achieve and, in many species, almost impossible due to missing technologies and techniques. In the past, genes of organisms have been changed by breeding, and new variability was added either by random mutagenesis due to treatment with radiation or chemicals or by the insertion of T-DNA randomly into the genomes. In the 1990s, a new method appeared using natural occurring nucleases from either bacteria, fungi, or archaea. This so-called Homing Endonucleases or Meganucleases were used for precise editing of genomes as they have a defined recognition site. By modifying the natural occurring ones and

T. Sprink (✉) · F. Hartung · J. Metje-Sprink
Institute for Biosafety in Plant Biotechnology, Julius Kühn-Institut, Quedlinburg, Germany
e-mail: thorben.sprink@julius-kuehn.de

© The Author(s), under exclusive license to Springer Nature   171
Switzerland AG 2022
S. H. Wani, G. Hensel (eds.), *Genome Editing*,
https://doi.org/10.1007/978-3-031-08072-2_8

by creating artificial Meganucleases with altered recognition sites could be altered and further regions of the genome were addressable. These Meganucleases founded the first era of genome editing tools, since the usage of engineered nucleases has made great progress and three other nucleases have been identified which in principle work the same way. Zinc finger nucleases are more flexible as the recognition domain (the zinc-finger) is a univalent present domain in all kinds of proteins, which interact with the DNA. These domains have been fused and the enzymatic part of a nuclease (FokI) has been added. Even though being more flexible, zinc-finger nucleases are still tricky in design and handling as one zinc-finger recognizes a base triplet and some domains even cross-react making them hard to design and apply (Beumer et al. 2013). The next generation of genome editing tools were based on Transcription Activator like effectors (TALEs), proteins used by bacteria to high-jack plant genes and use them for their own good. They use a one-by-one recognition code, where each repeated protein domain recognizes a single base pair. However, the recognition still depends on a protein. Things changed when CRISPR-Systems have been adapted for genome editing applications, as their recognition depends on RNAs, making the design and application much easier. To date, most genome editing applications in plants are based on CRISPR-Systems (see Menz et al. 2020) and new discoveries in the field of CRISPR-based GE are made nearly on a daily basis. Even though, all of the above mentioned techniques can theoretically be used in a DNA-free manner, but we will focus on the most commonly used ones TALENs and CRISPR-Systems in this chapter. DNA-free genome editing has gained more and more attention since many jurisdictions have released guidelines or amendments to their national biosafety laws (for an overview see Menz et al. 2020). In many of those guidelines, products of genome editing do not face GMO regulation in cases where a transgene has not been used or is not present in the final product. Nevertheless, the concept or method of proofing this is not further elaborated in the respective legislations and remain interpretable by the competent authorities. If the use of a transgene or any kind of DNA has been avoided, in the first place, the space for interpretation is rather limited and products should pass the legislation inspection without any further delays.

2 Targeted Nucleases

Organisms found multiple ways to interact with their own or foreign DNA due to many reasons, some of them are the regulatory control of genes or the defense against viruses or other invading nucleic acids. Transcription activator-like effectors (TALEs), for example, have its origin in *Xanthomonas spec.*, which infect plants. These bacteria use a segregation type III system to introduce the effector-like proteins into plant cells to manipulate the cellular processes of its host (Göhre and Robatzek 2008). Once reaching the nucleus, the TALE recognizes its target by "scanning" the DNA for its binding partner (Becker and Boch 2016). Once bound to the DNA, the TALE begins to modify the expression of its host gene, e.g., the

SWEET 14 gene in rice (Oliva et al. 2019), to provide itself with nutrients and enable the best conditions for the bacteria to live and start an infection such as bacterial leave blight. In 2010, the code of TALEs has been revealed paving the way to use them as a genome editor. The TALE protein consists of three functional domains: (i) the N-terminal region with the bacterial secretion signal and a nonspecific DNA-binding domain which enables general attachment of the protein to the DNA; (ii) the C-terminal domain with an interaction part for plant transcription factors IIA, an acidity activation domain, as well as two nuclear localization signals; and (iii) the center part of the TALE with its programmable DNA-binding region (for more details see Becker and Boch 2021).

The CRISPR system has a different origin, and most of the archaea and roughly 60% of all bacteria species encode for one or even more CRISPR systems. CRISPR-Cas formerly known as clustered regularly interspaced short palindromic repeats in combination with CRISPR-associated proteins (Cas) was first described in 1987 as "junk DNA" found in *E. coli* (Ishino et al. 1987). For a long time, this discovery had no huge impact on science; in 2002, two groups of scientist agreed on the term CRISPR, which is used since then (Jansen et al. 2002). They described that CRISPR is common in all kinds of bacteria and archaea. In 2005, the CRISPR spacer sequence was found to be highly homologous with exogenous sequences from bacterial plasmids and phages (Pourcel et al. 2005; Mojica et al. 2005; Bolotin et al. 2005). In addition, in 2007 the practical application of CRISPR had been recognized as a kind of adaptive immune system of bacteria against invading viruses and nucleic acids, which can be programmed to protect the host (Barrangou et al. 2007). However, since their discovery as DNA-interacting partners to a programmable DNA editing tool, it took additional years. CRISPR systems are very diverse and are currently classified into two main classes with five types and 16 subtypes. The types differ in the number of proteins and the distribution of the main domains, as well as the type of nucleic acid recognized (for detail see Makarova et al. 2020).

3 From Natural DNA Modifiers to Programmable Genome Editors

The main step on the way to the TALE nucleases (TALENs) was the decryption of the hidden code of these natural occurring navigation systems, the mode of interaction with the nucleic acid and fusing them to the already used enzymatic part of the FokI nuclease (Boch et al. 2009; Cermak et al. 2011). TALENs are always used as pairs because FokI is cutting DNA, as a dimer making them a precise tool for addressing and cutting DNA (Sprink et al. 2015). TALEs and TALEN offer a unique feature compared to other genome editing tools; TALENs are able to differentiate between methylated and non-methylated nucleotides. They can even be used to differentiate between two sequences which differ only in the methylation stage of a single nucleotide opening them for methylation-dependent DNA modification.

Furthermore, the mode of action for the target search is different compared to other genome editors. TALENs wrap around the DNA and perform a one-dimensional, non-rotational unbiased search for the target sequences. Many other genome editors are either hopping from one site to the other by frequent binding and release of DNA or by "rotating" around the DNA following its major groove. The concept of wrapping that is used by the TALENs is astonishingly faster, allowing a rapid identification of the target even in large genomes.

CRISPR systems have been formed to genome editors in 2012 when the two RNAs which are responsible for the target identification (CRISPR-RNA and trans-activating CRISPR-RNA) have been artificially fused to a single-guide RNA (Jinek et al. 2012). In its "original" form, CRISPR is still used to optimize and protect bacteria in dairy production from phage invasion when cheese is produced (Grens 2015). Since the first described editing of a plant in 2013 by Feng et al., hundreds of publications showing CRISPR editing in plants appeared. Besides "proof-of-principle" articles, basic research and gene function studies play an important role in CRISPR-based genome editing as the technique is easy to apply and enable scientist to address precise locations of the genome (Modrzejewski et al. 2019). Furthermore, companies and scientist are also addressing market-oriented traits with CRISPR as well as TALENs (see Menz et al. 2020; Metje-Sprink et al. 2019, 2020). First plants have entered the market and more are likely to be released in the near future (Menz et al. 2020).

The classical system that is used for CRISPR-based genome editing is based on the CRISPR-Cas9 System from *Streptococcus pyogenes*, which belongs to the Class II type II variants of CRISPRs. Its great advantage is that all functional domains are combined in a single protein and that many vector systems are available for research; one disadvantage is the relatively large size of the protein and the rather restrict intellectual property policy of using SpCas9 for industrial uses. Since the discovery of Cas9 from *S. pyogenes*, multiple variants of Cas9 have been identified in many species and modified to be used for genome editing in plants; the most frequently used one besides *S. pyogenes* is the one from *Staphylococcus aureus* as well as protein-engineered versions of those (see Huang and Puchta 2021). Another nuclease that is used in CRISPR-based genome editing is CRISPR/Cpf1 (from *Prevotella* and *Francisella*), recently named Cas12a. The system differs in some points from Cas9 as the PAM is different and the Cpf1 protein is even smaller compared to Cas9.

Recently, a new nuclease has been described which belongs to the same type of CRISPR systems as Cpf1, Mad7, which has the great advantage of being even smaller and licensing is easy and public available. Mad7 has been shown to function for genome editing applications in humans and recently also in plants (Lin et al. 2021).

Other CRISPR systems like Cas13 tend to alter RNA instead of DNA and can be used for transcriptome editing (see Ali et al. 2018). Recent work has shown the possibility of using Cas13 for mRNA editing as well as to combat invading viral RNA replication (Aman et al. 2018).

Besides the classical CRISPR systems and TALENs, many modifications exist which all can be used in a DNA-free matter. Some of those have been made for both systems and some are specific for either CRISPR or TALEN.

4 Base Editors

Base editors are a fusion of a deaminase either to a dead-Cas (dCas, both nuclease sites deactivated) or a TALE without a nuclease. They are precise tools to modify a chosen sequence by mutating it without the need of introducing a double-strand break, foreign DNA template or homologous recombination. Like in a classical CRISPR system, the guide RNA defines the site of modification and guides the dCas deaminase to the distinct region of the genome; the dCas is then generating single-stranded DNA R-loops which are accessible to the deaminase. Currently there are two types of deaminases, cytidine deaminase and adenine deaminase. Cytidine deaminases convert cytidine to uridine creating mismatches between the former C-position and guanine within an editing window. Most organisms have a uracil base excision repair pathway to excise U from its DNA as it is mutagenic. By blocking this repair pathway, researchers have increased the C to T transition rate in mammals as well as in plants (Zong et al. 2017). With the cytidine base editor, it is possible to mediate transition mutations of C to T and G to A. One further aim was to enable the other transition mutations of T to C and A to G. As no natural occurring adenine deaminase is known, researchers started to do targeted protein engineering with the *E. coli* tRNA adenine deaminase, receiving a version of the protein that is able to deaminase adenine in DNA rather than that in tRNA (Gaudelli et al. 2017). The adenine base editor is converting adenine to inosine, which is recognized as guanine by the polymerase converting AT to GC pairs also in plants (Kang et al. 2018). As many desirable traits in plants can be achieved by a single mutation of a nucleotide, base editors may have a great potential for future application in plants as well. Furthermore, they can be used in a DNA-free matter enabling the change of the genome without any break or introduced nucleic acid, as shown recently for wheat (Zhang et al. 2019). The typical base editing window is defined by the type of nuclease used and can be between one and five nucleotides. Furthermore, CRISPR base editors seem to have a low off-target rate (Modrzejewski et al. 2020).

To be used as base editors, it was necessary to modify CRISPR systems as the Cas9 proteins are unwinding double-stranded DNA and deaminases only work on single-stranded DNA. TALENs do not have the ability to unwind double-stranded DNA as the mechanism is different from CRISPR, so they were originally not suited for base editing. Nevertheless, recently also TALEs have been modified to be used for base editing by fusing them to the newly discovered bacterial double-stranded DNA deaminase toxin A (DddA). DddA is a bacterial toxin that serves as a deaminase facilitating C/G to T/A conversions. First approaches of fusing DddA to TALEs resulted in cell death, but using a split protein fusing each half to an opposite binding TALE solved this issue (Mok et al. 2020; for an overview see Becker and Boch

2021). A combination with uracil glycosylase inhibitor is enhancing the efficiency as shown for CRISPR-cytidine base editors. Nevertheless, adenine base editors for TALEs are not available so far.

5 Mito TALEs (Plastid Editing)

Another promising approach is the possibility of novel gene editing tools to alter plastids. Especially TALENs offer a great opportunity as they are purely protein based, making them in principle able to translocate in any part of an organism that is accessible by proteins. By fusing a mitochondrial targeting sequence to the TALENs, it was possible to edit mitochondrial DNA and cure cytoplasmic male sterility in rice and rapeseed (Kazama et al. 2019). Furthermore, TALE base editors are functional in mitochondria, thus making them the first functional base editing tools available for compartments, whereas CRISPR systems always struggle with the co-transfection of RNA and protein into the mitochondria. TALE base editors seem to be working also in chloroplasts as recent data let assume (Kim et al. 2021). Even though these approaches have still been generated using classical transformation methods, they are also suitable for a DNA-free approach and will be applied in the near future.

6 Delivery of Genome Editing Reagents into the Cells

Just like classical methods, DNA-free genome editing can only work if the reagents have been delivered to the point of action (mostly the nucleus). Since the first description of DNA-free delivery of Cas9 into plants, various methods have been tested, but two main ways of genome editing tool delivery in the cell are currently used to achieve DNA-free editing of various plant species. The first method is the delivery of RNP (ribonucleoprotein) complexes or proteins into plants, and the second method is based on the use of RNA-virus vector systems. Both methods are feasible for DNA-free genome editing, but there are some minor differences between the techniques, which are discussed further. For many jurisdictions, it is necessary to prove the absence of foreign DNA (or recombination of DNA) in the final product. This can be an easy task in some jurisdictions but can turn nasty in others as no general protocol or definition is defined how to prove the absence of something. Therefore, it would be of great advantage to show that DNA has ever been used during the production of the plant, which is only the case for the RNP delivery systems, making those the system of choice. Nevertheless, we will introduce all systems and also show the differences in the delivery methods.

7   Viral Vector Delivery

Viral-based delivery of reagents to cells is a routine application in animal cells and has been lately also transferred to plants (Butler et al. 2016; Ma et al. 2020). Most of the currently published methods are referring to the use of DNA viruses such as geminiviruses or to the introduction of guide RNAs into a plant that is constantly expressing Cas9 (Baltes et al. 2014; Cody et al. 2017). These applications are often described as DNA-free but should be handled with caution as DNA is used as an intermediate step and does not fulfill the sensu stricto definition of DNA-free genome editing. Modifying plant viruses as vectors for genome editing is somehow challenging due to the size of the genome editing tools. Many viruses, mostly DNA and RNA plus strand, are limited in the capacity of nucleic acid, which can be incorporated without losing their function. Furthermore, as viruses replicate rather fast, sequences in the viral genome can be unstable and change during replications including deletions. Lately two approaches have been successful in delivering guide RNA and Cas9 in parallel into the plants.

Ariga et al. constructed Potato virus X, a plus-stranded RNA virus for the delivery of Cas9 and guide RNA and introduced it into *N. benthamiana* plants (Table 1). They have achieved editing in two different genes of *N. benthamiana*, by using both *Agrobacterium*-based transfection and direct delivery of the virus into the plants by mechanical inoculation. Although the editing efficiency has been reduced (2.5% vs. 60%) in the DNA-free approach, plants with biallelic mutations were obtained. Furthermore, they also constructed a viral-based base editing system that can also be used in a DNA-free manner (Ariga et al. 2020).

In another approach Ma et al. used the *sonchus yellow net rhabdovirus*, a minus-stranded RNA virus for the delivery of Cas and guide RNA into the plants. The advantage of this approach is that minus-stranded RNA viruses are tolerating larger DNA inserts, but the disadvantage is that their transformation is tricky. Ma et al. also used *N. benthamiana* for the experiments targeting multiple genes in the plants reaching an efficiency of up to 90%. Just like Ariga et al., they also used agrobacterium delivery as well as mechanical inoculation. Both methods resulted in edited plants with a satisfying efficiency (Ma et al. 2020).

8   RNP or Protein Delivery into Plants

The delivery of preassembled RNP complexes or unassembled proteins and free guide RNA into plants is used more frequently compared to the viral-based systems. For the delivery of RNPs, the complex is assembled prior to transformation in vivo. This is only possible due to the ability of Cas proteins to interact with guide RNA without additional factors. For this, Cas9 (or TALENs) have to be produced using bacterial production systems followed by an extensive cleanup and concentration. Many vector systems for the production of various Cas variants are currently

Table 1 Recent publications using DNA-free genome editing approaches

Plant species	Trait	GE technique	Tissue	Delivery	Editing efficiency	Off targets	Reference
N. benthamiana	POC	CRISPR/ Cas9 RNPs	Leaves	Mechanical inoculation RNA(+)Virus	3% biallelic	n.d.	Ariga et al. (2020)
N. benthamiana	POC	CRISPR/ Cas9 RNPs	Leaves	Mechanical inoculation RNA(-)Virus	n.d.	0/13	Ma et al. (2020)
N. benthamiana	POC	CRISPR/ Cas9RNPs	Leaves	Agrobacterium type IV secretion system	2.5%	n.d.	Schmitz et al. (2020)
H. brasiliensis	Fast breeding, veg. growth	CRISPR/ Cas9RNPs	Protoplasts	PEG delivery	3.74– 20.1%	n.d.	Fan et al. (2020)
B. rapa	POC	CRISPR/ Cas9RNPs	Protoplasts	PEG delivery	1.2– 24.5%	n.d.	Murovec et al. (2018)
B. oleracea	POC	CRISPR/ Cas9RNPs	Protoplasts	PEG delivery	0.1–2.3%	n.d.	Murovec et al. (2018)
S. tuberosum	Reduced browning	CRISPR/ Cas9RNPs	Protoplasts	PEG delivery	68% mono allelic 24% tetra allelic	0/2	González et al. (2019)
C. arietinum	Drought tolerance	CRISPR/ Cas9RNPs	Protoplasts	PEG delivery	2–79%	0/0	Badhan et al. (2021)
G. max	Oil composition	CRISPR/ Cpf1 RNPs	Protoplasts	PEG delivery	4.8–15%	0/13	Kim and Choi (2021)
G. max	Oil composition	CRISPR/ Cpf1 RNPs	Protoplasts	PEG delivery	0–11%	0/13	Kim et al. (2017)
Musa × paradisiaca	POC	CRISPR/ Cas9RNPs	Protoplasts	PEG delivery	0.9%	0/1	Wu et al. (2020)
Petunia spec	Flower color	CRISPR/ Cas9RNPs	Protoplasts	PEG delivery	11.9– 26%	n.d.	Yu et al. (2021)
O. sativa	POC	CRISPR/ Cas9RNPs	Zygotes	PEG delivery	14–64%	n.d.	Toda et al. (2019)
N. tabacum	POC	CRISPR/ Cas9RNPs	Protoplasts	Lipofection	6%	n.d.	Liu et al. (2020)
N. tabacum	POC	CRISPR/ Cas9RNPs	Suspension culture cells	Biolistic bombardment	3%	n.d.	Liu et al. (2020)

(continued)

Table 1 (continued)

Plant species	Trait	GE technique	Tissue	Delivery	Editing efficiency	Off targets	Reference
T. aestivum	Herbicide resistance	Cas9 base editing mRNA	Immature embryos	Biolistic bombardment	0.5%	n.d.	Zhang et al. (2019)
B. oleracea	POC	CRISPR/ Cas9 RNPs	Protoplasts	Electroporation	3.4%	n.d.	Lee et al. (2020)
B. oleracea	POC	CRISPR/ Cas9 RNPs	Protoplasts	PEG delivery	1.8%	n.d.	Lee et al. (2020)
N. tabacum	POC, drought stress	CRISPR/ Cas9 RNPs	Leaf discs	Pulsed laser-induced shockwaves	5.6%; 8.7%	n.d.	Augustine et al. (2021)
C. reinhardtii	POC	CRISPR/ Cas9 RNPS	Cells	Cell-penetrating peptide	POC	n.d.	Kang et al. (2020)

GE technique genome editing technique, *POC* proof-of-concept, *n.d.* not determined, *RNP* ribonucleoprotein

available, but as the process is complicated and labor-intensive, not all labs can produce their own reagents. Due to this, many vendors offer purified Cas proteins. The guide RNAs can either be chemically synthesized (having the advantage that no DNA has been used as template) or be produced by using a T7 (or another) polymerase and a suitable template. One big advantage of using RNPs for the transformation of plants is the omission of codon optimization, making it, once established, a universal system for all kinds of plants and tissues. Currently, there are two main ways to introduce RNPs (or mRNA) into the cells and multiple minor ones. The main techniques of introduction are (i) protoplast transformation and (ii) biolistic delivery.

9 Protoplast Transformation

Production and transformation of protoplasts is a long-known method to introduce plasmid DNA into plant cells. For the isolation of protoplasts, different tissues can be used, in many cases meristem tissue is the most suitable one to lead to efficient regeneration. For the protoplast generation, the tissue is treated with an enzyme mixture that is degrading the cell wall making the cellular membrane accessible to reagents. Much care has to be taken during handling of protoplasts as they are fragile and each species may need its own media concerning enzyme concentration, pH, and sugar/salt content. Protoplast transformation has been the first method of choice to introduce RNPs into plants cells and has been used in many species to date. The use of protoplast for DNA-free editing is highly efficient but comes with some drawbacks. First, protoplastation and regeneration protocols are genotype dependent and not all species are amenable to regeneration from protoplasts. Furthermore, there is a long in vitro phase up to several months involved in the process of protoplastation to regeneration, which might enhance unwanted somaclonal variation. Different methods are available for the transformation of protoplasts; the most frequently used one is the polyethylene glycol (PEG)-mediated vesicle fusion, which has also been the first method that was adapted for RNP delivery. PEG-mediated transformation makes use of PEG vesicles that enclose cargo (e.g., RNPs or DNA), these vesicles fuse with the cell membrane, and the cargo is released into the cell. PEG-mediated protoplast transformation is still the most commonly used technique for the introduction of RNPs into plants. It was recently used in many species, including soybean (Kim et al. 2017; Kim and Choi 2021), cabbage and Chinese cabbage (Murovec et al. 2018), chickpea (Badhan et al. 2021), banana (Wu et al. 2020), rubber tree (Fan et al. 2020), and potato (González et al. 2019). Another upcoming approach for protoplast transformation is the use of lipofection reagents. The lipofection method is frequently used in animal cell culture transformation and has recently been adapted for protoplast transformation. The functional principle of lipofection-mediated protoplast transformation is similar to PEG-mediated transformation but has not been widely applied in plants so far. In a first proof-of-concept (POC) approach, tobacco protoplasts were transformed with RNPs mediated by two lipofection reagents (lipofectamine 3000 and RNAiMAX) which both worked well (Liu et al. 2020) for editing endogenous targets as well as an introduced transgene. Aside from the mentioned vesicle fusion-based transformation methods, scientists have tried to fuse protoplasts with (semi)synthetic nanoparticles, and the delivery of DNA by nanoparticles is already used for genetic engineering in plants (for a review see Wang et al. 2019). However, RNP delivery by nanoparticles was only shown for fungi protoplasts so far, but it is probably only a matter of time until plant protoplasts are addressed by nanoparticles as well. Li et al. delivered RNPs into protoplast of the phytopathogenic fungus *M. oryzae* by biomimetic mineralized RNP nanoparticles and reached an editing efficiency of up to 20 % (Li et al. 2019). Just like bacteria protoplast can also be transformed using electric voltage, and electroporation of protoplasts with RNPS was shown in a proof-of-concept study in cabbage and has been compared with PEG-mediated transformation. Lee et al. found a slightly higher editing efficiency after electroporation, but further studies need to be done before any recommendation can be made (Lee et al. 2020).

10   Biolistic Delivery

Just like protoplast transformation, bombardment of tissues with gold or tungsten particles is an established technology for the transformation of plants with DNA plasmids, which has been adapted to carry RNPs. The particles that breach the cell wall are used as a carrier for the RNPs. One main advantage of this technique is that there is no need for enzymatic digestion of the cells. Nevertheless, it still has some restrains as not all plants are suitable for biolistic delivery and the efficiency of the delivery is highly depended on physical parameters, e.g., helium pressure, particle size, and target distance, that need to be reevaluated for each tissue individually. Besides leaf tissue, which is subsequently regenerated into whole plants via tissue culture, also embryogenic material can be used for the delivery having the advantage of avoiding the regeneration phase, as embryos develop directly into plants. However, in most cases biolistic delivery results in chimeric plants that need to go through an additional generation before non-chimeric T1 plants can be generated. Although biolistic delivery is frequently used with plasmid delivery of Cas9, the studies using RNPs are still low in the last 2 years because only two studies in rice and one in potato have been published as previously reported (Banakar et al. 2019, 2020; Makhotenko et al. 2019; see Metje-Sprink et al. 2019).

11 Additional Methods

Besides biolistic delivery and the transformation of protoplasts, some other methods have been developed or adapted from animal cell culture that might have applications in RNP-based plant transformation. However, most of these techniques have only been applied in a single species so far and need more studies to prove their ability for an efficient editing. All of the techniques presented here have some advantages compared to the established ones but may also come with some challenges to be implemented in various species. One method that is comparable to the PEG-mediated transformation of protoplast is the PEG-mediated transformation of zygotes. Instead of enzymatic digestion of cells, zygotes are produced in vitro by electrofusion of egg and sperm cell. This technology has the big advantage that the zygotes directly develop into plants, without going through long in vitro culture phases reducing the chance of somaclonal variation and chimeric plants as one plant results from a single cell. Currently, this system has only been applied to rice but shows great potential also for other cultures in which in vitro production of zygotes is established (Toda et al. 2019).

A novel approach that also avoids the use of protoplasts and even of other isolated cells is the use of a pulsed laser-induced shockwave to generate cavitation bubbles on plant leaf discs. This cavitation bubbles introduce transient cell wall and membrane pores in intact cells for direct RNP uptake. Augustine et al. (2021) have shown in a proof-of-concept approach that this technique is functional in tobacco by editing PDS as well as a drought resistance gene. This technique shows the unique feature that no pre-preparation of cells is necessary which could have an advantage for some species, but the technique has to be tested on other species and also non-model organisms to prove its common ability (Augustine et al. 2021).

Another approach, for the delivery of RNPs, that seems to be promising is the use of cell-penetrating peptides (CPPs). However, to our knowledge to date, no publication on the editing of a higher plant, using CPPs, is published, but a lot of work has been put into this method, and protein delivery of CPPs into higher plants has been shown (e.g., Guo et al. 2019). A current problem might be the size of the cargo, which seems to be limited for some CPPs. Nevertheless, efficient editing of the green algae *Chlamydomonas reinhardtii* has been achieved using CPPs in a proof-of-concept study albeit with low efficiency (Kang et al. 2020). It might be only a matter of time until first editing of a higher plant using CPP-based RNP editing will be published.

Apart from using isolated RNPs for the transformation of plant material, there is an alternative approach using Cas9-producing *Agrobacteria* tested by Schmitz ct al. In this approach, *Agrobacterium tumefaciens* is transformed with a plasmid to produce Cas9, and guide RNA and RNPs do form in the bacteria. These *Agrobacteria* are subsequently used to infiltrate plant material, and the RNP complex is transported into the cells via the type IV secretion system, which is normally used to transfer the Vir genes into plants. Schmitz et al. succeeded to edit the PDS gene in *N. benthamiana* in a proof-of-concept study. However, the editing efficiency increased when the guide RNA is transferred to the plants on a separate plasmid and is being produced in the plant, but this approach is not fulfilling the requirement of being DNA-free (Schmitz et al. 2020).

12   Current Regulatory Views on DNA-Free Genome Editing

Products of genome editing are no longer only present in the lab and made the transition from research to application in a remarkable short time. The first products already entered national markets, and more and more will appear on national and probably on international markets in the coming months or years. Due to this, a clarification of the legal status of such crops is urgently needed. In some jurisdiction such as the European Union, products of genome editing are seen and handled just alike GMOs; this status has been strengthened with the ruling of the European court of justice in July 2018. We have recently published a study on this and the regulatory status of GE products (see Menz et al. 2020). However, since then the European Commission published a "Study on the status of new genomic techniques under Union law and in light of the Court of Justice ruling in Case C-528/16" accompanied by two reports from the Joint Research Center and one by the European group of Ethics in Science and new technologies in April 2021. This study is a first step in a policy action to clarify the status of products of new technologies including genome editing in the European Union. This is important for an appropriate handling of products via import from countries outside the Union, as many jurisdictions outside Europe don't regulate all products of GE as GMO (Menz et al. 2020). Especially in the Americas and in some Asian countries (particularly Japan), some products of genome editing are out of the regulatory scope of GMO legislation and special regulations are in place (Menz et al. 2020). In most of these regulations, a special focus is laid on the transgene status of the final and/or intermediate product. In most cases an absence of foreign DNA has to be shown by the developers, but a clear guidance how this has to be proven is lacking and has to be shown on a case-by-case basis. If a DNA-free approach is used, in the first place this analysis can be dismissed and may, depending on the regulation, spare developers' time and money and also come with additional benefits for international markets. However, for cases that are defined as a GMO, a clear identification and detection protocol that is needed for import and export is impossible to implement; due to this trade markets will face problems if regulation of GE products stays unharmonized (Grohmann et al. 2019).

References

Ali Z, Mahas A, Mahfouz M (2018) CRISPR/Cas13 as a Tool for RNA Interference. Trends Plant Sci 23:374–378. Available: http://dx.doi.org/10.1016/j.tplants.2018.03.003

Aman R, Ali Z, Butt H, Mahas A, Aljedaani F, Khan MZ et al (2018) RNA virus interference via CRISPR/Cas13a system in plants. Genome Biol 19(1):1–9

Ariga H, Toki S, Ishibashi K (2020) Potato virus X Vector-mediated DNA-Free genome editing in plants. Plant Cell Physiol 61(11):1946–1953

Augustine SM, Cherian AV, Seiling K, Di Fiore S, Raven N, Commandeur U, Schillberg S (2021) Targeted mutagenesis in Nicotiana tabacum ADF gene using shockwave-mediated ribonucleoprotein delivery increases osmotic stress tolerance. Physiol Plant 173(3):993–1007. https://doi.org/10.1111/ppl.13499. Epub 2021 Jul 28. PMID: 34265107

Badhan S, Ball AS, Mantri N (2021) First report of CRISPR/Cas9 Mediated DNA-Free Editing of 4CL and RVE7 genes in Chickpea protoplasts. Int J Mol Sci 22(1). https://doi.org/10.3390/ijms22010396

Baltes NJ, Gil-Humanes J, Cermak T, Atkins PA, Voytas DF (2014) DNA replicons for plant genome engineering. Plant Cell 26(1):151–163

Banakar R, Eggenberger AL, Lee K, Wright DA, Murugan K, Zarecor S et al (2019) High-frequency random DNA insertions upon co-delivery of CRISPR-Cas9 ribonucleoprotein and selectable marker plasmid in rice. Sci Rep 9(1):19902. https://doi.org/10.1038/s41598-019-55681-y

Banakar R, Schubert M, Collingwood M, Vakulskas C, Eggenberger AL, Wang K (2020) Comparison of CRISPR-Cas9/Cas12a Ribonucleoprotein complexes for genome editing efficiency in the Rice Phytoene Desaturase (OsPDS) gene. Rice (New York, N.Y.) 13(1):4. https://doi.org/10.1186/s12284-019-0365-z

Barrangou R, Fremaux C, Deveau H, Richards M, Boyaval P, Moineau S et al (2007) CRISPR provides acquired resistance against viruses in prokaryotes. Science 315(5819):1709–1712

Becker S, Boch J (2016) TALEs spin along, but not around. Nature Chem Biol 12(10):766–768

Becker S, Boch J (2021) TALE and TALEN genome editing technologies. Gene Genome Editin 2:100007

Beumer KJ, Trautman JK, Christian M, Dahlem TJ, Lake CM, Hawley RS et al (2013) Comparing zinc finger nucleases and transcription activator-like effector nucleases for gene targeting in Drosophila. G3 (Bethesda, Md.) 3(10):1717–1725. https://doi.org/10.1534/g3.113.007260

Boch J, Scholze H, Schornack S, Landgraf A, Hahn S, Kay S et al (2009) Breaking the code of DNA binding specificity of TAL-type III effectors. Science 326(5959):1509–1512

Bolotin A, Quinquis B, Sorokin A, Ehrlich SD (2005) Clustered regularly interspaced short palindrome repeats (CRISPRs) have spacers of extrachromosomal origin. Microbiology 151(8):2551–2561

Butler NM, Baltes NJ, Voytas DF, Douches DS (2016) Geminivirus-mediated genome editing in potato (Solanum tuberosum L.) using sequence-specific nucleases. Front Plant Sci 7:1045

Cermak T, Doyle EL, Christian M, Wang L, Zhang Y, Schmidt C et al (2011) Efficient design and assembly of custom TALEN and other TAL effector-based constructs for DNA targeting. Nucleic Acids Res 39(12):e82–e82

Cody WB, Scholthof HB, Mirkov TE (2017) Multiplexed gene editing and protein overexpression using a tobacco mosaic virus viral vector. Plant Physiol 175(1):23–35

Fan Y, Xin S, Dai X, Yang X, Huang H, Hua Y (2020) Efficient genome editing of rubber tree (hevea brasiliensis) protoplasts using CRISPR/Cas9 ribonucleoproteins. Ind Crops Prod 146:112146. https://doi.org/10.1016/j.indcrop.2020.112146

Fcng Z, Zhang B, Ding W, Liu X, Yang D-L, Wei P et al (2013) Efficient genome editing in plants using a CRISPR/Cas system. Cell Research 23(10):1229–1232. https://doi.org/10.1038/cr.2013.114

Gaudelli NM, Komor AC, Rees HA, Packer MS, Badran AH, Bryson DI, Liu DR (2017) Programmable base editing of A•T to G•C in genomic DNA without DNA cleavage. Nature 551(7681):464–471. https://doi.org/10.1038/nature24644

Göhre V, Robatzek S (2008) Breaking the barriers: microbial effector molecules subvert plant immunity. Annu Rev Phytopathol 46:189–215

González MN, Massa GA, Andersson M, Turesson H, Olsson N, Fält A-S et al (2019) Reduced enzymatic browning in potato tubers by specific editing of a Polyphenol Oxidase gene via Ribonucleoprotein complexes delivery of the CRISPR/Cas9 system. Front Plant Sci 10:1649. https://doi.org/10.3389/fpls.2019.01649

Grens K (2015) There's CRISPR in your yogurt: we've all been eating food enhanced by the genome-editing tool for years. Scientist 29(1):1–5

Grohmann L, Keilwagen J, Duensing N, Dagand E, Hartung F, Wilhelm R et al (2019) Detection and identification of genome editing in plants: challenges and opportunities. Front Plant Sci 10:236

Guo B, Itami J, Oikawa K, Motoda Y, Kigawa T, Numata K (2019) Native protein delivery into rice callus using ionic complexes of protein and cell-penetrating peptides. PloS One 14(7):e0214033. https://doi.org/10.1371/journal.pone.0214033

Huang T-K, Puchta H (2021) Novel CRISPR/Cas applications in plants: from prime editing to chromosome engineering. Trans Res. https://doi.org/10.1007/s11248-021-00238-x

Ishino Y, Shinagawa H, Makino K, Amemura M, Nakata A (1987) Nucleotide sequence of the iap gene, responsible for alkaline phosphatase isozyme conversion in Escherichia coli, and identification of the gene product. J Bacteriol 169(12):5429–5433

Jansen R, van Embden J, Gaastra W, Schouls LM (2002) Identification of genes that are associated with DNA repeats in prokaryotes. Mol Microbiol 43(6):1565–1575

Jinek M, Chylinski K, Fonfara I, Hauer M, Doudna JA, Charpentier E (2012) A programmable dual-RNA–guided DNA endonuclease in adaptive bacterial immunity. Science 337(6096):816–821

Kang B-C, Yun J-Y, Kim S-T, Shin YJ, Ryu J, Choi M et al (2018) Precision genome engineering through adenine base editing in plants. Nat Plant 4(7):427–431. https://doi.org/10.1038/s41477-018-0178-x

Kang S, Jeon S, Kim S, Chang YK, Kim Y-C (2020) Development of a pVEC peptide-based ribonucleoprotein (RNP) delivery system for genome editing using CRISPR/Cas9 in Chlamydomonas reinhardtii. Sci Rep 10(1):1–11

Kazama T, Okuno M, Watari Y, Yanase S, Koizuka C, Tsuruta Y et al (2019) Curing cytoplasmic male sterility via TALEN-mediated mitochondrial genome editing. Nat Plant 5(7):722–730

Kim H, Choi J (2021) A robust and practical CRISPR/crRNA screening system for soybean cultivar editing using LbCpf1 ribonucleoproteins. Plant Cell Rep 40(6):1059–1070. https://doi.org/10.1007/s00299-020-02597-x

Kim H, Kim S-T, Ryu J, Kang B-C, Kim J-S, Kim S-G (2017) CRISPR/Cpf1-mediated DNA-free plant genome editing. Nat Commun 8:14406. https://doi.org/10.1038/ncomms14406

Kim J-S, Kang B-C, Bae S-J, Lee S, Lee JS, Kim A et al (2021) Chloroplast and mitochondrial DNA editing in plants. Nat Plant 7:899–905

Lee MH, Lee J, Choi SA, Kim Y-S, Koo O, Choi SH et al (2020) Efficient genome editing using CRISPR–Cas9 RNP delivery into cabbage protoplasts via electro-transfection. Plant Biotechnol Rep 14(6):695–702. https://doi.org/10.1007/s11816-020-00645-2

Li S, Song Z, Liu C, Chen X-L, Han H (2019) Biomimetic mineralization-based CRISPR/Cas9 ribonucleoprotein nanoparticles for gene editing. ACS Appl Mater Interface 11(51):47762–47770. https://doi.org/10.1021/acsami.9b17598

Lin Q, Zhu Z, Liu G, Sun C, Lin D, Xue C et al (2021) Genome editing in plants with MAD7 nuclease. J Genet Genomic 48:444–451

Liu W, Rudis MR, Cheplick MH, Millwood RJ, Yang J-P, Ondzighi-Assoume CA et al (2020) Lipofection-mediated genome editing using DNA-free delivery of the Cas9/gRNA ribonucleoprotein into plant cells. Plant Cell Rep 39(2):245–257

Ma X, Zhang X, Liu H, Li Z (2020) Highly efficient DNA-free plant genome editing using virally delivered CRISPR–Cas9. Nat Plants 6(7):773–779

Makarova KS, Wolf YI, Iranzo J, Shmakov SA, Alkhnbashi OS, Brouns SJJ et al (2020) Evolutionary classification of CRISPR–Cas systems: a burst of class 2 and derived variants. Nat Rev Microbiol 18(2):67–83

Makhotenko AV, Khromov AV, Snigir EA, Makarova SS, Makarov VV, Suprunova TP et al (2019) Functional analysis of Coilin in virus resistance and stress tolerance of Potato Solanum tuberosum using CRISPR-Cas9 editing. Doklady Biochem Biophys 484(1):88–91. https://doi.org/10.1134/S1607672919010241

Menz J, Modrzejewski D, Hartung F, Wilhelm R, Sprink T (2020) Genome edited crops touch the market: a view on the global development and regulatory environment. Front Plant Sci 11:586027

Metje-Sprink J, Menz J, Modrzejewski D, Sprink T (2019) DNA-free genome editing: past, present and future. Front Plant Sci 9:1957

Metje-Sprink J, Sprink T, Hartung F (2020) Genome-edited plants in the field. Curr Opn Biotechnol 61:1–6

Modrzejewski D, Hartung F, Sprink T, Krause D, Kohl C, Wilhelm R (2019) What is the available evidence for the range of applications of genome-editing as a new tool for plant trait modification and the potential occurrence of associated off-target effects: a systematic map. Environ Eviden 8(1):1–33

Modrzejewski D, Hartung F, Lehnert H, Sprink T, Kohl C, Keilwagen J, Wilhelm R (2020) Which factors affect the occurrence of off-target effects caused by the use of CRISPR/Cas: a systematic review in plants. Front Plant Sci 11:1838

Mojica FJM, García-Martínez J, Soria E (2005) Intervening sequences of regularly spaced prokaryotic repeats derive from foreign genetic elements. J Mol Evol 60(2):174–182

Mok BY, de Moraes MH, Zeng J, Bosch DE, Kotrys AV, Raguram A et al (2020) A bacterial cytidine deaminase toxin enables CRISPR-free mitochondrial base editing. Nature 583(7817):631–637

Murovec J, Guček K, Bohanec B, Avbelj M, Jerala R (2018) DNA-free genome editing of Brassica oleracea and B. rapa protoplasts using CRISPR-Cas9 ribonucleoprotein complexes. Front Plant Sci 9:1594. https://doi.org/10.3389/fpls.2018.01594

Oliva R, Ji C, Atienza-Grande G, Huguet-Tapia JC, Perez-Quintero A, Li T et al (2019) Broad-spectrum resistance to bacterial blight in rice using genome editing. Nat Biotechnol 37(11):1344–1350

Pourcel C, Salvignol G, Vergnaud G (2005) CRISPR elements in Yersinia pestis acquire new repeats by preferential uptake of bacteriophage DNA, and provide additional tools for evolutionary studies. Microbiology 151(3):653–663

Schmitz DJ, Ali Z, Wang C, Aljedaani F, Hooykaas PJJ, Mahfouz M, de Pater S (2020) CRISPR/Cas9 mutagenesis by translocation of Cas9 protein into plant cells via the Agrobacterium Type IV secretion system. Front Genome Editin 2:6. https://doi.org/10.3389/fgeed.2020.00006

Sprink T, Metje J, Hartung F (2015) Plant genome editing by novel tools: TALEN and other sequence specific nucleases. Curr Opn Biotechnol 32:47–53

Toda E, Koiso N, Takebayashi A, Ichikawa M, Kiba T, Osakabe K et al (2019) An efficient DNA-and selectable-marker-free genome-editing system using zygotes in rice. Nat Plants 5(4):363–368

Wang JW, Grandio EG, Newkirk GM, Demirer GS, Butrus S, Giraldo JP, Landry MP (2019) Nanoparticle-mediated genetic engineering of plants. Mol Plant 12(8):1037–1040

Wu S, Zhu H, Liu J, Yang Q, Shao X, Bi F et al (2020) Establishment of a PEG-mediated protoplast transformation system based on DNA and CRISPR/Cas9 ribonucleoprotein complexes for banana. BMC Plant Biology 20(1):425. https://doi.org/10.1186/s12870-020-02609-8

Yu J, Tu L, Subburaj S, Bae S, Lee G-J (2021) Simultaneous targeting of duplicated genes in Petunia protoplasts for flower color modification via CRISPR-Cas9 ribonucleoproteins. Plant Cell Rep 40(6):1037–1045. https://doi.org/10.1007/s00299-020-02593-1

Zhang R, Liu J, Chai Z, Chen S, Bai Y, Zong Y et al (2019) Generation of herbicide tolerance traits and a new selectable marker in wheat using base editing. Nat Plants 5(5):480–485. https://doi.org/10.1038/s41477-019-0405-0

Zong Y, Wang Y, Li C, Zhang R, Chen K, Ran Y et al (2017) Precise base editing in rice, wheat and maize with a Cas9-cytidine deaminase fusion. Nat Biotechnol 35(5):438–440. https://doi.org/10.1038/nbt.3811

Genome Editing by Ribonucleoprotein Based Delivery of the Cas9 System in Plants

Karina Y. Morales and Michael J. Thomson

Abstract Clustered regularly interspaced palindromic repeat (CRISPR)-based gene editing technology has opened the doors for targeted mutagenesis in a variety of crop species. The sole requirements for CRISPR/Cas-based gene editing are a guide RNA and a Cas protein. As a result, DNA-free methods employing Cas9/gRNA ribonucleoprotein (RNP) delivery have been developed to genome edit a variety of crop species and tissue types. In plants, common RNP-mediated transformation techniques include protoplast transformation and biolistics (particle bombardment) into immature embryos and shoot apical meristems. These approaches show promise for enabling *in planta* transformation techniques that may ultimately bypass lengthy in vitro tissue culture and regeneration steps. This chapter gives an overview of the current status of editing with ribonucleoprotein complexes in plants while giving vision for future prospects in this field.

Keywords CRISPR · Genome editing · Plant transformation · Ribonucleoprotein delivery

1 Introduction

Over the coming decades, global agriculture will need to significantly increase the amount of food produced each year while facing a myriad of challenges, including climate change, decreased land availability, and new disease and pest pressures. Conventional breeding methods have created significant advances combatting these hurdles in recent years; however, the timeline these goals can be achieved with conventional methods is far too long. Recently clustered regularly interspaced palindromic repeat (CRISPR) systems have been proposed as a solution for improving

K. Y. Morales · M. J. Thomson (✉)
Department of Soil and Crop Sciences, Texas A&M University, College Station, TX, USA
e-mail: michael.thomson@ag.tamu.edu

© The Author(s), under exclusive license to Springer Nature
Switzerland AG 2022
S. H. Wani, G. Hensel (eds.), *Genome Editing*,
https://doi.org/10.1007/978-3-031-08072-2_9

plant genetics at a faster rate. CRISPR was originally discovered as a bacterial immunity system which targeted viruses and is composed of palindromic repeats that are transcribed into guide RNAs (gRNAs). These gRNAs then guide Cas proteins to recognize a viral sequence, and the Cas protein will then create a double-stranded break in the viral DNA (Doudna and Charpentier 2014). Subsequently, this system has been found to be perfectly suited to gene editing applications by inducing targeted double strand breaks in the genome, which are then repaired by the cells through non-homologous end joining (NHEJ) or homology-directed repair (HDR). By taking advantage of these controlled double-strand breaks to introduce deletions, insertions, and substitutions, CRISPR has since been demonstrated as a powerful tool for creating targeted edits within a variety of plant species, including rice (*Oryza sativa*), wheat (*Triticum aestivum*), Arabidopsis (*Arabidopsis thaliana*), sorghum (*Sorghum bicolor*), and tomato (*Solanum lycopersicum*) (Belhaj et al. 2015).

The sole requirement for the successful creation of CRISPR mutants is a CRISPR-associated protein (Cas) and its associated gRNAs specified for each gene target. Although the majority of gene editing studies in plant species to date have introduced the Cas protein (such as Cas9 or Cas12a) and gRNA into plant cells in the form of plasmids through *Agrobacterium*-mediated delivery or particle bombardment, gene editing can also occur by directly transferring the Cas9/gRNA ribonucleoprotein (RNP) complex into plant cells. As has been demonstrated in lettuce protoplasts, these Cas9/gRNA ribonucleoprotein complexes do not integrate foreign DNA into the plant genome and can also minimize potential off-target effects (Woo et al. 2015). While other countries are still setting a framework for how they will regulate genome edited crops, the USDA has decided that plants developed using genome editing techniques that could have otherwise been produced through conventional breeding and do not incorporate foreign genetic material do not fall under their regulatory jurisdiction (USDA 2018). Although the recent implementation of the SECURE rule by USDA APHIS allows for *Agrobacterium*-mediated delivery due to a focus on the plant pest risk presented by the product, rather than the process (USDA 2020), there are a number of advantages to using gene editing techniques that avoid DNA integration altogether. For example, as no genetic material needs to be incorporated through a bacterial or viral vector, RNP methods may circumvent previous transformation limitations through which only specific species and genotypes are viable for genetic modification or editing. As a result, interest in the development of RNP methods has increased over the past few years. This chapter explores current methods and applications of RNP delivery methods in plants, along with the possibility for improved delivery methods in the future.

2 Delivery Methods

2.1 Protoplast Transformation Methods

Protoplasts, which are isolated plant cells that have had the cell wall digested away, are often used as the initial tissue type to test genome editing methods due to its ease of delivery of reagents into hundreds of cells for testing the editing efficiency of a gRNA. Common methods for transforming protoplasts include polyethylene glycol (PEG)-mediated transformation and electroporation. While the mechanisms behind PEG-mediated transformation are not completely understood, it is believed that both methods influence membrane dynamics to allow DNA and other materials to passively cross the membrane of a cell (Shillito 1999). While each of these methods is relatively easy to perform, they are not commonly used to create commercial varieties or breeding material as this requires the ability to perform protoplast culture and whole plant regeneration, a species- and genotype-dependent procedure that is technically challenging and time intensive for most plant species. Despite this, protoplast transformation using RNPs has been demonstrated in a variety of plant species including Arabidopsis, tobacco, rice, lettuce (Woo et al. 2015), potato (Andersson et al. 2018), wheat (Brandt et al. 2020; Liang et al. 2017), soybean (Kim et al. 2017), grapevine, apple (Malnoy et al. 2016), and various *Brassica* species (Murovec et al. 2018). Due to the limitations of protoplast culture, protoplasts are often best for in vivo validation of a gRNA across a full genome as many cells can be screened at the same time to test for efficiency. For plant species where protoplast culture can be easily performed, such as lettuce, protoplasts are also advantageous in that the plants created from protoplast culture are derived from a single cell. As a result, there is less of a worry of chimeras being formed after regenerating a whole plant from the initial protoplast.

Novel methods for protoplast transformation are still being developed. Currently the newest method shown to deliver RNPs is lipofection. In lipofection, RNP complexes are associated with a lipid to change the charge of the complex and allow for interaction with and passage across the cell membrane. First optimized in mammalian cell cultures, this method is still in its infancy as applied to plants; however, it has been shown effective in tobacco using both Lipofectamine 3000 and RNAiMAX as lipid sources. Furthermore, even without extensive optimization, this method is showing improved efficiency over PEG-mediated transformation and is a fairly low cost transformation option (Liu et al. 2019).

2.2   Biolistics

In biolistics, otherwise known as particle bombardment, gold particles are coated with DNA, RNA, and/or proteins. These particles are then shot at the material being transformed at a high velocity using a gene gun, such as the biolistic PDS-1000/He Particle Delivery System produced by Bio-Rad, in order to allow for penetration of the cell wall. Particle bombardment is not species- or genotype-dependent and has been shown to create efficient edits when delivering RNPs in wheat (Hamada et al. 2018; Liang et al. 2019), rice (Banakar et al. 2019, 2020), and maize (Svitashev et al. 2016). While delivery via biolistics can occur with or without a selectable marker, it is important to note that there is a higher frequency of the incorporation of plasmid DNA when using biolistics with RNPs than would be observed in delivering plasmid DNA or performing *Agrobacterium* transformation (Banakar et al. 2019). Furthermore, the velocity at which the gold particles are delivered can do significant damage to the tissue the reagents are aimed at. Chimeras can also be a common occurrence with biolistics as the material used for transformation is generally multicellular with only a small percentage of the cells receiving the reagents needed for transformation.

3 Validation of gRNAs and Target Edits

Before investing time and resources into plant transformation and regeneration, it can be beneficial to validate the gRNA design using an in vitro RNP assay. For this procedure, purified Cas protein (either purchased commercially or expressed and purified in the lab) is combined with the custom gRNA (either purchased commercially as a synthetic gRNA or expressed through in vitro transcription) to create RNPs that can be used to cut PCR amplicons containing the wild-type target sequence. After incubating the PCR products and RNPs, the results can be seen through gel electrophoresis: cut bands indicate successful gRNA activity, while uncut bands show a lack of gRNA efficacy. Once the gRNAs are validated in vitro, they can be further tested in vivo with a protoplast system or used directly for gene editing.

Upon transforming a plant, one of the most important steps is validating the presence of a mutation. A myriad of strategies exist to perform this essential step. The first approach follows a similar approach to the in vitro RNP validation described above. However, instead of using the native target gene to amplify by PCR and test the gRNA activity, DNA from the gene edited plants is used as the PCR template. In this case, since a mutation generally forms within 3–5 bp of the PAM site, if the gene has been edited, it is unlikely the specific 20 bp Cas recognition site will still be present, thereby leading to cut bands indicating non-mutated targets while the uncut bands indicating a mutated target (Liang et al. 2018). Similar to performing an in vitro cut by RNP, mutations can also be validated using the T7E1 assay. The T7E1 enzyme will identify bulges, as will occur when one strand has a mismatch or an insertion/deletion (INDEL) while its complementary strand does not, and therefore will cleave this region, which is then validated using gel electrophoresis (Vouillot et al. 2015). Cleavage assays can also be performed using restriction endonucleases. In this case the gRNA is designed to have a restriction enzyme recognition site near the PAM site. An edited plant would then likely have a mutation to the cut site and, as a result, would not be cleaved by the enzyme, in contrast to the wild-type sequence that would be cleaved.

Mutations caused by Cas proteins often result in INDELs, meaning size separation can also be used as a method to detect edits. One method which relies on this is high-resolution fragment analysis (HRFA). In this method an amplicon of the target region is generated using a fluorescently labeled primer. Products are then run through capillary electrophoresis where differences in size can be detected to a 1 bp resolution (Andersson et al. 2017). Likewise, high-resolution melting (HRM) curve analysis can be used, which uses slight changes in the melting temperature of double-strand DNA products to identify insertion/deletions down to a single base pair in length (Denbow et al. 2018).

Finally, edits can be validated through sequencing. This can be performed through Sanger sequencing or next-generation sequencing. However, sequencing can be cost prohibitive, depending on how many samples need to be screened. As a result, other validation methods are often used as a preliminary screen to ensure an edit is present prior to sequencing. After narrowing down the number of samples, sequencing is the best method for identifying the precise mutations, especially for substitutions, as many of these methods focus instead on whether bases have been inserted or deleted.

4 Tissue Types to Edit

4.1 Protoplasts

Protoplasts are the most optimized option for delivering RNPs into plants due to their widespread use as a rapid tool for in vivo validation of successful gRNA design. Approximately one dozen plant species have successfully been transformed with RNP at the protoplast stage including Arabidopsis, tobacco, rice, lettuce (Woo et al. 2015), potato (Andersson et al. 2018), wheat (Brandt et al. 2020; Liang et al. 2017), soybean (Kim et al. 2017), grapevine, apple (Malnoy et al. 2016), and various *Brassica* species (Murovec et al. 2018). Protoplasts have had their cell walls digested and, as a result, are easier to deliver genome editing reagents. Furthermore, protoplasts are single cells which means that any plant generated from protoplast culture will hold the desired mutation in all tissues, avoiding the problem of chimeric plants. However, protoplast culture is often difficult to perform and is severely limited to a small range of species and genotypes within these species. This often makes protoplasts better to use for validation of RNP efficacy across a genome as they are relatively easy to generate, many cells can be tested at the same time, and next-generation sequencing can be performed to identify the most common edits occurring with specific RNP combinations.

4.2 Immature Embryos

As an alternative to regenerating plants from protoplasts, tissue culture can be performed on transformed immature embryos, significantly limiting the hurdles needed to overcome to create genome edited plants. At this time, editing in immature embryos has been demonstrated in rice (Banakar et al. 2019), wheat (Liang et al. 2017), and maize (Svitashev et al. 2016). All methods developed for RNP transformation in immature embryos currently use biolistics. Although the process for regenerating plants from immature embryos is easier than from protoplasts, this is still a large time investment with the need for strong technical skills. Tissue culture is also incredibly dependent on the species and variety being used, which limits the genetic material an experiment can be started with.

4.3 Shoot Apical Meristem

Shoot apical meristems (SAM) are the actively growing part of a germinated seedling which will form the aboveground portion of a plant. SAMs do not require tissue culture and can easily form roots after growth on rooting media. At this time transformation of SAMs has only been demonstrated using biolistics in wheat (Hamada et al. 2018). RNP transformation of SAMs allows for *in planta* transformation and significantly cuts down the time to generate plants to phenotype as the lengthy tissue culture process is skipped. Although transformation of SAMs offers many advantages, ensuring inheritance of the targeted mutation can be difficult to accomplish. SAMs are multicellular and must have the reagents penetrate down to the L2 layer as pollen and eggs are developed from this layer. Furthermore, due to the relatively large surface area of tissue being transformed, chimeras can be fairly common in SAM transformations with only some cells being transformed instead of the full plant receiving the same mutation.

5   Benefits of RNP Delivery

At this time there are not many crops edited with RNP delivery systems which are in the process of being commercialized. Despite this, RNP delivery systems offer a vast number of advantages over comparable delivery systems. Plants transformed via RNP delivery systems generally display fewer off-target effects than those transformed via integration of DNA as the RNP complex is easily degraded within the cell. This transient nature of the delivery method means the Cas enzyme used is not actively produced throughout the lifetime of the plant, limiting the number of cuts that can be made (Woo et al. 2015). Furthermore, as the RNP complexes are already active upon delivery, there is no need to optimize codons or to identify the best promoters for transcribing desired products. Even more importantly, DNA is not randomly integrated into the genome as occurs with *Agrobacterium*-mediated transformation and biolistics with plasmids (Banakar et al. 2019). This allows for better assurance that only the targeted gene is modified while all other genes are left intact as plasmid DNA is not randomly incorporated at multiple points throughout the genome. Likewise, it bypasses the cumbersome process of identifying how many copies are in the genome and removing the DNA integration sites through segregation before commercialization.

Beyond the precision of edits made, RNP edited crops will likely face less scrutiny in the regulatory phase of releasing new varieties to the market. At this time, USDA has decided that as long as no foreign DNA is integrated and the mutation made could occur naturally, CRISPR edited crops do not fall under their jurisdiction to regulate (USDA 2018). This allows RNP edited plants to reach market faster than those edited by *Agrobacterium* or biolistics as there is no need to breed multiple generations to segregate out plasmid DNA, and without any DNA integration there is less burden to show that all foreign DNA sequences have been completely removed from the genome. Australia, New Zealand, and the EU are all regulating genome edited crops on the basis of how the edit was created. As a result, due to the integration of foreign DNA, plants transformed to incorporate plasmid DNA are automatically flagged as GMOs (Friedrichs et al. 2019). Clarity surrounding the use of RNPs in these countries has not been released yet; however, RNP-edited lines have a greater chance that they may not fall under the same regulations as traditionally transformed crops, as no foreign genetic material is incorporated into the genome during the gene editing process.

6 Future Prospects

Although considerable advances have been made in the field of DNA-free editing in recent years, there is still a strong need to improve the efficacy of delivering genome editing reagents in materials that do not need to pass through tissue culture. A summary of current methods readily available for use in plants can be found in Fig. 1.

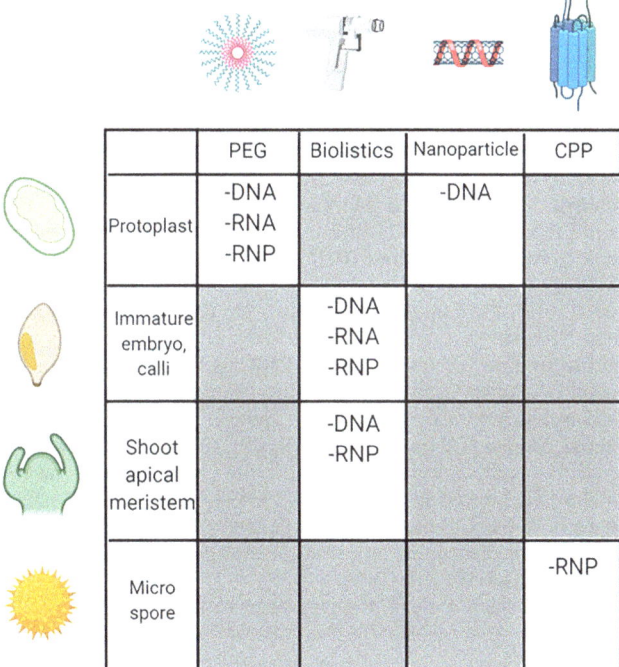

		PEG	Biolistics	Nanoparticle	CPP
	Protoplast	-DNA -RNA -RNP		-DNA	
	Immature embryo, calli		-DNA -RNA -RNP		
	Shoot apical meristem		-DNA -RNP		
	Micro spore				-RNP

Fig. 1 An outline of methods and their corresponding reagents that are currently available for use in genome editing for specific plant tissue types. Combinations which are blocked out in grey have no proven methods at this time

Recently, nanoparticles have been discovered to allow for passive passage of DNA into plant cells; although this has yet to be tested with RNPs, this method could allow for genome editing without the need for subsequent tissue culture (Demirer et al. 2019). Alternatively, cell-penetrating peptides have been shown to deliver RNPs into wheat microspores (Bilichak et al. 2020). With better characterization of cell-penetrating peptides, this method has the potential to become commonplace for editing plants. At this time, the available tissues for performing genome editing with RNPs are fairly limited. It is possible that in the future this can be expanded to include pollen and egg cells allowing for tissue culture and in vitro regeneration to be completely bypassed and for edits to be naturally introgressed into already existing populations. RNP-based editing is very much still in its infancy; however, this technology has the potential to radically change the methods by which genome edited crops are created, which can accelerate progress in crop improvement and, ultimately, may prove to have greater acceptance by consumers.

References

Andersson M, Turesson H, Nicolia A, Fält A-S, Samuelsson M, Hofvander P (2017) Efficient targeted multiallelic mutagenesis in tetraploid potato (Solanum tuberosum) by transient CRISPR-Cas9 expression in protoplasts. Plant Cell Rep 36(1):117–128

Andersson M, Turesson H, Olsson N, Fält A-S, Ohlsson P, Gonzalez MN et al (2018) Genome editing in potato via CRISPR-Cas9 ribonucleoprotein delivery. Physiologia Plantarum 164(4):378–384. https://doi.org/10.1111/ppl.12731

Banakar R, Eggenberger AL, Lee K, Wright DA, Murugan K, Zarecor S et al (2019) High-frequency random DNA insertions upon co-delivery of CRISPR-Cas9 ribonucleoprotein and selectable marker plasmid in rice. Sci Rep 9(1):19902. https://doi.org/10.1038/s41598-019-55681-y

Banakar R, Schubert M, Collingwood M, Vakulskas C, Eggenberger AL, Wang K (2020) Comparison of CRISPR-Cas9/Cas12a Ribonucleoprotein complexes for genome editing efficiency in the Rice Phytoene Desaturase (OsPDS) gene. Rice 13(1):4. https://doi.org/10.1186/s12284-019-0365-z

Belhaj K, et al. (2015). Editing plant genomes with CRISPR/Cas9. Current Opinion in Biotechnology 32: 76–84

Bilichak A, et al. (2020). Delivery of Cas9/sgRNA RNP into Wheat Microspores Using Synthetic CPP for Genome Editing and Gene Expression Modulation. CRISPR-Cas Methods. M. T. Islam, P. K. Bhowmik and K. A. Molla. New York, NY, Springer US: 191–202

Brandt KM, Gunn H, Moretti N, Zemetra RS (2020) A streamlined protocol for Wheat (Triticum aestivum) protoplast isolation and transformation with CRISPR-Cas ribonucleoprotein complexes. Front Plant Sci 11(769). https://doi.org/10.3389/fpls.2020.00769

Demirer GS, Zhang H, Goh NS, González-Grandío E, Landry MP (2019) Carbon nanotube–mediated DNA delivery without transgene integration in intact plants. Nat Protoc 14(10):2954–2971. https://doi.org/10.1038/s41596-019-0208-9

Denbow C, Ehivet SC, Okumoto S (2018) High resolution melting temperature analysis to identify CRISPR/Cas9 mutants from Arabidopsis. Bio-protocol 8(14):e2944. https://doi.org/10.21769/BioProtoc.2944

Doudna JA, Charpentier E (2014) The new frontier of genome engineering with CRISPR-Cas9. Science 346(6213):1258091–1258099. https://doi.org/10.1126/science.1258096

Friedrichs S, Takasu Y, Kearns P, Dagallier B, Oshima R, Schofield J, Moreddu C (2019) An overview of regulatory approaches to genome editing in agriculture. Biotechnol Res Innov 3(2):208–220. https://doi.org/10.1016/j.biori.2019.07.001

Hamada H, Liu Y, Nagira Y, Miki R, Taoka N, Imai R (2018) Biolistic-delivery-based transient CRISPR/Cas9 expression enables in planta genome editing in wheat. Sci Rep 8(1):1–7

Kim H, Kim S-T, Ryu J, Kang B-C, Kim J-S, Kim S-G (2017) CRISPR/Cpf1-mediated DNA-free plant genome editing. Nat Commun 8:14406

Liang Z, Chen K, Li T, Zhang Y, Wang Y, Zhao Q et al (2017) Efficient DNA-free genome editing of bread wheat using CRISPR/Cas9 ribonucleoprotein complexes. Nat Commun 8:14261. https://doi.org/10.1038/ncomms14261. https://www.nature.com/articles/ncomms14261#supplementary-information

Liang Z, Chen K, Yan Y, Zhang Y, Gao C (2018) Genotyping genome-edited mutations in plants using CRISPR ribonucleoprotein complexes. Plant Biotechnol J 16(12):2053–2062

Liang Z, Chen K, Gao C (2019) Biolistic delivery of CRISPR/Cas9 with ribonucleoprotein complex in wheat. In: Qi Y (ed) Plant genome editing with CRISPR systems: methods and protocols. Springer, New York, pp 327–335

Liu W, Rudis MR, Cheplick MH, Millwood RJ, Yang J-P, Ondzighi-Assoume CA et al (2019) Lipofection-mediated genome editing using DNA-free delivery of the Cas9/gRNA ribonucleoprotein into plant cells. Plant Cell Rep 39(2):245–257. https://doi.org/10.1007/s00299-019-02488-w

Malnoy M, Viola R, Jung M-H, Koo O-J, Kim S, Kim J-S et al (2016) DNA-free genetically edited grapevine and apple protoplast using CRISPR/Cas9 Ribonucleoproteins. Front Plant Sci 7(1904). https://doi.org/10.3389/fpls.2016.01904

Murovec J, Guček K, Bohanec B, Avbelj M, Jerala R (2018) DNA-free genome editing of Brassica oleracea and B. rapa protoplasts using CRISPR-Cas9 ribonucleoprotein complexes. Front Plant Sci 9(1594). https://doi.org/10.3389/fpls.2018.01594

Shillito R (1999) Methods of genetic transformation: electroporation and polyethylene glycol treatment. In: Vasil IK (ed) Molecular improvement of cereal crops. Springer, Dordrecht, pp 9–20

Svitashev S, Schwartz C, Lenderts B, Young JK, Cigan AM (2016) Genome editing in maize directed by CRISPR–Cas9 ribonucleoprotein complexes. Nat Commun 7:13274

USDA (2018) Secretary Perdue issues USDA statement on plant breeding innovation. (Release No. 0070.18).

USDA (2020) About the SECURE Rule. Retrieved from https://www.aphis.usda.gov/aphis/ourfocus/biotechnology/biotech-rule-revision

Vouillot L, Thélie A, Pollet N (2015) Comparison of T7E1 and surveyor mismatch cleavage assays to detect mutations triggered by engineered nucleases. G3 Genes Genome Genetic 5(3):407–415. https://doi.org/10.1534/g3.114.015834

Woo JW, Kim J, Kwon SI, Corvalan C, Cho SW, Kim H et al (2015) DNA-free genome editing in plants with preassembled CRISPR-Cas9 ribonucleoproteins. Nat Biotech 33(11):1162–1164. https://doi.org/10.1038/nbt.3389. http://www.nature.com/nbt/journal/v33/n11/abs/nbt.3389.html#supplementary-information

Virus-Mediated Delivery of CRISPR/CAS9 System in Plants

Monika Bansal and Shabir Hussain Wani

Abstract Plant virus research had resulted in precise understanding of viral replication, motility, and interactions with host after more than a century of study. A huge number of viral genes have been found and their functions have been determined. With a better knowledge of plant virus interactions with host, a variety of viruses have been produced as vectors for gene expression for functional investigations and biotechnology applications. These methods are useful for molecular breeding and functional studies of genomes for important agronomic features but also for the manufacture of health-promoting proteins. This overview outlines the most recent development in these areas, also as the viral vectors that are accessible for commercially significant crops and much further.

Keywords Guide RNAs (gRNA) · Viral genomes · Coat protein · Mobility protein · Virus-induced gene silencing

1 Introduction

Plant genome editing is an effective tool for researching and developing biological systems; as a result new key agricultural traits could be discovered. Genome editing is a technique in that uses designer endonucleases to edit the target genome and achieve the knockout and addition of specified DNA sequences within a cell or organism. Early gene editing relied on homologous recombination targeting technology, which was prone to off-target consequences. This situation was changed by the subsequent invention of designer endonucleases like clustered regularly

M. Bansal (✉)
SIILAS Campus, Jaipur National University, Jaipur, Rajasthan, India

S. H. Wani
Mountain Research Centre for Field Crops, SKUAST-Kashmir, Anantnag,
Jammu and Kashmir, India

© The Author(s), under exclusive license to Springer Nature
Switzerland AG 2022
S. H. Wani, G. Hensel (eds.), *Genome Editing*,
https://doi.org/10.1007/978-3-031-08072-2_10

interspaced short palindromic repeats-associated (CRISPR/Cas) as endonucleases utilized for genome editing. The Cas9 endonuclease had acquired broad usage as a tool for genome editing for a variety of model plants. This Cas9 protein and guide RNAs(gRNA) is delivered into target cells for effective genome editing. The Cas9 endonuclease is used in conjunction with gRNA, for directing the Cas9 protein to the target DNA sequence before the protospacer-associated motif (PAM). In this chapter we will discuss recent breakthroughs in plant genome editing technique utilizing viral vectors, their present constraints, and future directions.

2   Plant Viruses as Vectors

Plant viruses had been employed as vectors for a variety of applications, which includes the formation of commercially valuable proteins (Rybicki 2009). Viral genomes are an ideal choice as vectors because of efficient machinery and extensive structure of genome. Virus-based vectors provide successful means for delivering GE components into plant cells. These include the RNA viruses in case of monocots which are wheat streak mosaic virus and barley stripe mosaic virus (Lee et al. 2012) and *Tobacco rattle virus* for dicots. The importance of using RNA virus-based vectors over DNA virus-based vectors is that RNA virus does not give rise to undesired integration into plant genome. Therefore, plants modified by use of RNA viruses are well thought out to be transgene-free edited plants. Although these viruses have a limited cargo capacity, they can be transformed into non-infectious replicons by substitution of genes involved with infection into host cells and cell-to-cell migration like SSN expression cassettes. The coding sequences of mobility protein (MP) and coat protein (CP) were deleted for achieving this purpose, because of which cell-to-cell and insect-mediated transmission from one plant to another plant are no longer possible. The absence of CP enhances the quantity of replicon intermediates, mostly because CP isn't available for sequestering and packaging of ssDNA into virions and the loss of CP/Rep interactions inhibits viral replication. Tobacco rattle virus (TRV) is an example of such type of viruses which are popularly exploited for effective virus-induced gene silencing in different plants species. The small size of its genome is suitable for its cloning, agroinfection, and multiplexing which shows its great ability to be used as a vector for delivery of genome editing components (Ellison et al. 2020). When the recombinant TRV vector is transformed into plants, the expression system of virus results into formation of the recombinant viral RNA in the infected host cells (Senthil-Kumar and Mysore 2014). These cells then prove to be a source for replication of viral RNAs and their systemic spreading to a number of different tissues. Geminiviruses belong to the family *Geminiviridae*, found all over the world and infects a wide range of plants which includes wheat, cotton, cucurbits, maize, fruits, and legumes (Briddon 2015; Rey et al. 2012). Geminiviruses in most of the cases serve as efficient vector for genome editing because of their capability to infect a wide host range from different families; only one protein, Rep (replication-associated protein) is needed for initiation of replication within the host cell. These viruses can efficiently replicate within host cell and makes large numbers of replicons, producing of SSNs for target sequence significantly improving the efficacy for transformation (Hanley-Bowdoin et al. 2013).

3   Virus as Vectors for Their Use in Genome Editing

TRV1 and TRV2 are the two genomic components of TRV. TRV1 contains genes for replicase proteins, as well as a 29-kDa MP, which are necessary for viral movement. Resistance against tomato yellow leaf curl virus genome can be delivered with the help of *Tobacco rattle virus* (TRV) into *Nicotiana benthamiana* plants and expresses Cas9, in plants that result in less buildup of viral DNA and reduced symptoms of infection (Tashakandi et al. 2018). The Cas9 system targets a conserved region in various begomoviruses which demonstrated that targeting viral intergenic, non-coding sequences was found to be effective than targeting coding sequences of virus, and reverting the generated mutations through NHEJ limits the production of recovered viral types that circumvent CRISPR-mediated immunity (Ali et al. 2016). Ma and Li (2020) worked on delivery of CRISPR/Cas9 cassette *in Nicotiana benthamiana* and obtained genetically edited plants with greater efficacy. This experiment was performed by use of negative-strand sonchus yellow net rhabdovirus as a vector. This development opened possibilities for genome editing of plants because of its larger cargo capacity that negative-strand RNA viruses have for foreign sequences.

CRISPR/Cas9 technology had been used for engineering geminivirus resistance (Ali et al. 2015, 2016; Baltes et al. 2015; Ji et al. 2015) and potyvirus resistance (Chandrasekaran et al. 2016) by targeting and cleaving virus genome (Ali et al. 2015, 2016; Ji et al. 2015) or by editing plant genome for triggering immunity for invading virus (Chandrasekaran et al. 2016; Zaidi et al. 2016; Mahas et al. 2019; Pyott et al. 2016).

In maize, *Nicotiana benthamiana*, and *Setaria viridis*, a collection of foxtail mosaic virus (FoMV) vectors was created for transient expression and delivery of gRNA for CRISPR-based genome editing. This was achieved by duplication of the FoMV capsid protein subgenomic promoter, deleting the redundant open reading frame 5A, and insertion of a cloning site just after the duplicated promoter. In the leaves of infected maize seedlings, the modified FoMV vectors transiently produce GFP and BAR protein. Ali et al. (2015) studied single gRNA expressed under regulation of duplicated promoter edits for the *Phytoene desaturase* gene in case of *N. benthamiana*, *Carbonic anhydrase 2* gene in *S. viridis*, and the *HKT1* gene in maize which codes for a potassium transporter. FoMV's value in monocots for virus-induced gene silencing, virus-mediated overexpression, and virus-enabled gene editing has been expanded in this study. Both DNA viruses and RNA viruses have shown effective gene editing frequency in several model plants.

In plant cells editing of tobacco rattle virus and pea early browning virus (PEBV) was done for delivery of multiple sgRNAs into tobacco and *Arabidopsis* plants.

TRV and PEBV deliver sgRNAs into systemic leaves and result in mutation of the target genomic loci. TRV and PEBV can enable genome editing and will be used for producing mutations at target site for functional analysis across various plant species (Ali et al. 2018).

The apple latent spherical virus (ALSV) is a strongly infectious virus that can be used to transmit genome editing elements. In the case of ALSV-based gRNA delivery, the Cas9-based Csy4-processed ALSV Carry (CCAC) system was designed (Luo et al. 2021). In this case soybean-infecting ALSV was engineered for delivering gRNA. The endoribonuclease Csy4 efficiently releases gRNAs which functions competently in Cas9-mediated editing. Confirmation of editing for phytoene desaturase loci and exogenous 5-enol-pyruvylshikimate-3-phosphate synthase in *Nicotiana benthamiana* was performed by sequencing.

The first reports of use of viruses for gene editing involved studies on geminiviruses, which are widely distributed, spread by insects, and infect a diverse range of plant hosts. These viruses have SS circular DNA with monopartite or bipartite genomes and four ORFs. When it enters plants, its single-stranded genome turns into intermediate that is double stranded and finds its use as template for transcription and replication. Single-stranded genomes are transformed into double-stranded intermediates for starting a new replication cycle, or else they are coated by the coat protein to make virions that can spread to neighboring cells with the help of plasmodesmata. Their small size makes them easier to manage, but it also restricts their cargo capacity; as a result, they can't carry large DNA pieces like the genes that code for Cas nucleases (~4.2 kb). Geminiviruses are not able to carry long DNA fragments like in the case of genes which code for Cas nucleases and produce large amounts of sgRNA. Geminiviruses had been deployed into non-infectious replicons (GVRs) by eliminating sequence for movement protein and coat protein and stop movement of virus from cell to cell and its transmission. Viral vectors are not infectious themselves and need to be transformed into plant cells by use of *Agrobacterium*-mediated transformation.

Bean yellow dwarf virus replicons were used for delivering a Cas9 nuclease and a repair template into tobacco for editing. Baltes et al. (2014) show significant cargo capacity and targeting efficiency with twofold high compared with *Agrobacterium* transformation. BeYDV replicons were also used to facilitate genome editing in potatoes (Butler et al. 2016) creating mutations that supported a reduced herbicide resistance phenotype (Cermark et al. (2015). BeYDV replicons were utilized for inserting a strong promotor upstream of a gene which regulates anthocyanin synthesis in tomato, resulting in a 12-fold greater delivery as compared to Agrobacterium T-DNA delivery. Yin et al. (2015) used cabbage leaf curl virus for genome editing by replacement of viral CP by sgRNA, for editing genes (namely, NbPDS3 and NbIspH) in *Nicotiana benthamiana*. Wheat dwarf virus replicons were also transferred for genome editing in wheat and rice and show high gene targeting efficiency for targeting multiple genes (Wang et al. 2017; Gil-Humanes et al. 2017). TRV1 is required for replication and movement of virus and TRV2 genome codes for the CP and nonstructural proteins involved with nematode transmission. TRV was originally used as a vector for genetic engineering in plants for delivery of zinc-finger nucleases (ZFN) by replacement of RNA2 with the Zif268: FokI ZFN. Targeted genome alterations were observed with help of reporter gene into the cells of tobacco and petunia in this approach, and inheritance of mutations to the following generation validated the ZFN's stability. Through agroinfection, a TRV vector encoding sgRNA for phytoene desaturase gene was injected into the leaves of tobacco that overexpress Cas9, causing PDS gene changes. TRV was initially used as a CRISPR vector as a vehicle for delivering sgRNAs into *N. benthamiana* and *A. thaliana*. TMV was used for delivery of sgRNA by substitution of the CP gene with a sgRNA. TMV also shows its capability for promoting gene editing by delivering a large number of sgRNA and efficiently editing the gene in *N. benthamiana* plants. Ali et al. showed that PEBV had the ability for delivering sgRNAs and cause mutagenesis at the target loci into plants of *Nicotiana benthamiana*. PEBV can infect meristematic tissues that allow the recovery of seeds with the given mutations (Constantin et al. 2004). Barley stripe mosaic virus (BSMV) had been edited for delivery of Cas9-mediated mutation at target site in maize and wheat (Hu et al. 2019). Beet necrotic yellow vein virus-based vectors were designed for synchronized expression of foreign proteins and used for effective sgRNA delivery into the genome in *Nicotiana benthamiana* plants. FoMV had been shown to express sgRNAs in *Nicotiana benthamiana*, *Setaria viridis*, and maize that express Cas9 and proves that FoMV can be used for successful genome editing (Mei et al. 2019).

Barley yellow striate mosaic virus (BYSMV) and sonchus yellow net rhabdovirus (SYNV), both negative-strand viruses, were used for transporting CRISPR/Cas components into plant. SYNV has been found to be capable of knocking out multiple genes in plants, resulting in increased efficiency for DNA-free genome editing (Ma and Li 2020). This work proves that genome-edited plants carry the genome change to the following generations by creating sgRNAs for distinct genes

4   Conclusions and Future Prospects

Plant virus vectors are being developed in terms of ease of use and range of use in the field of genetic engineering studies in plants. Several viral vectors must be modified to accommodate more nucleic acid while also being aggressive in the infection process. Plant virus vectors and components produced provide critical benefits for the quick and cost-effective studies, due to advancements in strategies to allow their introduction and transient expression in plants.

References

Ali Z, Abulfaraj A, Idris A, Ali S, Tashkandi M, Mahfouz MM (2015) CRISPR/Cas9-mediated viral interference in plants. Genome Biol 16:238. https://doi.org/10.1186/s13059-015-0799-6

Ali Z, Ali S, Tashkandi M, Zaidi SSEA, Mahfouz MM (2016) CRISPR/Cas9-Mediated Immunity to Geminiviruses: Differential Interference and Evasion. Sci Rep 6:26912

Baltes NJ, Gil-Humanes J, Cermak T, Atkins PA, Voytas DF (2014) DNA replicons for plant genome engineering. Plant Cell 26:151–163. https://doi.org/10.1105/tpc.113.119792

Baltes NJ, Hummel AW, Konecna E, Cegan R, Bruns AN, Bisaro DM et al (2015) Conferring resistance to geminiviruses with the CRISPR–Cas prokaryotic immune system. Nat Plants 1:15145. https://doi.org/10.1038/nplants.2015.145

Briddon RW (2015) Geminiviridae. In: eLS. Wiley, Chichester. https://doi.org/10.1002/9780470015902.a0000750.pub

Butler NM, Baltes NJ, Voytas DF, Douches DS (2016) Geminivirus-mediated genome editing in potato (Solanum tuberosum L.) using sequence-specific nucleases. Front Plant Sci 7:1045. https://doi.org/10.3389/fpls.2016.01045

Constantin GD, Krath BN, MacFarlane SA, Nicolaisen M, Johansen IE, Lund OS (2004) Virus-induced gene silencing as a tool for functional genomics in a legume species. Plant J 40(4):622–631

Čermák T, Baltes NJ, Čegan R, Zhang Y, Voytas DF (2015) High-frequency, precise modification of the tomato genome. Genome Biol 16:232

Chandrasekaran J, Brumin M, Wolf D, Leibman D, Klap C, Pearlsman M et al (2016) Development of broad virus resistance in non-transgenic cucumber using CRISPR/Cas9 technology. Mol Plant Pathol 17:1140–1153. https://doi.org/10.1111/mpp.12375

Ellison et al (2020) Multiplexed heritable gene editing using RNA viruses and mobile single guide RNAs

Gil-Humanes J, Wang Y, Liang Z, Shan Q, Ozuna CV, Sánchez-León S, Baltes NJ, Starker C, Barro F, Gao C, Voytas DF (2017) High-efficiency gene targeting in hexaploid wheat using DNA replicons and CRISPR/Cas9. Plant J 89(6):1251–1262

Hanley-Bowdoin L, Bejarano ER, Robertson D, Mansoor S (2013) Geminiviruses: masters at redirecting and reprogramming plant processes. Nat Rev Microbiol 11:777–788. https://doi.org/10.1038/nrmicro3117

Hu J, Li S, Li Z, Li H, Song W, Zhao H, Lai J, Xia L, Li D, Zhang Y (2019) A barley stripe mosaic virus-based guide RNA delivery system for targeted mutagenesis in wheat and maize. Mol Plant Pathol 20(10):1463–1474

Ji X, Zhang H, Zhang Y, Wang Y, Gao C (2015) Establishing a CRISPR-Cas-like immune system conferring DNA virus resistance in plants. Nat Plants 1:15144. https://doi.org/10.1038/nplants.2015.144

Luo Y, Na R, Nowak JS, Qiu Y, Lu QS, Yang C, Marsolais F, Tian L 2021) Development of a Csy4-processed guide RNA delivery system with soybean-infecting virus ALSV for genome editing. BMC Plant Biol 13;21(1):419

Lee WS, Hammond-Kosack KE, Kanyuka K (2012) Barley stripe mosaic virus-mediated tools for investigating gene function in cereal plants and their pathogens: virus-induced gene silencing, host-mediated gene silencing, and virus-mediated overexpression of heterologous protein. Plant Physiol 160:582–590

Ma X, Li Z (2020) Significantly improved recovery of recombinant Sonchus yellow net rhabdovirus by expressing the negative-strand genomic RNA. Viruses 12:1459

Mahas A, Ali Z, Tashkandi M, Mahfouz MM (2019) Virus-Mediated Genome Editing in Plants Using the CRISPR/Cas9 System. Methods Mol Biol 1917:311–326. https://doi.org/10.1007/978-1-4939-8991-1_23. PMID: 30610646

Mei Y, Beernink BM, Ellison EE, Konečná E, Neelakandan AK, Voytas DF, Whitham SA (2019) Protein expression and gene editing in monocots using foxtail mosaic virus vectors. Plant Direct 3(11):e00181

Pyott DE, Sheehan E, Molnar A (2016) Engineering of CRISPR/Cas9-mediated potyvirus resistance in transgene-free *Arabidopsis* plants. Mol Plant Pathol 17:1276–1288. https://doi.org/10.1111/mpp.12417

Rey ME, Ndunguru J, Berrie LC, Paximadis M, Berry S, Cossa N et al (2012) Diversity of dicotyledenous-infecting geminiviruses and their associated DNA molecules in southern Africa, including the South-west Indian ocean islands. Viruses 4:1753–1791. https://doi.org/10.3390/v4091753

Rybicki, E.P. (2009) Plant-produced vaccines: promise and reality. Drug Discov. Today 14, 16–24

Senthil-Kumar M, Mysore KS (2014) Tobacco rattle virus-based virus-induced gene silencing in *Nicotiana benthamiana*. Nat. Protoc 9:1549–1562. https://doi.org/10.1038/nprot.2014.092

Tashkandi M, Ali Z, Aljedaani F, Shami A, Mahfouz MM (2018) Engineering resistance against Tomato yellow leaf curl virus via the CRISPR/Cas9 system in tomato. Plant Signal Behav 13(10):e1525996

Wang M, Lu Y, Botella JR, Mao Y, Hua K, Zhu JK (2017) Gene targeting by homology-directed repair in rice using a geminivirus-based CRISPR/Cas9 system. Mol Plant 10(7):1007–1010

Yin K, Han T, Liu G, Chen T, Wang Y, Yu AY, Liu Y (2015) A geminivirus-based guide RNA delivery system for CRISPR/Cas9 mediated plant genome editing. Sci Rep 5:14926

Zaidi SS, Tashkandi M, Mansoor S, Mahfouz MM (2016) Engineering plant immunity: using CRISPR/Cas9 to generate virus resistance. Front Plant Sci 7:1673. https://doi.org/10.3389/fpls.2016.01673

Characterization of Gene Edited Crops via Metabolomics

Muhammad Qudrat Ullah Farooqi, Sanathanee Sachchithananthan, Muhammad Afzal, and Zahra Zahra

Abstract Metabolomics is a new branch of "omics" that entails identifying and quantifying both the metabolites and chemical impression of the processes that are regulated by the cells in different types of species. The metabolome is a type of metabolite reservoir of an organism that can be analyzed to define the variations on the basis of genetically and environmental traits. This approach is very important while studying the interaction that happens between the environment and the gene of the crops, characterization of the mutant, assessment of the expressed traits, recognition of clinical assessment, and discovery of the drugs. Metabolomics is an approach which is very promising when it comes to deciphering the metabolic networks of the crops that are linked with the tolerance capacity of the biotic and abiotic stress. The breeding that is metabolomic-assisted helps the breeders in providing a screening that is very efficient when it comes to crop yielding and tolerating the stress at the metabolic level. The process of metabolic profiling has been supported by the metabolomic analytical tools which are very advanced. These tools are nondestructive nuclear magnetic resonance spectroscopy (NMR) and mass spectrometry. Linking metabolomics to genomic tools makes it possible to have an efficient dissection of the genetic association with the phenotypes of crop plants. This book chapter provides an overview of cutting-edge plant metabolomic tools for crop improvement.

M. Q. U. Farooqi (✉)
UWA School of Agriculture and Environment, The University of Western Australia, Perth, WA, Australia

S. Sachchithananthan
School of Molecular Science, The University of Western Australia, Perth, WA, Australia

M. Afzal
College of Food and Agricultural Sciences, King Saud University, Riyadh, Saudi Arabia

Z. Zahra
Department of Civil & Environmental Engineering, University of California-Irvine, Irvine, CA, USA
e-mail: nzahra@uci.edu

© The Author(s), under exclusive license to Springer Nature
Switzerland AG 2022
S. H. Wani, G. Hensel (eds.), *Genome Editing*,
https://doi.org/10.1007/978-3-031-08072-2_11

1 Introduction

The past decade has been a witness of the development in the omics sector. These omics are genomics, transcriptomics, epigenomics, proteomics, metabolomics, and phenomics. The approaches or platforms of omics create information about the improvement and enhancement in the speed and precision of the breeding programs that are currently in progress. These breeding programs aim at developing germplasm which is climate smart and rich in nutrition to ensure food safety (Muthamilarasan et al. 2019). The omics approach has provided great scope in improving the present understanding of the important traits so that new strategies to improve the plants can be developed. Among all the approaches of omics, metabolomics is the one branch which is the most difficult and complex. It has never received proper attention in the crop science especially in the context of trait mapping and selecting the plants (Van Emon 2016). This essay will discuss how the gene-edited crops have been characterized by the metabolomics.

2 Metabolomics

In the past 10 years, the approach of metabolomics has been proved as one of the major developments in the field of science, which has paved a way for correct profiling of the metabolites in the microorganisms, plants, and animals. This approach is capable of recognizing a large number of metabolites from just one extraction and has the potential to recognize a wide variety of metabolites from just a single sample. This enhances the fast and proper evaluation of the metabolites. It can also be said that the metabolomics gives a detailed view of the cellular metabolites and describes it as natural compounds that are small and actively participate in the various events that occur at the cellular level, thus representing the cell's full physiological state. In the context of advancement of metabolomics, the study of metabolite recognition of the mutants and the transgenic lines has the capacity to comprehend metabolic networks and genetic markers (Sousa Silva et al. 2019). Metabolomics helps in expressing the function of gene and the impact of a particular gene on the pathway of the metabolites and shows different stages of regulation and interception between the connected pathways which gets difficult to analyze with the other methods such as microarray. The metabolomics allow the researchers to catalogue and prioritize the genes that can improve the important characteristics in the crops (Kumar et al. 2017).

The abovementioned studies of omics approaches have been expanded to examine regulatory steps that are associated with them, such as control of the epigenetic and changes at the post-transcriptional and post-translational level. Regardless of whether or not a transgenic system is present, the use of metabolomics in crop science has increased. Metabolomics has the ability to help in the development of improved traits, allowing for superior breeding materials (Razzaq et al. 2019). The presence of whole sequence of the genome, huge genetic variants of the genome, and genotyping assays which are of very low cost, combined with developments in

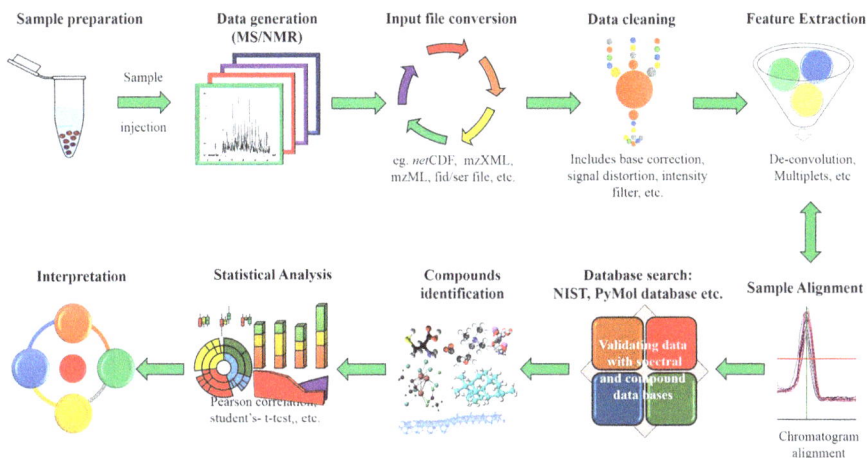

Fig. 1 Scheme for crop improvement using metabolomics. (Source: Kumar et al. 2017)

metabolomics, provides an exciting opportunity to link the metabolomics with the crop breeding programs very efficiently. In metabolomics, the methods and tools that are used include the mass spectrometry (MS) and nuclear magnetic resonance (NMR) spectroscopy. These approaches and tools have resulted in significant progress. Metabolomic approaches have the potential to conduct metabolite surveys on a wide scale (Alseekh et al. 2018). Figure 1 depicts the improvement of crop through metabolomics.

3   Analytical Tools for Metabolomic Studies

The generation of metabolomics data on modern metabolomic platforms involves the use of two crucial techniques, namely, NMR and MS. Metabolites present in the crop can be detected using NMR which is dependent on the magnetization of the atom nuclei present in the magnetic field. The system of nuclear magnetic resonance (NMR) is non-destructive which has been widely used to recognize the metabolites that have lower molecular weight and is used for a variety of applications such as fingerprinting of the metabolites, profiling, metabolic flux, and extracting the information from the atomic structure of the biological samples. The technique's poor sensitivity, which is due to the lack of coverage provided by low-abundance biomarkers, is a major drawback that prevents it from being widely used. In contrast to NMR, MS has greater sensitivity and it allows researchers to obtain a wider coverage of metabolomics data. This has prompted the researchers to find out new metabolic biomarkers as well as molecules that can be of assistance in the redevelopment of metabolic systems and processes. With the development in the methods of ionization such as the atmospheric pressure chemical ionization (APCI), electrospray ionization (ESI), and MALDI-TOF, MS has been successful in achieving the accuracy. To improve throughput, MS is commonly used in conjunction with chromatography techniques including gas chromatography (GC), liquid chromatography (LC), capillary electrophoresis (CE), Fourier transform ion-cyclotron resonance (FT-ICR), and field asymmetric waveform ion mobility spectrometry (FAIMS). NMR is favored over MS for identifying physical properties of ligands, binding sites present on the proteins, uncovering structures of protein ligand complexes, and direct binding of target protein despite its low sensitivity and broad sample requirement (Rubert et al. 2015).

4 The Advent and Adoption of Genome Editing in Plants

Meganucleases were used to carry out the procedure of genome editing. The target site was as long as 18 bp. It is undeniable that the specificities of these enzymes that exist naturally were the ones that determined the double-stranded breaks that they make. These specificities made it difficult to work with these enzymes. As a consequence, the designer nucleases that consist of the features such as zinc finger nucleases and transcription activators like effector nucleases have been established to address this problem. The most recent technology that has been developed is based on the adaptive immunity of the bacteria. In this technique the invasive DNA that has been previously encountered is imprinted in the molecular memory and is the subject of the target to the future challenges by expressing the CRISPRs. The resulting CRISPR RNAs are the guides for the CRISPR-associated (CAS) nucleases, which target pathogens after they've been infected. The advantages and applications of genome editing include increased precision and the absence of foreign DNA in the edited genome (Fraser et al. 2020).

The advanced metabolomics plays an important role in the deciding the changes of the global metabolome. It also decides how these changes are manifested to the chemical structure of currently available validated crops. The classic mutagenesis that takes place is the only mutagenesis that has continued to be the only exception to the procedures that regulate GM technologies. This determination is based on a safety track record. As a result, actively participating in genetic manipulation exploration and viability studies is crucial in order to produce a wealth of reliable scientific proof to substantiate these technological innovations (Nadakuduti and Enciso-Rodríguez 2020).

5 Metabolomics for Improvement of Fruits

Metabolomic research has revealed new information about fruit genetics, particularly in relation to ripening and quality of food. Tomato is a very rich source of carotenoids, antioxidants, and flavonoids (Tohge and Fernie 2015). The patterns of metabolite segregation of 50 cultivars were found to be very similar to the segregation of the fruit size. By plotting a correlation with the fruit transcriptome, the metabolome can be used to dissect the event of ripening. This metabolome can be a very good explanation of the diverse and distinct biochemical pathways found in tomato ILs and ecotypes' fruits (Upadhyaya et al. 2017). Apple peel and flesh contain beneficial nutrients such as antioxidants, which helps in reducing the risk of chronic diseases such as asthma, cancer, cardiovascular disease, and diabetes. The metabolite content of the apple is used to differentiate the cultivars that are crucial for commercial use. If examples of this are considered, then the cultivar "Golden Delicious" shows a high concentration of myo-inositol, sugars, and succinic acid, whereas the cultivars "Red Delicious" and "Fuji" have a higher concentration of sterols, flavonoids, phenolic acids, stearic acid, anthocyanin, and carbohydrates (Eisenmann et al. 2016). Fuji's peel extract contains very high concentration of carbohydrates such as the glucose and sorbitol, which can be differentiated from the Red Delicious that contains very high level of unsaturated fatty acids such as oleic and linoleic acid. The spatial variability of sugars and organic acids between the layers of the apple was revealed in a recent report on metabolic characterization of the fruit (Cuthbertson et al. 2012). Citrus Huanglongbing causes infection in the *Candidatus* Liberibacter asiaticus and deteriorates the quality of its juice. This infection causes a significant decrease in the amount of glucose, fructose, sucrose, and amino acids such as alanine, arginine, isoleucine, leucine, proline, threonine, and valine while increasing the level of citrate and phenylalanine (Wang et al. 2016).

In the study of metabolomics, fruit heat treatment is widely used to avoid any kind of fruit infection during the process of post-harvest storage. Heat treatment very efficiently reduces the organic and amino acid content and also promotes the accumulation of certain metabolites such as 2-keto-D-gluconic acid, tetradecanoic acid, oleic acid, ornithine, succinic acid, myo-inositol, glucose, fructose, sucrose, and turanose. These metabolites reduce the possibility of infection in the crops post-harvest (Yun et al. 2013).

6   Metabolomics for Improvement of Legume Crops

Despite the extensive research on model legumes, metabolomic studies in other legumes are still restricted. In terms of model legumes, research into the impact of rhizobia node factor (Nod) in Medicago showed a decrease in oxylipins. Through metabolic characterization of salt-resistant lotus plants, another finding revealed a variety of reforms involving metabolic modifications of shoot constituents for sustainability (Ramalingam et al. 2015).

7   Metabolomics for Improvement of Cereal Crops

The amount of metabolites in cereals has been thoroughly investigated, as has their association with gene sequences. Different rice research projects have harnessed the power of metabolomics to uncover the variety of metabolites found in different varieties and natural variants (Hamany Djande et al. 2020). In the same way, the study of metabolomics in maize has made it easier for the researchers to differentiate and then select the genotypes that are superior and consist of the improved nutritional composition. Presently, a metabolomic technique is being used to investigate the chemical concentration of different maize and rice crops, as well as their natural varieties (Gálvez Ranilla 2020).

8   Effect of Biotic and Abiotic Stress on Plant

Crop productivity is negatively impacted by biotic and abiotic stresses, which result in a huge reduction of the annual yield of crops globally. To address abiotic and biotic stresses in plants, metabolomic tools can be combined with other omics tools such as genomics, transcriptomics, and proteomics. It aids in the analysis of various plant metabolites, both exogenous and endogenous, under the extreme stress related to the climate change and is very important to understand the biology of the plant system. Figure 2 illustrates the use of omics-based approaches to explain the stress regulation system from the genome to the phenome (Atkinson et al. 2015). There are two types of metabolites throughout the plant metabolome: primary metabolites and secondary metabolites. Primary and secondary metabolite metabolic profiling provides accurate and comprehensive knowledge about the biochemical processes that occur during plant metabolism. The primary and secondary metabolites of the pants are also associated with the metabolic pathways which are very complex. Advanced metabolomics tools, such as gas chromatography-mass spectrometry (GC-MS), liquid chromatography, mass spectroscopy (LC-MS), and non-destructive nuclear magnetic resonance spectroscopy (ND-NMR), can successfully detect, identify, assess, and evaluate these metabolites (Dresselhaus and Hückelhoven 2018).

9   Metabolomics: An Integral Part of Knowledge-Based Plant Breeding

In the last 10 years, tremendous technological advancements have been made, which have been used to fully understand various fields of biology in order to obtain a better insight into the underlying dynamics of required gene coding characteristics. Knowledge-based plant breeding (KPB) incorporates all relevant data derived from large-scale data analysis pertaining to the genome, epigenome, transcriptome, metabolome, and proteome, all of which contribute to a particular phenotype. The data that has been recorded by the "omics" platforms has allowed the researchers to have a better understanding of the system of traits, which until now has been mainly contributed by the systems genetics and genomics and to a lesser extent by the transcriptomics (Dawid and Hille 2018). Other "omics" approaches, such as

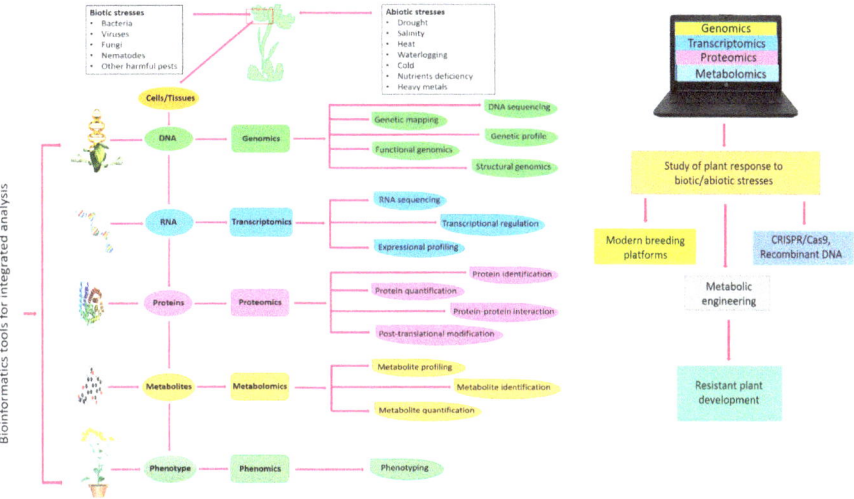

Fig. 2 Plant stress mechanism. (Source: Razzaq et al. 2019)

metabolomics and proteomics, have also started their contribution to initiate important information so that it can become an integral part of KPB, thus strengthening the approach even more so that they can achieve much better genetic advantage (Liu and Locasale 2017).

Technological progress has helped to improve the efficiency of the techniques of the plant breeding by allowing for the proper selection of desired plants. Access to various "omics" approaches will result in a paradigm shift in the overall process of breeding by promoting the selection of the plants on the basis of the information on the genome-scale which are generated at various biological levels. The breeders of the plants will slowly accept these developments and changes which will help them in making proper decisions (Hong et al. 2016).

10 Concluding Remarks and Future Perspectives

Recent advances in plant metabolomics have made it possible to select the desirable traits along with the possibility of developing metabolically engineered plants. The transition from single metabolite analysis to high throughput assays that initiates the footprints of multiple metabolites has immediately made ways for constructing better models for the networks of the metabolites as well as in recognition of biomarkers. In the last 10 years, the use of metabolomics along with the other omics approaches has revealed many metabolites which were novel and known to the researchers. It has also allowed the recognition of the particular contribution of these metabolites to improve the attributes of the plants such as quality of the crops, yields, shelf life, and so on. To that end, high throughput genotyping and sequencing platforms have been a huge help to provide a technique which is a cost-efficient, high-throughput method of elucidating the design of metabolic traits (Pinu et al. 2019).

In conclusion, it can be assumed that the use of the metabolomics approach has very efficiently improved the ability of plant breeders so that they can create and develop highly superior plants and therefore can improve the crop genotypes and make it highly productive so that they can meet the challenges of the twenty-first-century agriculture.

References

Alseekh S, Bermudez L, De Haro LA, Fernie AR, Carrari F (2018) Crop metabolomics: from diagnostics to assisted breeding. Metabolomics 14(11):1–13

Atkinson NJ, Jain R, Urwin PE (2015) The response of plants to simultaneous biotic and abiotic stress. In: Combined stresses in plants. Springer, Cham, pp 181–201

Cuthbertson D, Andrews PK, Reganold JP, Davies NM, Lange BM (2012) Utility of metabolomics toward assessing the metabolic basis of quality traits in apple fruit with an emphasis on anti-oxidants. J Agric Food Chem 60(35):8552–8560

Dawid C, Hille K (2018) Functional metabolomics—a useful tool to characterize stress-induced metabolome alterations opening new avenues towards tailoring food crop quality. Agronomy 8(8):138

Dresselhaus T, Hückelhoven R (2018) Biotic and abiotic stress responses in crop plants

Eisenmann P, Ehlers M, Weinert CH, Tzvetkova P, Silber M, Rist MJ et al (2016) Untargeted NMR spectroscopic analysis of the metabolic variety of new apple cultivars. Metabolites 6(3):29

Fraser PD, Aharoni A, Hall RD, Huang S, Giovannoni JJ, Sonnewald U, Fernie AR (2020) Metabolomics should be deployed in the identification and characterization of gene-edited crops. Plant J 102(5):897–902

Gálvez Ranilla L (2020) The application of metabolomics for the study of cereal corn (Zea mays L.). Metabolites 10(8):300

Hamany Djande CY, Pretorius C, Tugizimana F, Piater LA, Dubery IA (2020) Metabolomics: a tool for cultivar phenotyping and investigation of grain crops. Agronomy 10(6):831

Hong J, Yang L, Zhang D, Shi J (2016) Plant metabolomics: an indispensable system biology tool for plant science. Int J Mol Sci 17(6):767

Kumar R, Bohra A, Pandey AK, Pandey MK, Kumar A (2017) Metabolomics for plant improvement: status and prospects. Front Plant Sci 8:1302

Liu X, Locasale JW (2017) Metabolomics: a primer. Trends Biochem Sci 42(4):274–284

Muthamilarasan M, Singh NK, Prasad M (2019) Multi-omics approaches for strategic improvement of stress tolerance in underutilized crop species: a climate change perspective. Adv Genet 103:1–38

Nadakuduti SS, Enciso-Rodríguez F (2020) Advances in genome editing with CRISPR systems and transformation technologies for plant DNA manipulation. Front Plant Sci 11

Pinu FR, Goldansaz SA, Jaine J (2019) Translational metabolomics: current challenges and future opportunities. Metabolites 9(6):108

Ramalingam A, Kudapa H, Pazhamala LT, Weckwerth W, Varshney RK (2015) Proteomics and metabolomics: two emerging areas for legume improvement. Front Plant Sci 6:1116

Razzaq A, Sadia B, Raza A, Khalid Hameed M, Saleem F (2019) Metabolomics: a way forward for crop improvement. Metabolites 9(12):303

Rubert J, Zachariasova M, Hajslova J (2015) Advances in high-resolution mass spectrometry based on metabolomics studies for food–a review. Food Addit Contam Part A 32(10):1685–1708

Sousa Silva M, Cordeiro C, Roessner U, Figueiredo A (2019) Metabolomics in crop research—current and emerging methodologies. Front Plant Sci 10:1013

Tohge T, Fernie AR (2015) Metabolomics-inspired insight into developmental, environmental and genetic aspects of tomato fruit chemical composition and quality. Plant Cell Physiol 56(9):1681–1696

Upadhyaya P, Tyagi K, Sarma S, Tamboli V, Sreelakshmi Y, Sharma R (2017) Natural variation in folate levels among tomato (Solanum lycopersicum) accessions. Food Chem 217:610–619

Van Emon JM (2016) The omics revolution in agricultural research. J Agric Food Chem 64(1):36–44

Wang J, Sun L, Xie L, He Y, Luo T, Sheng L et al (2016) Regulation of cuticle formation during fruit development and ripening in 'Newhall' navel orange (Citrus sinensis Osbeck) revealed by transcriptomic and metabolomic profiling. Plant Sci 243:131–144

Yun Z, Gao H, Liu P, Liu S, Luo T, Jin S et al (2013) Comparative proteomic and metabolomic profiling of citrus fruit with enhancement of disease resistance by post-harvest heat treatment. BMC Plant Biol 13:44. https://doi.org/10.1186/1471-2229-13-44

Part III
Applications of CRISPR/Cas Technology for Biotic and Abiotic Stress Tolerance

Genome Editing in Plants for Resistance Against Bacterial Pathogens

Kalpesh Yajnik, Rajesh Mehrotra, and Purva Bhalothia

Abstract Agricultural sector serves the ground for global food security, which prioritizes the need to enhance the yield of crops and protect them from phyto-pathogens. These pathogens pose a threat by damaging the plantations and reducing their produce. CRISPR/Cas9, a foremost genome editing tool, corresponding to other techniques, is capable of inducing insertions or deletions (indels) and alters the genomic sequence by specific nucleases. Also, it has proven to be successful in providing resistance from phytopathogens by editing site-specific genes – susceptible or resistant. The technique employs nucleases at targeted sites, inducing cuts in DNA which are then repaired by non-homologous end-joining (NHEJ) or homology-directed repair (HDR) repair pathway, leading to the production of genetically modified plants. In spite of being highly efficient, CRISPR/Cas9 system needs to overcome certain uncertainties such as off-target effects, approval, and regulation of genetically altered plants. The chapter provides a detailed rundown toward the application of CRISPR/Cas9 in increasing the resistance in plants against bacterial pathogens and dealing with restrictions associated with the technique.

Keywords Phytopathogens · CRISPR/Cas9 · Indels · HR · NHEJ · Off-targets

K. Yajnik
Dr. B. Lal Institute of Biotechnology, Jaipur, India

R. Mehrotra
Department of Biological Sciences, BITS Pilani, Goa, India

P. Bhalothia (✉)
Birla Institute of Scientific Research (BISR), Jaipur, India

© The Author(s), under exclusive license to Springer Nature
Switzerland AG 2022
S. H. Wani, G. Hensel (eds.), *Genome Editing*,
https://doi.org/10.1007/978-3-031-08072-2_12

217

1 Introduction

The field of agriculture is the major source of food required to support the increasing population. The global population is estimated to reach around 9.5 billion by 2050 (Clarke and Zhang 2013). An increment in the production of food became an evident step to cope with the rising population. The evaluation done by Godfray et al. (2010) showed the need of the food production to be increased by a minimum of 70%, to cope with the needs of increasing population and ensure global food security. Crops have to adapt to varying climatic changes and gain tolerance against biotic and abiotic stress for a good yield (Haque et al. 2018). Pathogenic diseases have proven to be the major reasons for the loss in the yield of crops by reducing the plant growth, resulting in deteriorating quality of the products (Yin and Qiu 2019). The use of pesticides has shown detrimental effects to the habitat, thus disturbing the natural ecosystem. Also, an increase in the phytopathogen resistance of pesticide demands an alternative need to reduce the agricultural and economic loses (Damalas and Eleftherohorinos 2011).

In response to pathogenic attack, plants carry out numerous defense networks, consisting of two-layer immune response (Jones et al. 2016). The cellular receptors recognize pathogens and counteract their activity by resistance (R) gene-arbitrated immunity. Previous studies (Lapin and Van den Ackerveken 2013; Pavan et al. 2010) reported the role of susceptibility (S) genes in combating the resistance mediated by R genes, by either negatively regulating the immunity or encoding proteins in host; when monitored by pathogen, it can restrain the immune response (Langner et al. 2018). In due course, pathogens have evolved in their effector secretions which enabled them to adapt in the host environment (Dodds and Rathjen 2010; Vleeshouwers et al. 2011; Win et al. 2012). The scientific community aims to address this issue by engineering variants of nucleotide-binding oligomerization domain (NOD)-like receptors (NLR), encoded by R genes. The NLRs designed will have the ability to perceive broader spectrum of pathogen effectors (Harris et al. 2013; Giannakopoulou et al. 2015; Kim et al. 2016). A comprehensive study of NLRs and its role in plant defense responses can provide an opportunity to exploit the functioning of NLRs, opening new possibilities in development of synthetic immune receptors (Langner et al. 2018).

Developing mutant cell lines are essential for strain improvement and higher yield in crops. Conventional practices involving screening of libraries were extensive but tedious. Advancement in the field of recombinant DNA technology brought a transient change in the crop varieties (Langner et al. 2018). Genome modification done by zinc-finger nucleases (ZFN) (Maeder et al. 2008), transcription activator-like effector nucleases (TALENs) (Boch et al. 2009; Bogdanove and Voytas 2011), and clustered regularly interspaced short palindromic repeat (CRISPR)/CRISPR-associated protein 9 (Cas9) (Jinek et al. 2012) has proven to be successful in developing targeted mutations. CRIPSR/

Cas9, being the most efficient gene modification tool, has its current application in plant pathology, that is, to develop disease resistance in plants by either targeting the desired gene directly or by regulating the immune response indirectly.

The current chapter comprises an application of CRISPR/Cas9 genome editing in the field of phytopathology, for enhancing the resistance in plants against bacterial pathogen. It also focuses on interaction of genes in plants during pathogen infection and provides approaches to improve resistance in plants.

2   CRISPR/Cas9 Genome Editing

CRISPR/Cas9 genomic editing tool has proven to be the most successful innovation for editing the genomic sequence and recognizing the role of each gene. This technique is adapted from the immune system of bacteria and certain archaebacteria (Jinek et al. 2012). It performs breakage of foreign DNA/RNA by acquiring DNA fragments of invading plasmids and bacteriophages and transcribes them into CRISPR RNA (crRNA), which base pairs with foreign DNA/RNA (Garneau et al. 2010; Makarova et al. 2015). The genome editing machine requires two major components for definite editing: DNA-binding domain and an effector domain which allows breaks in DNA (Wang et al. 2016b). Double-stranded breaks (DSBs) are generated by sequence-specific nucleases (SSN), which further activate DNA repair system and increase the chance of gene alteration in the target sequence. Cas9 is one such nuclease, which interacts and hybridizes with guide RNA (gRNA)-DNA sequence along with protospacer adjacent motif (PAM) in the DNA, making Cas9 an ideal tool for site-specific genome editing (Gasiunas et al. 2012; Jinek et al. 2012).

CRISPR/Cas is classified into three major classes, I, II, and III, based on their components and working mechanism. CRISPR editing system follows three steps to initiate an immune response against foreign DNA. The first step, termed as acquisition, involves the fragments of foreign DNA (called protospacers) introduced in the locus of CRISPR as spacers, between CRISPR-RNAs (crRNAs) of the host. Except in type III, type I and II Crispr/Cas systems depend on protospacer adjacent motif (PAM) for introduction of protospacers (Marraffini and Sontheimer 2010). The second step involves expression of Cas proteins. The CRISPR sequence accommodating protospacers is then transcribed to pre-crRNA and then processed to form crRNA. Mature crRNA now contains spacer sequence, which targets foreign genome, and crRNA sequence, which interacts with Cas proteins and other components of RNA (Deltcheva et al. 2011).

In type I class, CRISPR-associated complex for antiviral defense (Cascade) forms a complex with pre-crRNA and is then processed into crRNA via Cas6. The matured crRNA guides the Cascade complex toward the target site and cleaves the foreign DNA by recruiting Cas3 (Makarova et al. 2011). The type III Crispr/Cas system also gets processed by Cas6, and matured crRNA then forms a complex with Csm and Cmr in type III A and III B systems, respectively. The subtypes type III A cleave the target DNA via Cas6 nuclease, and type III B cleaves the foreign RNA by Cas-crRNA ribonucleoprotein complex (Marraffini and Sontheimer 2008; Hale et al. 2009).

The chapter discusses the use of type II CRISPR system. It involves hybridization of trans-activating CRISPR RNA (tracrRNA) with crRNA sequence, to initiate the processing of crRNA, Cas9 binding, and cleavage of desired sequence via Cas9. Finally, third step involves targeting of Cas9 proteins via crRNA, causing destruction of foreign genome. The specificity of CRSIPR/Cas9 system is due to the presence of protospacer adjacent motif (PAM), found next to the target sequence in the foreign genome, to which crRNA binds (Mojica et al. 2009; Shah et al. 2013). The technique described is now manipulated for gene editing in numerous plant species too (Liang et al. 2014; Brooks et al. 2014; Zhang et al. 2016; Chilcoat et al. 2017). Initially, sgRNA targets Cas9 at a specific sequence forming a DSB and initiates DNA repair process via two different pathways: nonhomologous end joining (NHEJ) pathway, which causes gene knockout by inducing insertions or deletions (indels) at DSB, and homology-directed repair (HDR) pathway, which provides an opportunity of gene replacement at DSB through homologous recombination using donor template, thus creating gene specific mutation, deletion, or insertion (Rudin et al. 1989; Choulika et al. 1995).

3 Plant-Pathogen Interaction

A series of multiple events initiates during plant-pathogen interaction (Silva et al. 2018). Here, the type of interacting pathogen determines the recruitment of biomolecules. Bacteria involve II, III, and IV types of secretion system and promote the infection (Kamber et al. 2017). Generally, the defense system of plants recruits certain receptors which recognize target pathogens in extracellular regions and in cell cytoplasm (Dodds and Rathjen 2010). Trans-membrane pattern recognition receptors (PRRs) present at the surface of the cell recognize specific microbes named as pathogen-associated molecular patterns (PAMPs) and pathogenic proteins delivered via apoplast, thus initiating PRR-triggered immunity (PTI) (Win et al. 2012). For the infection to spread, microbes have to make a suitable interaction to feed on nutrients of the hosts. The susceptibility (S) gene then supports infection by suppressing PTI, and the process is referred as effector triggered susceptibility (ETS) (Andersen et al. 2018).

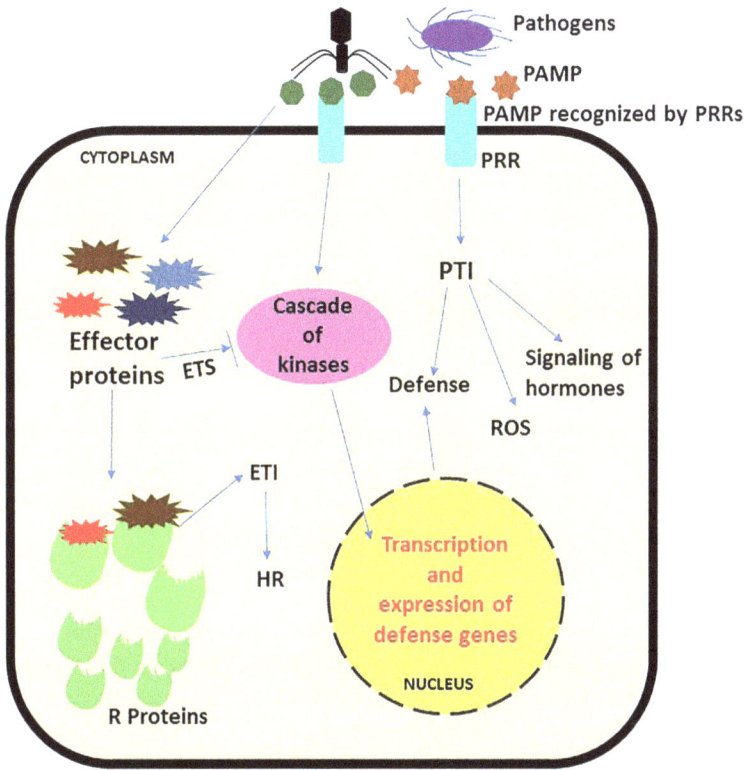

Fig. 1 A model presenting the defense mechanism in plants during infection. The defense signaling and PAMP/PTI responses are initiated by PAMP association via PRRs. Also, activation of ETI begins through *R* proteins. Expression of *R* and defense genes in response to PTI and ETI are regulated by upregulation or downregulation of certain transcription factors, hormones, reactive oxidation species (ROS), and hypersensitive reactions (HR). (The figure has been adapted and modified from Nejat et al. (2017) with the permission from Taylor & Francis with license number- 4980790305745)

To combat with effectors, plants commence second round of defense mechanism by resistance (*R*) genes. These genes in cytoplasm encode intracellular leucine-rich repeat-containing/NOD-like receptors (NLRs) along with nucleotide-binding domain and perceive pathogen effectors. This, in turn, triggers NLR/effector-triggered immunity (NTI/ETI) and activates plant immune system (Dodds and Rathjen 2010; Thomma et al. 2011). Figure 1 depicts a model of plant defense mechanism during infection. Therefore, by modifying the functional regions of *S* genes via genome editing approaches, desired mutants conferring pathogenic resistance can be obtained (Yin and Qiu 2019). Similarly, *R* genes can be engineered to other susceptible plant varieties.

4   CRISPR/Cas9-Mediated Editing of *S* Genes

Jiang et al. (2013) reported a favorable expression of CRISPR/Cas9 in plants. The report presented three versions of Cas9/sgRNA system, delivered by polyethylene glycol (PEG) or *Agrobacterium* in *Arabidopsis*, tobacco, rice, and sorghum. On targeting the *S* genes, *OsSWEET14* and *OsSWEET11*, at the promoter sites of bacterial blight in rice via CRISPR/Cas9, showed mutagenesis. Thus, the experiment demonstrated the use of CRISPR/Cas9 genome editing tool in plants and other crop varieties.

Similar experiments employing CRISPR/Cas9 technology in wheat to induce mutations in *S* gene, *TaMLO-A1*, encoding *Mildew Locus O* (*MLO*) proteins, presented a surge in resistance to counter powdery mildew (Wang et al. 2014). Malnoy et al. (2016) also targeted *S* gene, *MLO-7*, via CRISPR/Cas9 ribonucleoproteins in protoplast cells of apple to increase resistance against fire blight.

Biotic factors possess a great impact in the production of citrus. Diseases caused by *Xanthomonas citri* ssp. citri (*Xcc*) and *Candidatus* Liberibacter asiaticus are the main causes of citrus canker and Huanglongbing (HLB), respectively (Jia et al. 2017). An *S* gene, *Cs*LOB1, is responsible for citrus disease and is a part of lateral organ boundaries domain (LBD) family of transcription factors in plants. *Xcc* encodes transcription activator-like (TAL) effectors which interacts effector binding element (EBE) in the promoter region of *Cs*LOB1 and induces the expression of susceptibility genes (Hu et al. 2014). Thus, EBE may act as a target gene for genetic engineering in inducing resistance against citrus canker. Jia et al. (2017) conducted the study incorporating CRISPR/Cas9 to alter *Cs*LOB1 gene in Duncan grapefruit. Out of six transgenic lines, $D_{LOB}9$ and $D_{LOB}10$ resulted in the highest mutation rate of 89.36% and 88.79%, respectively. The results were then verified when pustules produced by *Xcc* did not show symptoms of canker in $D_{LOB}9$ and $D_{LOB}10$ transgenic lines. The study opened a new gateway to develop disease-resistant varieties of citrus via CRISPR/Cas9 technology. Similar data presented in Table 1 comprises the list of CRISPR-mediated editing of genes associated with disease resistance against bacterial pathogen.

5   Adoption of Wider Approaches to Improve Resistance in Plants

The development of new techniques made identification of desired genes and their modification easier, without harming the yield of the crops. Currently, advancement in the next-generation sequencing (NGS) has simplified the procedures of differential expressions of gene during plant-pathogen interaction (Buermans and Den Dunnen 2014). In silico studies involving transcriptome analysis of pathogen and plants provide a detailed interpretation of their pathogenic and immune responses (Mushtaq et al. 2019). The section elaborates role of phytohormones, RNAi, and

Table 1 Genome editing of target genes involved in disease resistance against bacterial pathogen via CRISPR/Cas9

Crop	Gene targeted	Pathogen involved	Mutation induced	Remarks	References
Apple cultivar	DIPM-1, DIPM-2, DIPM-4	Erwinia amylovora	Indels	Direct transfer of CRISPR/Cas9 ribonucleoproteins (RNPs) in apple protoplast was achieved and targeted mutations obtained might provide resistance against fire blight	Malnoy et al. (2016)
Citrus sinensis	CsLOB1	Xanthomonas citri subspecies citri	Indels	Mutation in the promoter region of CsLOB1 is necessary to develop resistance in plants	Jia et al. (2016)
Citrus sinensis	CsLOB1	Xanthomonas citri subspecies citri	Indels	When inoculated with pustules, $D_{LOB}9$ and $D_{LOB}10$ transgenic lines did not show symptoms of citrus canker	Jia et al. (2017)
Citrus sinensis	CsLOB1	Xanthomonas citri subspecies citri	Indels	By editing promoter region of CsLOB1, four mutant lines obtained revealed an increase in resistance to canker disease	Peng et al. (2017)
Citrus sinensis	CsWRKY22	Xanthomonas citri subspecies citri	Indels	Three mutant lines, W1, W2, and W3, showed mutation efficiency of 85.7%, 79.2%, and 68.2%, respectively. On evaluation of resistance, the susceptibility against bacterial canker declined in these plants	Wang et al. (2019a, b)
Oryza sativa	OsSWEET11, OsSWEET14	Xanthomonas oryzae	Indels	Disruption of promoter of S gene was achieved	Jiang et al. (2013)
Oryza sativa	OsMPK5	Burkholderia glumae	Indels	Mutation efficiency was estimated to be low. Also, off-target effects were imperfectly matched. These results did not bring light about resistance in rice	Xie and Yang (2013)
Oryza sativa	OsSWEET13	Xanthomonas oryzae pv. oryzae	Indels	Gene profiling of strains revealed the induction of OsSWEET13 by X. oryzae and therefore support the basis that PthX02 targets OsSWEET13	Zhou et al. (2015)
Oryza sativa	Os8N3	Xanthomonas oryzae	Indels	The homozygous mutants developed displayed enhanced resistance against Xoo	Kim et al. (2019)

(continued)

Table 1 (continued)

Crop	Gene targeted	Pathogen involved	Mutation induced	Remarks	References
Oryza sativa	*OsSWEET11, OsSWEET13, OsSWEET14*	*Xanthomonas oryzae* pv. *oryzae*	Indels	Editing was confirmed by analyzing sequence of TALe genes in 63 different strains of *Xoo*, revealing multiple variants for SWEET13. Five mutations at promoter region were also introduced in rice lines. Paddy trails confirmed the broad-spectrum resistance in edited promoter region of SWEET	Oliva et al. (2019)

other transgenic approaches of genome editing to enhance disease resistance in plants.

CRISPR/Cas9 system can be employed in increasing resistance by HDR-mediated transfer of *R* genes to stress-responsive host plants. This approach results in gain-of-function mutation via CRISPR/Cas9 (Schenke and Cai 2020). The reports of Bonardi et al. (2011) and Wu et al. (2017) discussed that NLR genes work in pairs as helper NLR and sensor NLR in perceiving pathogen effectors. To improve resistance, transferring and expressing *R* genes in response to pathogen attack could give positive results. But due to limited knowledge of *R* genes, identification of helper NLR using Illumina sequencing was carried out, which provided detailed information of its function (Wang et al. 2019a, b). Another alternative involves transfer of PRRs, as seen in *Arabidopsis* and tomato (Lacombe et al. 2010), *Arabidopsis* and wheat (Schoonbeek et al. 2015), and rice and banana (Tripathi et al. 2014). It is because of the conserved pathways of immune signaling between dicots and monocots (Holton et al. 2015).

6   RNA Silencing

RNA silencing or RNA interference (RNAi) is mediated by a set of noncoding RNAs which protect plants from pathogenic infection by controlling the expression of genes. The mechanism of RNAi initiates with the production of small RNAs (sRNAs), with the help of proteins – Dicer-like (DCL), Argonaute (AGO), and RNA-dependent RNA polymerase (RDRs) enzyme (Baulcombe 2004; Vaucheret 2006). Here, DCL proteins produce sRNAs and integrate it into RISCs (Vaucheret 2008). AGO protein, than from a complex with RISCs, interacts with similar RNAs, regulates DNA methylation, and inhibits translational activities of mRNAs (Voinnet 2009). With the mentioned properties of RNAi, researchers proposed diverse roles of its genes, DCLs, AGOs, and RDRs in plant defenses

(Xie et al. 2004; Cao et al. 2016; Wang et al. 2011; Katiyar-Agarwal et al. 2007; Jauvion et al. 2012).

The primary targets of RNAi system are viruses. Studies have shown a correlation in transcription of RNAi genes and virus inoculation in plants. The examples include upregulation of DCLs, RDRs, and AGOs by tomato yellow leaf curl virus (TYLCV) (Bai et al. 2012) and increased expression of RDR1, RDR2, and RDR3 by cucumber mosaic virus (CMV) in *Salvia miltiorrhiza* (Shao and Lu 2014). Findings revealed the role of RNAi in response to bacterial infections too. Reports submitted by Pumplin and Voinnet (2013) and Staiger et al. (2013) presented that during pathogen infection, multiple siRNAs and miRNAs were in tune with PTI and ETI. miR393 is one such transcription factor which confers antibacterial immunity with the help of AGO1 and AGO2, and both the proteins show some correlation with PTI and ETI (Zhang et al. 2011).

Wagh et al. (2016) tested an RNA silencing factor, *RDR6*, to understand its role in providing defense against bacterial pathogens. They conducted a study to test the expression of *OsRDR6* during pathogen attack, and whether *OsRDR6* affects the susceptibility of disease when inoculated with bacterial pathogens using *shl2-rol* mutant. The result showed that the expression levels of *OsRDR6*, when induced by *shl2-rol* mutant, had more susceptibility than wild ones, thus suggesting its roles in defense responses. But some findings also revealed that certain plant viruses and bacterial effector proteins have evolved many counter-defense systems to suppress RNAi machinery (Voinnet 2005; Navarro et al. 2008). Therefore, a constitutive expression of CRISPR/Cas9 system, targeting desired genes, can build an advanced immune system (Ji et al. 2015; Mushtaq et al. 2020).

7 Phytohormones

We classify plant hormones as auxins, gibberellins (GA), abscisic acid (ABA), cytokinin (CK), ethylene (ET), jasmonates (JA), salicylic acid (SA), strigolactones, and brassinosteroids. Among them, ABA is responsible in monitoring abiotic stress responses in plants (Lata and Prasad 2011), while SA, JA, and ET show response against pathogen infection (Bari and Jones 2009). Studies conducted by Bari and Jones (2009), Navarro et al. (2008), and Nishiyama et al. (2013) provided the evidence for molecular crosstalk of these hormones with GAs, auxins, and CKs.

SA is believed to be associated during initial defense responses against biotrophic and hemi-biotrophic microbes (Loake and Grant 2007), while JA and ethylene show defense response against necrotrophic pathogens and insects eating herbivores (Wasternack and Hause 2013; Gamalero and Glick 2012). In response to pathogen attack, SA induces broad-spectrum resistance defined as systemic acquired resistance (SAR) at the site of infection. The increasing levels of SA in infected tissues induce *pathogenesis-related* (*PR*) genes. These genes encode numerous proteins having antimicrobial properties, thus increasing resistance against several pathogens (van Loon et al. 2006).

Similar changes in the level of JA were also observed during pathogen invasion (Wasternack and Hause 2013). Data from the reports by Lee et al. (2010) and Mizoi et al. (2012) revealed that the members of the family *APETALA2/ETHYLENE-RESPONSIVE-FACTOR* (*AP2/ERF*) are known to participate in JA-mediated stress responses. Further, a protein, *JASMONATE-JIM-DOMAIN* (*JAZ*), is known to play an important role in biotic and abiotic stresses. In its inactive form, it forms a complex with *JIN1/MYC2* and blocks regulation of JA-responsive genes, while, in the presence of JA, JA-Ile interacts with an F-box protein and degrades *JAZ*, permitting upregulation of JA-associated genes by myelocytomatosis (*MYC36*).

ET also plays distinct roles in activating defense responses (Bari and Jones 2009; Gamalero and Glick 2012). It can interact with the pathways of SA and JA and can upregulate or downregulate their expression to attain customized defense responses. *ETHYLENE-RESPONSIVE-FACTOR1* (*ERF1*) is one such modulator of ethylene responses during infection (Berrocal-Lobo et al. 2002). Solano et al., in 1998, reported the regulation of *ERF1* by *EIN3*, which consecutively monitors the expression of pathogen-related genes. Therefore, the experiment was conducted in which the ethylene-responsive genes were overexpressed in *Arabidopsis*. An enhanced expression of *ERF1* was observed in *Arabidopsis*, when infected by *B. cinerea*, thus conferring resistance against necrotrophic pathogens. These findings suggest that manipulation of the mentioned genes by CRISPR/Cas9 editing tool can increase resistance against pathogens.

Rice suffers severe damage by fungus *Magnaporthe oryzae* (Dean et al. 2012; Liu et al. 2014). As mentioned that *ERF* performs a very imperative role in stress responses, CRISPR/Cas9-mediated knockout of *ERF* gene, *OsERF922*, was performed by Wang and colleagues in 2016a, b. Out of 50 T_0 transgenic plants, 21 mutations were identified. The results showed a considerable decrease in the lesions, when homozygous mutant culture lines were compared with the wild ones. Consecutively, multiple sites in *OsERF922* were also targeted by Cas9/multi-target sgRNAs to achieve desired mutations. The above results served as a base for further investigation. *Solanum lycopersicum* belongs to the family of Solanaceae and has one of the biggest turnovers in the market. Regardless of such economic value, pathogens such as *Pseudomonas syringae* and *Xanthomonas* spp. possess a major threat in their production (Schwartz et al. 2015). Zeilmaker et al. (2015) published in their paper that a single gene mutation in *DMR6* (*Downy Mildew Resistance 6*) linked with SA homeostasis resulted in production of resistant *Arabidopsis* plants. Therefore, the study employing CRISPR/Cas9 system was conducted to modify *SlDMR6-1* orthologue *Solanum lycopersicum* 03g080190. Frameshift deletions and premature truncation introduced in *SlDMR6-1* showed resistance against pathogens as *P. syringae*, *P. capsica*, and some species of *Xanthomonas* (de Toledo Thomazella et al. 2016).

Genomic studies showed upregulation of many transcription factors (TFs) during leaf senescence (Woo et al. 2016). WRKY is one such TF which plays an important role in senescence in leaves (Li et al. 2012). Guo et al. (2017) took *WRKY75* for investigation to identify its role in regulating SA production and resistance against diseases. The CRISPR/Cas9-mediated genome editing was performed to generate mutant *Arabidopsis wrky75* alleles, *wrky75-c1*, and *wrky75-c2*. This mutation created a premature stop codon. The results displayed a decrease in the level of SA, downregulation of expression in SA *induction-deficient 2* (*SID2*), and increased disease severity in the mutant cell lines, with respect to the wild ones, when infected with *P. syringae* pv *tomato*. Moreover, overexpression of *Arabidopsis WRKY75* (*WRKY75ox*) using 35S promoter showed an increase in the levels of SA, upregulation in the expression of *SID2*, and enhanced resistance against *P. syringae*. The analysis also revealed a positive correlation of *WRKY75* in disease resistance in *Arabidopsis* by altering SA levels (Guo, et al. 2017).

Among the *Pseudomonas* species, *P. syringae* pv *tomato* DC3000 (*Pto* DC3000) is considered as the main biological pathogen of bacterial speck in tomato (Blancard 2012). The early process of infection depends on either natural opening as stomata or any accidental wounds (Melotto et al. 2017). The opening/closure of stomata is regulated by SA and ABA. Hence, they inhibit the entry of pathogens (Zeng and He 2010; Zhang et al. 2008; Melotto et al. 2006). As mentioned by Robert-Seilaniantz et al. (2011), SA positively regulates the resistance against biotrophic and hemi-biotrophic pathogens, whereas JA in association with ET provides resistance contrary to necrotrophic pathogens. *Pto* DC3000 produces a mimic of JA-Ile (Fonseca et al. 2009), coronatine (COR), which stimulates the opening of stomatal aperture and promotes bacterial colonization (Brooks et al. 2005; Laurie-Berry et al. 2006; Zheng et al. 2012; Melotto et al. 2006). The study was conducted, reporting a solution by uncoupling the hostility between SA-JA pathways and producing tomatoes resistant to *Pto* DC3000. The functional analog of *AtJAZ2* in tomato, *SlJAZ2*, was truncated via CRISPR/Cas9 editing tool. Therefore, this gain-of-function mutation completely inhibited the reopening of stomata by COR, thus enhancing resistance against infection. Also, stomatal apertures were found to be normal at the time of transpiration in these edited plants (Ortigosa et al. 2019).

An experimental analysis done by Zhou et al. (2017) showed that the changes in level of expression of *CsWRKY22* brought about by *Xanthomonas citri* subsp. citri (*Xcc*) in sensitive Newhall nucellar navel orange were higher when compared to resistive calamondin, indicating a negative correlation of *CsWRKY22* with respect to citrus resistance from canker. Based on this research, CRISPR/Cas9-mediated genome editing was performed, which targeted *CsWRKY22* in Wanjincheng orange to enhance resistance against citrus canker. Four sgRNAs, targeting the first exon of *CsWRKY22*, were selected, and out of them, two sgRNAs revealed cleavage activities. Three out of seven transgenic lines, W-1,

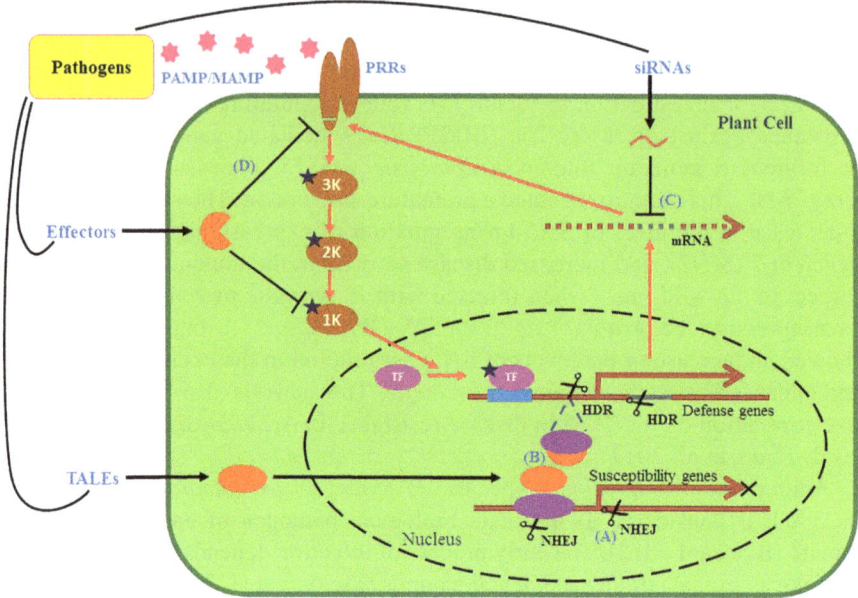

Fig. 2 Development of non-transgenic plants by mutating target sites via CRISPR/Cas9 to engineer resistance against pathogens. (**a–d**) During pathogen infection, plant induces defense responses, for example, cascade of MAP kinases, which causes activation of transcription factors and certain defense-responsive genes. (**a**) Frame shift mutation of *S* genes via NHEJ, during pathogen infection; (**b**) elimination of cis-element in the promoter region using NHEJ to inhibit the triggering of *S* genes in cis-element/TALE introduction via HDR, thus triggering the activation of defense genes; (**c**) blocking of cross-kingdom RNAi via HDR-mediated editing of pathogenic siRNA target sequence; (**d**) HDR-mediated editing of sequences to change amino acid residues needed by effectors for interaction, thus hindering further modifications. (The figure has been adapted and modified from Schenke and Cai (2020) with the permission from Elsevier with license number- 4980790799252)

W-2, and W-3, have mutation efficiencies of 85.7%, 79.2%, and 68.2% respectively. Complete evaluation of mutated plants displayed a lower susceptibility toward citrus canker, thus indicating that genome modification via CRISPR/Cas9 is proven to be the more effective tool for increasing resistance against bacterial pathogen (Wang et al. 2019a, b).

8 Conclusions and Future Perspectives

With the advancement in the high-throughput sequencing mechanisms, the molecular pathways and their interaction can be easily elucidated. Therefore, the genome editing methods for increasing the resistance in plants against bacterial pathogens can be precisely applied. CRISPR/Cas9 editing tool has proven to be the current and most successful genome-modifying tool, by inducing targeted mutations. The summarized work includes targeting the desired genes directly by involving *S* genes, indirectly by the phytohormones, and other methods of genome editing to increase resistance against biotic stress in plants.

Production of genetically modified (GM) crops has increased in Asia-Pacific countries. Data presented by Gupta et al. (2014) summarized an increase in the use GM events in Asia-Pacific regions, especially for animal feed and food purposes. However, funding and showing the level of interest in research and development by public sectors is comparatively low than certain private institutions, which automatically reduces the competitive skills of public institutions to produce and commercialize GM crops. Also, the safety issues concerning the regulation of genome edited crops are still a subject of discussion and under evaluation in many European countries (Callaway 2018). The similar concern of regulation of GM products is also under evaluation by USDA regulatory partners (Jaganathan et al. 2018). A shift toward non-transgenic mutations induced by CRISPR/Cas9 system can be a better approach for increasing resistance. The deletion or introduction of cis-elements in promoter regions of *S* or *R* genes, respectively, can lead to the increase in resistance, as presented in Fig. 2.

Works done by Oliva et al. (2019) and Jia et al. (2016) presented non-transgenic approach by targeting promoter regions of *OsSWEET* genes. The method also ensures that regular spacing of cis-elements in the promoter region will not get interrupted by indels. However, introducing new cis-elements in the promoter region of defense-responsive genes could be effective in initiating a response against new pathogens, as mentioned by Rivas and Genin (2011). With these advancements, the gap in the research for increasing resistance against bacterial pathogens in plants can be covered and non-transgenic methods might also fix the issues concerned with fitness costs in addition to GMOs.

References

Andersen EJ, Al S, Byamukama E, Yen Y, Nepal MP (2018) Disease resistance mechanisms in plants. Genes 9:339

Bai M, Yang GS, Chen WT, Mao ZC, Kang HX, Chen GH et al (2012) Genome-wide identification of Dicer-like, Argonaute and RNA-dependent RNA polymerase gene families and their expression analyses in response to viral infection and abiotic stresses in *Solanum lycopersicum*. Gene 501:52–62

Bari R, Jones JD (2009) Role of plant hormones in plant defense responses. Plant Mol Biol 69:473–488

Baulcombe D (2004) RNA silencing in plants. Nature 431:356–363

Berrocal-Lobo M, Molina A, Solano R (2002) Constitutive expression of ETHYLENE-RESPONSE-FACTOR1 in *Arabidopsis* confers resistance to several necrotrophic fungi. Plant J 29:23–32

Blancard D (2012) Tomato diseases: identification, biology and control: a colour handbook. CRC Press

Boch J, Scholze H, Schornack S, Landgraf A, Hahn S, Kay S et al (2009) Breaking the code of DNA binding specificity of TAL-type III effectors. Science 326:1509–1512

Bogdanove AJ, Voytas DF (2011) TAL effectors: customizable proteins for DNA targeting. Science 333:1843–1846

Bonardi V, Tang S, Stallmann A, Roberts M, Cherkis K, Dangl JL (2011) Expanded functions for a family of plant intracellular immune receptors beyond specific recognition of pathogen effectors. Proc Natl Acad Sci 108:16463–16468

Brooks DM, Bender CL, Kunkel BN (2005) The *Pseudomonas syringae* phytotoxin coronatine promotes virulence by overcoming salicylic acid-dependent defenses in *Arabidopsis thaliana*. Mol Plant Pathol 6:629–639

Brooks C, Nekrasov V, Lippman ZB, Van Eck J (2014) Efficient gene editing in tomato in the first generation using the clustered regularly interspaced short palindromic repeats/CRISPR-associated 9 system. Plant Physiol 166:1292–1297

Buermans HPJ, Den Dunnen JT (2014) Next generation sequencing technology: advances and applications. BBA-Mol Basis Dis 1842:1932–1941

Callaway E (2018) EU law deals blow to CRISPR crops. Nature 560:16

Cao JY, Xu YP, Li W, Li SS, Rahman H, Cai XZ (2016) Genome-wide identification of Dicer-like, Argonaute, and RNA-dependent RNA polymerase gene families in Brassica species and functional analyses of their Arabidopsis homologs in resistance to *Sclerotinia sclerotiorum*. Front Plant Sci 7:1614

Chilcoat D, Liu ZB, Sander J (2017) Use of CRISPR/Cas9 for crop improvement in maize and soybean. Progress in molecular biology and translational science. Academic Press, pp 27–46

Choulika A, Perrin A, Dujon B, Nicolas JF (1995) Induction of homologous recombination in mammalian chromosomes by using the I-SceI system of Saccharomyces cerevisiae. Mol Cell Biol 15:1968–1973

Clarke JL, Zhang P (2013) Plant biotechnology for food security and bioeconomy. Plant Mol Biol 83:1–3

Damalas CA, Eleftherohorinos IG (2011) Pesticide exposure, safety issues, and risk assessment indicators. Int J Environ Res Public Health 8:1402–1419

de Toledo Thomazella DP, Brail Q, Dahlbeck D, Staskawicz B (2016) CRISPR-Cas9 mediated mutagenesis of a DMR6 ortholog in tomato confers broad-spectrum disease resistance. BioRxiv 064824

Dean R, Van Kan JA, Pretorius ZA, Hammond-Kosack KE, Di Pietro A, Spanu PD et al (2012) The Top 10 fungal pathogens in molecular plant pathology. Mol Plant Pathol 13:414–430

Deltcheva E, Chylinski K, Sharma CM, Gonzales K, Chao Y, Pirzada ZA et al (2011) CRISPR RNA maturation by trans-encoded small RNA and host factor RNase III. Nature 471:602–607

Dodds PN, Rathjen JP (2010) Plant immunity: towards an integrated view of plant–pathogen interactions. Nat Rev Genet 11:539–548

Fonseca S, Chico JM, Solano R (2009) The jasmonate pathway: the ligand, the receptor and the core signalling module. Curr Opin Plant Biol 12:539–547

Gamalero E, Glick BR (2012) Ethylene and abiotic stress tolerance in plants. In: Environmental adaptations and stress tolerance of plants in the era of climate change. Springer, New York, NY, pp 395–412

Garneau JE, Dupuis MÈ, Villion M, Romero DA, Barrangou R, Boyaval P et al (2010) The CRISPR/Cas bacterial immune system cleaves bacteriophage and plasmid DNA. Nature 468:67–71

Gasiunas G, Barrangou R, Horvath P, Siksnys V (2012) Cas9–crRNA ribonucleoprotein complex mediates specific DNA cleavage for adaptive immunity in bacteria. Proc Natl Acad Sci 109:E2579–E2586

Giannakopoulou A, Steele JF, Segretin ME, Bozkurt TO, Zhou J, Robatzek S et al (2015) Tomato I2 immune receptor can be engineered to confer partial resistance to the oomycete *Phytophthora infestans* in addition to the fungus *Fusarium oxysporum*. Mol Plant-Microbe Interact 28:1316–1329

Godfray HCJ, Beddington JR, Crute IR, Haddad L, Lawrence D, Muir JF et al (2010) Food security: the challenge of feeding 9 billion people. Science 327:812–818

Guo P, Li Z, Huang P, Li B, Fang S, Chu J, Guo H (2017) A tripartite amplification loop involving the transcription factor *WRKY75*, salicylic acid, and reactive oxygen species accelerates leaf senescence. Plant Cell 29:2854–2870

Gupta K, Karihaloo JL, Khetarpal RK (2014) Biosafety regulations for GM crops in Asia-Pacific. Asia-Pacific Consortium on Agricultural Biotechnology, New Delhi and Asia-Pacific Association of Agricultural Research Institutions, Bangkok, pp 1–160

Hale CR, Zhao P, Olson S, Duff MO, Graveley BR, Wells L et al (2009) RNA-guided RNA cleavage by a CRISPR RNA-Cas protein complex. Cell 139:945–956

Haque E, Taniguchi H, Hassan M, Bhowmik P, Karim MR, Śmiech M et al (2018) Application of CRISPR/Cas9 genome editing technology for the improvement of crops cultivated in tropical climates: recent progress, prospects, and challenges. Front Plant Sci 9:617

Harris CJ, Slootweg EJ, Goverse A, Baulcombe DC (2013) Stepwise artificial evolution of a plant disease resistance gene. Proc Natl Acad Sci 110:21189–21194

Holton N, Nekrasov V, Ronald PC, Zipfel C (2015) The phylogenetically-related pattern recognition receptors EFR and XA21 recruit similar immune signaling components in monocots and dicots. PLoS Pathog 11:e1004602

Hu Y, Zhang J, Jia H, Sosso D, Li T, Frommer WB et al (2014) Lateral organ boundaries 1 is a disease susceptibility gene for citrus bacterial canker disease. Proc Natl Acad Sci 111:E521–E529

Jaganathan D, Ramasamy K, Sellamuthu G, Jayabalan S, Venkataraman G (2018) CRISPR for crop improvement: an update review. Front Plant Sci 9:985

Jauvion V, Rivard M, Bouteiller N, Elmayan T, Vaucheret H (2012) RDR2 partially antagonizes the production of RDR6-dependent siRNA in sense transgene-mediated PTGS. PLoS One 7:e29785

Ji X, Zhang H, Zhang Y, Wang Y, Gao C (2015) Establishing a CRISPR–Cas-like immune system conferring DNA virus resistance in plants. Nat Plants 1:1–4

Jia H, Orbovic V, Jones JB, Wang N (2016) Modification of the PthA4 effector binding elements in Type I *Cs LOB 1* promoter using Cas9/sg RNA to produce transgenic Duncan grapefruit alleviating XccΔpthA4: dCs LOB 1.3 infection. Plant Biotechnol J 14:1291–1301

Jia H, Zhang Y, Orbović V, Xu J, White FF, Jones JB, Wang N (2017) Genome editing of the disease susceptibility gene *Cs LOB 1* in citrus confers resistance to citrus canker. Plant Biotechnol J 15:817–823

Jiang W, Zhou H, Bi H, Fromm M, Yang B, Weeks DP (2013) Demonstration of CRISPR/Cas9/sgRNA-mediated targeted gene modification in *Arabidopsis*, tobacco, sorghum and rice. Nucleic Acids Res 41:e188–e188

Jinek M, Chylinski K, Fonfara I, Hauer M, Doudna JA, Charpentier E (2012) A programmable dual-RNA–guided DNA endonuclease in adaptive bacterial immunity. Science 337:816–821

Jones JD, Vance RE, Dangl JL (2016) Intracellular innate immune surveillance devices in plants and animals. Science 354

Kamber T, Pothier JF, Pelludat C, Rezzonico F, Duffy B, Smits TH (2017) Role of the type VI secretion systems during disease interactions of *Erwinia amylovora* with its plant host. BMC Genomics 18:628

Katiyar-Agarwal S, Gao S, Vivian-Smith A, Jin H (2007) A novel class of bacteria-induced small RNAs in Arabidopsis. Genes Dev 21:3123–3134

Kim SH, Qi D, Ashfield T, Helm M, Innes RW (2016) Using decoys to expand the recognition specificity of a plant disease resistance protein. Science 351:684–687

Kim YA, Moon H, Park CJ (2019) CRISPR/Cas9-targeted mutagenesis of *Os8N3* in rice to confer resistance to *Xanthomonas oryzae* pv. *oryzae*. Rice 12:1–13

Lacombe S, Rougon-Cardoso A, Sherwood E, Peeters N, Dahlbeck D, Van Esse HP et al (2010) Interfamily transfer of a plant pattern-recognition receptor confers broad-spectrum bacterial resistance. Nat Biotechnol 28:365–369

Langner T, Kamoun S, Belhaj K (2018) CRISPR crops: plant genome editing toward disease resistance. Annu Rev Phytopathol 56:479–512

Lapin D, Van den Ackerveken G (2013) Susceptibility to plant disease: more than a failure of host immunity. Trends Plant Sci 18:546–554

Lata C, Prasad M (2011) Role of DREBs in regulation of abiotic stress responses in plants. J Exp Bot 62:4731–4748

Laurie-Berry N, Joardar V, Street IH, Kunkel BN (2006) The *Arabidopsis thaliana JASMONATE INSENSITIVE 1* gene is required for suppression of salicylic acid-dependent defenses during infection by *Pseudomonas syringae*. Mol Plant Microbe Interact 19:789–800

Lee SJ, Kang JY, Park HJ, Kim MD, Bae MS, Choi HI, Kim SY (2010) *DREB2C* interacts with *ABF2*, a bZIP protein regulating abscisic acid-responsive gene expression, and its overexpression affects abscisic acid sensitivity. Plant Physiol 153:716–727

Li Z, Peng J, Wen X, Guo H (2012) Gene Network Analysis and Functional Studies of Senescence-associated Genes Reveal Novel Regulators of Arabidopsis Leaf Senescence F. J Integr Plant Biol 54:526–539

Liang Z, Zhang K, Chen K, Gao C (2014) Targeted mutagenesis in *Zea mays* using TALENs and the CRISPR/Cas system. J Genet Genomics 41:63–68

Liu W, Liu J, Triplett L, Leach JE, Wang GL (2014) Novel insights into rice innate immunity against bacterial and fungal pathogens. Annu Rev Phytopathol 52:213–241

Loake G, Grant M (2007) Salicylic acid in plant defence—the players and protagonists. Curr Opin Plant Biol 10:466–472

Maeder ML, Thibodeau-Beganny S, Osia A, Wright DA, Anthony RM, Eichtinger M et al (2008) Rapid "open-source" engineering of customized zinc-finger nucleases for highly efficient gene modification. Mol Cell 31:294–301

Makarova KS, Haft DH, Barrangou R, Brouns SJ, Charpentier E, Horvath P et al (2011) Evolution and classification of the CRISPR–Cas systems. Nat Rev Microbiol 9:467–477

Makarova KS, Wolf YI, Alkhnbashi OS, Costa F, Shah SA, Saunders SJ et al (2015) An updated evolutionary classification of CRISPR–Cas systems. Nat Rev Microbiol 13:722–736

Malnoy M, Viola R, Jung MH, Koo OJ, Kim S, Kim JS et al (2016) DNA-free genetically edited grapevine and apple protoplast using CRISPR/Cas9 ribonucleoproteins. Front Plant Sci 7:1904

Marraffini LA, Sontheimer EJ (2008) CRISPR interference limits horizontal gene transfer in staphylococci by targeting DNA. Science 322:1843–1845

Marraffini LA, Sontheimer EJ (2010) Self versus non-self discrimination during CRISPR RNA-directed immunity. Nature 463:568–571

Melotto M, Underwood W, Koczan J, Nomura K, He SY (2006) Plant stomata function in innate immunity against bacterial invasion. Cell 126:969–980

Melotto M, Zhang L, Oblessuc PR, He SY (2017) Stomatal defense a decade later. Plant Physiol 174:561–571

Mizoi J, Shinozaki K, Yamaguchi-Shinozaki K (2012) AP2/ERF family transcription factors in plant abiotic stress responses. Biochim Biophys Acta Gene Regul Mech 1819:86–96

Mojica FJ, Díez-Villaseñor C, García-Martínez J, Almendros C (2009) Short motif sequences determine the targets of the prokaryotic CRISPR defence system. Microbiology 155:733–740

Mushtaq M, Sakina A, Wani SH, Shikari AB, Tripathi P, Zaid A et al (2019) Harnessing genome editing techniques to engineer disease resistance in plants. Front Plant Sci 10:550

Mushtaq M, Mukhtar S, Sakina A, Dar AA, Bhat R, Deshmukh R et al (2020) Tweaking genome-editing approaches for virus interference in crop plants. Plant Physiol Biochem 147:242–250

Navarro L, Bari R, Achard P, Lisón P, Nemri A, Harberd NP, Jones JD (2008) DELLAs control plant immune responses by modulating the balance of jasmonic acid and salicylic acid signaling. Curr Biol 18:650–655

Nejat N, Rookes J, Mantri NL, Cahill DM (2017) Plant–pathogen interactions: toward development of next-generation disease-resistant plants. Crit Rev Biotechnol 37:229–237

Nishiyama R, Watanabe Y, Leyva-Gonzalez MA, Van Ha C, Fujita Y, Tanaka M et al (2013) *Arabidopsis* AHP2, AHP3, and AHP5 histidine phosphotransfer proteins function as redundant negative regulators of drought stress response. Proc Natl Acad Sci 110:4840–4845

Oliva R, Ji C, Atienza-Grande G, Huguet-Tapia JC, Perez-Quintero A, Li T et al (2019) Broad-spectrum resistance to bacterial blight in rice using genome editing. Nat Biotechnol 37:1344–1350

Ortigosa A, Gimenez-Ibanez S, Leonhardt N, Solano R (2019) Design of a bacterial speck resistant tomato by CRISPR/Cas9-mediated editing of Sl JAZ 2. Plant Biotechnol J 17:665–673

Pavan S, Jacobsen E, Visser RG, Bai Y (2010) Loss of susceptibility as a novel breeding strategy for durable and broad-spectrum resistance. Mol Breed 25:1

Peng A, Chen S, Lei T, Xu L, He Y, Wu L et al (2017) Engineering canker-resistant plants through CRISPR/Cas9-targeted editing of the susceptibility gene *Cs LOB 1* promoter in *citrus*. Plant Biotechnol J 15:1509–1519

Pumplin N, Voinnet O (2013) RNA silencing suppression by plant pathogens: defence, counter-defence and counter-counter-defence. Nat Rev Microbiol 11:745–760

Rivas S, Genin S (2011) A plethora of virulence strategies hidden behind nuclear targeting of microbial effectors. Front Plant Sci 2:104

Robert-Seilaniantz A, Grant M, Jones JD (2011) Hormone crosstalk in plant disease and defense: more than just jasmonate-salicylate antagonism. Ann Rev Phytopathol 49:317–343

Rudin N, Sugarman E, Haber JE (1989) Genetic and physical analysis of double-strand break repair and recombination in *Saccharomyces cerevisiae*. Genetics 122:519–534

Schenke D, Cai D (2020) Applications of CRISPR/Cas to improve crop disease resistance–beyond inactivation of susceptibility factors. Iscience 101478

Schoonbeek HJ, Wang HH, Stefanato FL, Craze M, Bowden S, Wallington E et al (2015) *Arabidopsis* EF-Tu receptor enhances bacterial disease resistance in transgenic wheat. New Phytol 206:606–613

Schwartz AR, Potnis N, Timilsina S, Wilson M, Patané J, Martins J Jr et al (2015) Phylogenomics of *Xanthomonas* field strains infecting pepper and tomato reveals diversity in effector repertoires and identifies determinants of host specificity. Front Microbiol 6:535

Shah SA, Erdmann S, Mojica FJ, Garrett RA (2013) Protospacer recognition motifs: mixed identities and functional diversity. RNA Biol 10:891–899

Shao F, Lu S (2014) Identification, molecular cloning and expression analysis of five RNA-dependent RNA polymerase genes in *Salvia miltiorrhiza*. PloS One 9:e95117

Silva MS, Arraes FBM, de Araújo CM, Grossi-de-Sa M, Fernandez D, de Souza CE et al (2018) Potential biotechnological assets related to plant immunity modulation applicable in engineering disease-resistant crops. Plant Sci 270:72–84

Solano R, Stepanova A, Chao Q, Ecker JR (1998) Nuclear events in ethylene signaling: a transcriptional cascade mediated by *ETHYLENE-INSENSITIVE3* and *ETHYLENE-RESPONSE-FACTOR1*. Genes Dev 12:3703–3714

Staiger D, Korneli C, Lummer M, Navarro L (2013) Emerging role for RNA-based regulation in plant immunity. New Phytol 197:394–404

Thomma BP, Nürnberger T, Joosten MH (2011) Of PAMPs and effectors: the blurred PTI-ETI dichotomy. Plant Cell 23:4–15

Tripathi JN, Lorenzen J, Bahar O, Ronald P, Tripathi L (2014) Transgenic expression of the rice Xa21 pattern-recognition receptor in banana (*Musa* sp.) confers resistance to *Xanthomonas campestris* pv. *musacearum*. Plant Biotechnol J 12:663–673

van Loon LC, Rep M, Pieterse CM (2006) Significance of inducible defense-related proteins in infected plants. Annu Rev Phytopathol 44:135–162

Vaucheret H (2006) Post-transcriptional small RNA pathways in plants: mechanisms and regulations. Genes Dev 20:759–771

Vaucheret H (2008) Plant argonautes. Trends Plant Sci 13:350–358

Vleeshouwers VG, Raffaele S, Vossen JH, Champouret N, Oliva R, Segretin ME et al (2011) Understanding and exploiting late blight resistance in the age of effectors. Annu Rev Phytopathol 49:507–531

Voinnet O (2005) Induction and suppression of RNA silencing: insights from viral infections. Nat Rev Genet 6:206-220

Voinnet O (2009) Origin, biogenesis, and activity of plant microRNAs. Cell 136:669–687

Wagh SG, Alam MM, Kobayashi K, Yaeno T, Yamaoka N, Toriba T et al (2016) Analysis of rice RNA-dependent RNA polymerase 6 (*OsRDR6*) gene in response to viral, bacterial and fungal pathogens. J Gen Plant Pathol 82:12–17

Wang XB, Jovel J, Udomporn P, Wang Y, Wu Q, Li WX et al (2011) The 21-nucleotide, but not 22-nucleotide, viral secondary small interfering RNAs direct potent antiviral defense by two cooperative argonautes in Arabidopsis thaliana. Plant Cell 23:1625–1638

Wang Y, Cheng X, Shan Q, Zhang Y, Liu J, Gao C, Qiu JL (2014) Simultaneous editing of three homoeoalleles in hexaploid bread wheat confers heritable resistance to powdery mildew. Nat Biotechnol 32:947–951

Wang F, Wang C, Liu P, Lei C, Hao W, Gao Y et al (2016a) Enhanced rice blast resistance by CRISPR/Cas9-targeted mutagenesis of the ERF transcription factor gene *OsERF922*. PloS One 11:e0154027

Wang H, La Russa M, Qi LS (2016b) CRISPR/Cas9 in genome editing and beyond. Annu Rev Biochem 85:227–264

Wang L, Chen S, Peng A, Xie Z, He Y, Zou X (2019a) CRISPR/Cas9-mediated editing of *CsWRKY22* reduces susceptibility to *Xanthomonas citri* subsp. *citri* in Wanjincheng orange (*Citrus sinensis* (L.) Osbeck). Plant Biotechnol Rep 13:501–510

Wang L, Zhao L, Zhang X, Zhang Q, Jia Y, Wang G et al (2019b) Large-scale identification and functional analysis of NLR genes in blast resistance in the Tetep rice genome sequence. Proc Natl Acad Sci 116:18479–18487

Wasternack C, Hause B (2013) Jasmonates: biosynthesis, perception, signal transduction and action in plant stress response, growth and development. An update to the 2007 review in Annals of Botany. Ann Bot 111:1021–1058

Win J, Chaparro-Garcia A, Belhaj K, Saunders DGO, Yoshida K, Dong S et al (2012) Effector biology of plant-associated organisms: concepts and perspectives. Cold Spring Harbor Laboratory Press, In Cold Spring Harbor symposia on quantitative biology, pp 235–247

Woo HR, Koo HJ, Kim J, Jeong H, Yang JO, Lee IH et al (2016) Programming of plant leaf senescence with temporal and inter-organellar coordination of transcriptome in *Arabidopsis*. Plant Physiol 171:452–467

Wu CH, Abd-El-Haliem A, Bozkurt TO, Belhaj K, Terauchi R, Vossen JH, Kamoun S (2017) NLR network mediates immunity to diverse plant pathogens. Proc Natl Acad Sci 114:8113–8118

Xie K, Yang Y (2013) RNA-guided genome editing in plants using a CRISPR–Cas system. Mol Plant 6:1975–1983

Xie Z, Johansen LK, Gustafson AM, Kasschau KD, Lellis AD, Zilberman D et al (2004) Genetic and functional diversification of small RNA pathways in plants. PLoS Biol:e104

Yin K, Qiu JL (2019) Genome editing for plant disease resistance: applications and perspectives. Philos Tran R Soc B 374:20180322

Zeilmaker T, Ludwig NR, Elberse J, Seidl MF, Berke L, Van Doorn A et al (2015) *DOWNY MILDEW RESISTANT 6* and *DMR 6-LIKE OXYGENASE 1* are partially redundant but distinct suppressors of immunity in *Arabidopsis*. Plant J 81:210–222

Zeng W, He SY (2010) A prominent role of the flagellin receptor FLAGELLIN-SENSING2 in mediating stomatal response to *Pseudomonas syringae* pv tomato DC3000 in *Arabidopsis*. Plant Physiol 153:1188–1198

Zhang W, He SY, Assmann SM (2008) The plant innate immunity response in stomatal guard cells invokes G-protein-dependent ion channel regulation. Plant J 56:984–996

Zhang X, Zhao H, Gao S, Wang WC, Katiyar-Agarwal S, Huang HD et al (2011) *Arabidopsis* Argonaute 2 regulates innate immunity via miRNA393∗-mediated silencing of a Golgi-localized SNARE gene, *MEMB12*. Mol Cell 42:356–366

Zhang Z, Mao Y, Ha S, Liu W, Botella JR, Zhu JK (2016) A multiplex CRISPR/Cas9 platform for fast and efficient editing of multiple genes in Arabidopsis. Plant Cell Rep 35:1519–1533

Zheng XY, Spivey NW, Zeng W, Liu PP, Fu ZQ, Klessig DF et al (2012) Coronatine promotes *Pseudomonas syringae* virulence in plants by activating a signaling cascade that inhibits salicylic acid accumulation. Cell Host Microbe 11:587–596

Zhou J, Peng Z, Long J, Sosso D, Liu B, Eom JS et al (2015) Gene targeting by the TAL effector *PthXo2* reveals cryptic resistance gene for bacterial blight of rice. Plant J 82:632–643

Zhou P, Jia R, Chen S, Xu L, Peng A, Lei T et al (2017) Cloning and expression analysis of four citrus *WRKY* genes responding to *Xanthomonas axonopodis* pv. *citri*. Acta Hortic Sin 44:452–462

Improvement of Resistance in Plants Against Insect-Pests Using Genome Editing Tools

Sunaullah Bhat and Sandeep Kumar

Abstract During growth period plants are subjected to both biotic and abiotic stresses. Like other biotic stresses, insect-pests are the most serious challenge for the plants particularly in yield losses. Genome editing techniques are becoming an emerging technology bringing real revolution in genetic engineering and biotechnology. Editing of targeted gene provides ways to elucidate extensive ranges of aims for the improvement, protection, and increased yield of various crops.

Researchers all over the world have unraveled the usage of numerous gene editing methods from endonuclease to CRISPR/Cas in various aspects of plants like plant growth and development, insect-pest control, and other biotic stresses. The key goal of this chapter is to highlight various techniques of genome editing approaches which can be used to develop resistance in plants against insect-pests. New crop-based methods that reiterate the effective utilization of these techniques in insect-pest management as well as plant in resistance against pests are highlighted. This chapter also highlights the implication of genome editing as well as framework for its specific regulation.

Keywords Genome editing · CRISPR technology · Gene drive · Insect-pest management

1   Introduction

In the present state where population is accelerating at high level and climate is fluctuating, the improvement in crop productivity is very much essential. In the twenty-first century, the agriculture faces several challenges from abiotic as well as biotic challenges. Among the biotic factors, insects are the prime cause of crop loss. Insects are accountable for the reduced yield resulting in low productivity. Insects cause widespread yield losses in economically important crops all over the world, despite control measures (Dhaliwal et al. 2010; Tyagi et al. 2020). Insect-pests, along with other biotic stressors such as weeds, are estimated to reduce crop yield

S. Bhat · S. Kumar (✉)
Department of Zoology, Kumaun University, Almora, Uttarakhand, India
e-mail: bhatsunaullahijt@gmail.com; sandeepento@gmail.com

© The Author(s), under exclusive license to Springer Nature   237
Switzerland AG 2022
S. H. Wani, G. Hensel (eds.), *Genome Editing*,
https://doi.org/10.1007/978-3-031-08072-2_13

by 28 percent per year on a global scale (Ferry and Gatehouse 2010); therefore, development of insect-resistant varieties is the prime goal for breeders.

With the arrival of genetic engineering (Singer 1979) and plant biotechnology, pest management techniques have reached new horizons attaining numerous achievements. Various molecular techniques involving genetic mechanisms have been reported, and such studies have been reproduced to conduct in vitro experiments. Long investigation in the molecular genetics and biochemistry of organisms like bacteria and viruses allowed scientist to formulate new techniques of manipulating DNA with the help of various tools like vector for delivering DNA into the cell. Due to these techniques, not only transgenic microorganisms but genetic modifications of higher organisms like plants become possible (Nemudryi et al. 2014). Crops with modified gene having insect resistance like (*Bacillus thuringiensis*, Bt-ICPs) and its implementation (Bt gene) through introgressive hybridization have shown a reliable impact on productivity of crops and the sustainability of these techniques (Singh et al. 2018). However, development of resistance in insect against the Cry-toxins has been a major challenge (Wang et al. 2016).

Mutations in the genes encoding the receptor molecules that distract with the interaction between the insect and the toxin molecule result in insect-pest resistance. For prolonged efficacy against the insect-pests and to prevent resistance, chimeric toxins developed with the help of domain substitution and use of combinatorial ICPs (insecticidal crystal proteins) are also being exploited (Wang et al. 2016). An alternative to techniques is strategies aimed in silencing the selected gene like RNA interference (RNAi) involved in insect-pest feeding development which have also been formulated and evaluated for incorporation in crop improvement programs (Mamta & Rajam 2018).

Another approach is the cross-breeding of resistant (Arora and Sandhu 2017) and susceptible germplasm for the prevention of insect-pest. However, the lack of resistant gene pool and the incompatibility to breed in many crops minimize the possibility of breeding approaches for insect resistance. With the arrival of genomics, the plant genome has been sequenced which opens new horizons for the crop improvement (Jackson et al. 2011). With the advancement of genomic knowledge and various cellular functions, the essentiality variations in sequences for particular traits were revealed (Rathinam et al. 2019). This leads to the development of next-level biotechnological inventions called "genome editing" or "genome engineering." This technique emerged as an interesting contribution that allows the alteration in the genome by editing (adding or deleting or modifying) a particular part of the DNA, providing chances for its utility in plants (Fig. 1) and animals (including humans). In the present scenario, where population is growing at an alarming rate and agriculture land is constricted day by day, increased load of insect-pest on crop genome editing would serve as the key technique to combat insect-pests (Razzaq et al. 2019; Vats et al. 2019). For this reason genome editing is called "new breeding technology."

Presently several techniques are used for genome editing. Kim et al. (1996) revealed for the first time that the protein domains like "zinc fingers" coupled with

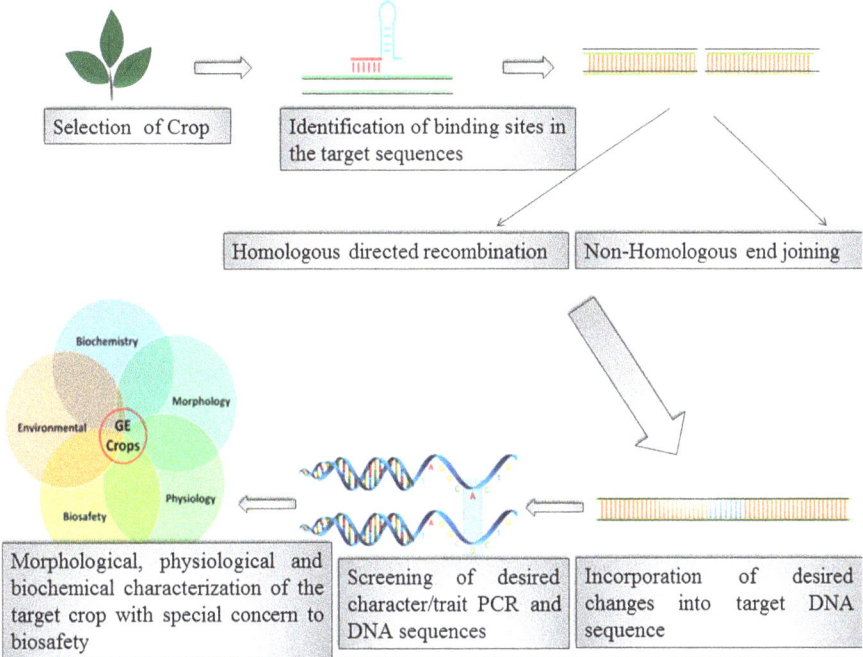

Fig. 1 The workflow diagram for gene editing in plants

the FokI endonuclease domains act as the site-specific nucleases, zinc finger nucleases (ZFNs), which divide the DNA in vitro study in firmly definite areas. These chimeric proteins have a linked structure, because the "zinc finger" domains recognize one triplet codon. To edit genome of cell culture, this method becomes the base including model as well as non-model plants (Weeks et al. 2016).

Regular efforts of researchers in this field lead to the advancement of novel genome editing tools like TALENs (transcription activator-like effector nucleases) and CRISPR/Cas (clustered regularly interspaced short palindromic repeats). To design TALENs new protein for every target needs to be reengineered. However, the design technique has been simplified by creating the modules of repeat combinations available that reduces importance of the cloning.

In contrast to TALENs, designing and usage of CRISPR are very simple. Such editing methodology when used properly for the manipulation of the genomes could resolve complex problems including the creation of mutant and transgenic plants (Hsu et al. 2013; Zhang and Guo 2014). Further, chimeric protein containing zinc finger and activation domains of additional protein or having TALE DNA-binding domain and Cas9 nuclease were used to study the regulation of transcription, epigenomes, and behavior of chromosome during the cell cycle (Cong et al. 2013; Petolino and Davies 2013; Lowder et al. 2016). This chapter was written to highlight the important gene editing tools to be used in insect-pest control.

2   Genome Editing for the Management of Insect-Pests

There are enormous scopes of biotechnology for the management of insect-pests to protect crops (Sun et al. 2017). These technologies vary from insect-pest resistant to transgenic hybridization (introgression) of new genes. The usage of genome editing in the development of insect-resistant crops has also been scarce. The most interesting aspect of genome editing for the management of insect-pest has the benefit of modifying plants as well as insects. Some of the genome editing tools used in the field of pest management are as follows.

2.1   Genome Editing to Diminish The Insect-Pests

In insect-pest management, the genome editing can be successfully done involving the strategy of modification of target insects and subsequently their release in the natural environment (McFarlane et al. 2018). Though Bt-technology has been used globally, the insect resistance developed against Bt insecticidal proteins (ICPs) has been the chief anxiety. To escape such problems, researches are being done to engineer the receptor protein in such a way that resistance management can be effective. Successful evidence of genome editing in *Helicoverpa armigera* shows that cadherin receptors were knock down which were genetically associated with Cry1Ac toxin resistance (Wang et al. 2016). This technique can be employed to modify the target sites in the midgut receptors which were responsible for the development of resistance against Bt ICPs. As insects possess specialized enzymes responsible for the detoxification (overcome chemical defense of plant), this technique can be employed for polyphagous insect-pest targeting their detoxification genes. Suppression of insecticidal detoxification genes such as gossypol-inducing cytochrome P450 can be utilized in producing the susceptibility in harmful insects (Hafeez et al. 2019); likewise, the knockdown of CYP6AE gene cluster (having a role to detoxify various phytochemicals) with the help of CRISPR/Cas9 (Fig. 2) in *H. armigera* revealed its reduced survival when exposed to botanical and insecticides (Wang et al. 2018) (Table 1).

Another technique to control the insect-pests utilizing genome editing techniques is to disrupt the genes that facilitate the insects in *chemoreception* and identification of mating partners. Olfactory receptors (ORs) in insects are essential for the recognition of host plants as well as their mating partners. For example, in *Drosophila* having mutation in Or83b gene interrupts the capacity of site selection for egg laying and reduced its olfactory detection (Larsson et al. 2004). In a similar way, the Orco (olfactory receptor co-receptor) gene knockdown in *Spodoptera litura* utilizing CRISPR/Cas9 demonstrated interruption in identifying host plants and also disrupts its mating selection

Fig. 2 The Cas9 nuclease of DNA and sgRNA complex. (https://commons.wikimedia.org/w/index.php?curid=62766587)

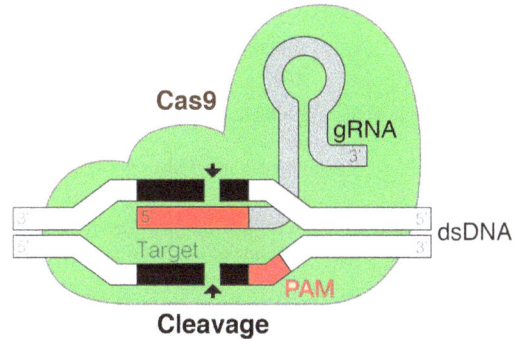

(Koutroumpa et al. 2016). In insects, female releases chemicals (sex phero-mones) to attract the male counterpart (Sun et al. 2017). *Using CRISPR/Cas9-based knockdown of the OR16 genes (odorant receptor 16) in H. armigera, males are unable to receive the pheromone signal form mature females, result-ing in mating with immature females subsequently producing sterile eggs* (Sun et al. 2017). Knockdown of OR16 receptor genes in lepidopteron pests may therefore be a new and effective approach to control mating time for pest man-agement in the agricultural crops. Injecting Cas9 messenger RNA and Slabd-A-specific single-guide RNA (sgRNA) into *S. litura* embryos efficiently caused the abd-A defective phenotype, which includes abnormal segmentation and *ectopic coloration* throughout the larval stage. The Cas9/sgRNA complex efficiently caused targeted mutagenesis in *S. litura* (Bi et al. 2016). Some other examples of genome editing in insect are given in Table 2. *Bemisia tabaci*, an invasive agricultural pest, was investigated that using efficient ReMOT-based CRISPR-Cas9 transduces Cas9 RNP into the ovaries of females, leading to the development of innovative controlling switch economically devastating pest insect (Heu et al. 2020). *Aedes aegypti*, one of the devastating vectors for spreading viruses like dengue, can be controlled through the release of sterile males. Chen et al. (2021) utilized CRISPR/Cas9 to produce a null mutation in β2-tubulin (B2t) gene of *A. aegypti*, which abolishes male fertility. Zhu et al. (2020) utilized CRISPR/Cas9 in *Spodoptera frugiperda* (fall armyworm), a serious polyphagous insect-pest. During this research they performed site-specific mutagenesis of three genes, two marker genes, and one developmental gene (biogenesis of lysosome-related organelles complex 1 subunit 2, tryptophan 2, 3-dioxygenase) and E93 (key ecdysone-induced transcription factor that promotes adult development). Knockout mutation of α-6-nicotinic acetylcholine receptor (nAchR) with the help of CRISPR/Cas9 in *H. armigera* and *S. exigua* that revealed α-6-nicotinic acetylcholine recep-tor is the target site of insecticide like spinosyn (Zuo et al. 2020; Wang et al. 2020). Some other such examples of genome editing in insect-pest *manage-ment* are given in Table 2.

Table 1 Comparison of various genome editing strategies

S.no	Description	Genome editing strategy used			References
		ZFN	TALEN	CRISPR/Cas9	
1	Components	Zn finger domains, nonspecific, FokI nuclease domain	TALE DNA-binding domains, nonspecific, FokI nuclease domain	crRNA, Cas9 proteins	Kumar and Jain (2015) and Sauer et al. (2016)
2	Catalytic Domain (enzyme)	Restriction endonuclease FokI, there is no catalytic domain	Restriction endonuclease FokI	RUVC and HNH	Chen et al. (2016)
3	Structural protein	Dimeric protein	Dimeric protein in nature	Monomeric in nature	Sauer et al. (2016)
4	Length of target sequence (bp)	24–36 bp	24–59 bp	20–22 bp	Chen et al. (2016)
5	Protein engineering steps	Required	Required	Should not be complex to test gRNA having 20 nucleotides	Sauer et al. (2016)
6	Gene cloning	Necessary	Necessary	Not necessary	Weeks et al. (2016) and Sauer et al. (2016)
7	gRNA production	Not applicable	Not applicable	Easy to produce	Cho et al. (2013), Noman et al. (2016) and Sauer et al. (2016)
8	Mode of action	Double-strand breaks in target	DNA double-strand breaks in target DNA	Double-strand breaks or single-strand nicks in target DNA	Sauer et al. (2016)
9	Target recognition efficiency	High	High	High	Gaj et al. (2013), Kumar and Jain (2015) and Sauer et al. (2016)
10	Mutation rate	High	Middle	Low	Li et al. (2016), Gaj et al. (2013) and Sauer et al. (2016)
11	Creation of large scale libraries	Impossible	Technically difficult	Possible	Hsu (2013), Cho et al. (2013) and Sauer et al. (2016)

(continued)

Table 1 (continued)

| S.no | Description | Genome editing strategy used | | | References |
		ZFN	TALEN	CRISPR/Cas9	
12	Multiplexing	Difficult	Difficult	Possible	Li et al. (2016), Mao et al. (2013), Noman et al. (2016) and Sauer et al. 2016)

2.2   Gene Drive Based on CRISPR in Insect for Crop Protection

CRISPR gene editing furnishes a "gene drive" that is powerful enough to spread the edited genes to transgenerational. In sexual reproducing organism, when one set of chromosomes has a gene drive, it dominates the partner genotype which drives the gene flow to be inherited with greater frequencies than expected by Mendelian inheritance (McFarlane et al. 2018). Such "super Mendelian inheritance" enables these genes to be passed from one generation to another even though these diminish the fitness of the particular organism. This technique may completely eradicate a particular species after 15–20 generations if only few individuals with CRISPR-based gene drive are released into the environment for mating (McFarlane et al. 2018). This may lead to catastrophic change in the environment due to the absence of a particular species by human intervention. An environmentally sustainable approach in this direction has to be followed to make genome editing techniques an effective way to target non-Mendelian genes and genes that do not undergo gene drive. Therefore, managing the resistance in insect-pest population by gene knockdown would be a more positive approach than those techniques aimed to target characters (traits) related to reproduction and developmental process in insect-pest, which might lead to ecological havoc (Tyagi et al. 2020).

2.3   Genome Editing in Plant for Insect-Pest Management

Genome editing in plant for the pest control has been less explored. In this chapter, we discuss the possibilities of targets that can be edited for sustainable insect-pest management (Vats et al. 2019). Most insect-pests (especially polyphagous insects) identify host plants using plants' visual appearance, volatile blends of plant, gustatory signs, and sites for the oviposition as these interactions are co-evolved (Larsson et al. 2004). Insect-pest selects host plant for egg laying for the availability of preferred food to the young ones. Insect-pests recognize few volatile blends of the host plant which direct it for egg laying. Researchers have

Table 2 Genome editing to diminish insect-pests

S. No	Target insect-pest	Target gene	Editing technique	Trait	References
1	*H. armigera*	OR16 (odorant receptor 16)	Gene knockout	Mating time and partner recognition	Sun et al. (2017)
2	*H. armigera*	α-6-nicotinic acetylcholine receptor (nAchR)	Gene knockout	Resistance to an insecticide, spinosyn	Wang et al. (2020)
3	*H. armigera*	CYP6AE gene cluster (P450 genes)	Gene knockdown	Regulation of detoxification enzymes	Wang et al. (2018)
4	*Bemisia tabaci*	Ovary-targeting peptide ligand "BtKV"	CRISPR-Cas9	Transduced the ribonucleoprotein complex to the germline	Heu et al. (2020)
5	*Spodoptera frugiperda*	Biogenesis of lysosome-related organelles complex 1 subunit 2	Knockout	Mosaic translucent integument phenotype	Zhu et al. (2020)
		Tryptophan 2,3-dioxygenase		Olive eye color phenotype	
		E93 (multiple sgRNA method used to knock out the E93 gene in one generation)		Larval-pupal intermediate phenotype	
6	*S. litura*	Abdominal-A (Slabd-A) gene	Knockout	Body segmentation and pigmentation	Bi et al. (2016)
7	*S. littoralis*	Olfactory receptor co-receptor (Orco)	Gene knockout	Mating time and partner recognition	Koutroumpa et al. (2016)
8	*S. exigua*	Ryanodine receptor	Substitution	Regulation of cellular processes like muscle contraction, hormone secretion, gene transcription, cell proliferation, and neurotransmitter release	Zuo et al. (2017) and Zuo et al. (2020)
9	*S. exigua*	α-6-Nicotinic acetylcholine receptor (nAchR)	Gene knockout	Resistance to an insecticide, spinosyn	Zuo et al. (2020)
10	*Plutella xylostella*	Abdominal-A	Deletions & insertions	Identify and functionality of abdomen segments	Huang et al. (2016)
11	*Drosophila melanogaster*	Chitin synthase-1	Substitution	Chitin biosynthesis	Douris et al. (2016)

(continued)

Table 2 (continued)

S. No	Target insect-pest	Target gene	Editing technique	Trait	References
12	*Drosophila melanogaster*	Nicotinic acetylcholine receptor α 6	Substitution	Nervous system	Zimmer et al. (2016)
13	*Drosophila suzukii*	White (w) and sex lethal (Sxl) genes	Knockout	Eye color and female development as well as X chromosome dosage compensation respectively	Scott et al. (2018)
14	*Aedes aegypti*	Male-determining factor (M factor, Nix)	Knockout	Sex determination	Hall et al. (2015)
15	*Aedes aegypti*	β2-tubulin (B2t) gene	CRISPR/ Cas9	Male fertility	Chen et al. (2021)
16	*Culex quinquefasciatus*	CYP9M10 gene	Knockout	Cytochrome P450 enzyme	Itokawa et al. (2016)
17	*Anopheles gambiae*	rDNA	Knockout	Sex-ratio distortion trait	Galizi et al. (2016)
18	*Anopheles gambiae*	*AGAP005958, AGAP011377, AGAP007280*	Knockout, Knock-in	Recessive female sterility phenotype upon disruption	Hammond et al. (2016)
19	*Anopheles stephensi*	Kynurenine hydroxylase	Knock-in	Eye color phenotype	Gantz et al. (2015)
20	*Locusta migratoria*	Odorant receptor co-receptor (Orco)gene	Knockout	Roles of the odorant receptor pathway	Zhang et al. (2014)
21	*Tribolium castaneum*	E-cadherin gene, EGFP	Knock-out, Knock-in	CRISPR-mediated gene knockout of the Tribolium E-cadherin gene causes defects in dorsal closure	Gilles et al. (2015)

demonstrated that change in these volatile blends retracts the pest from host plant. Aphid infestation in plant leads to the release of Eβf (sesquiterpene hydrocarbon (E)-β-farnesene) which attracts a parasitic wasp *Diaeretiella rapae* to control the population of aphids (Beale et al. 2006). Therefore, changing the volatile blends of host plant through genome editing is an alternative technique to control insect-pest population. However, care should be taken as these modifications would not lead to catastrophic effect toward the beneficial insects such as pollinators.

Plants' visual appearance plays an essential role to be recognized by the insect-pest. Changing the pigmentation of plant has been found to modify the insects' preferences for the host plant. This phenomenon was studied in red leaf tobacco, a transgenic plant in which anthocyanin pathway was modified (Malone et al. 2009). This report illustrated that overproduction of anthocyanin pigmentation (red coloration) of leaves acts as a deterrent to *S. litura* and *H. armigera*. These investigations demonstrate that engineering pathway of anthocyanin is a possible method for CRISPR-based editing for insect-pest management.

Plants have devised numerous strategies to respond to the attack of biotic stress like insect-pest. The resistance gene (R gene) decodes the ability of plants to resist pest/disease (Rathinam et al. 2019); the susceptible gene (S gene) invites the stress. Editing genome of susceptible genes for developing plant resistance against insect-pest is an emerging strategy. Insects rely on essential chemical communication from plants for their development, immunity, and behavior. This has been efficiently illustrated in rice. CRISPR/Cas9-based knockout of CYP71A1 gene encoding tryptamine 5-hydroxylase which is involved in the conversion of tryptamine to serotonin resulted in reduced growth of plant hopper. The research was hypothesized that serotonin, a neurotransmitter from plants, is vital for larval immunity and behavior (Lu et al. 2018).

3   Future Studies

- Despite the effectiveness of genome editing in combating pathogens in economically valuable crops, their use in insect control has not been fully utilized. Researchers must exercise techniques for effective pest resistance in plants.
- Globally scientists need to give emphasis for the identification of resistance source (genes) which will serve as platform for insect-pest management.
- Widespread phenotyping and enhancement of the resistance gene pool are essential intimations that need to be addressed in the path-breaking manner.

4   Conclusion

Genome editing techniques are becoming popular molecular methods for genomics, crop improvement, and crop protection. This tool is emerging as a method of choice in laboratories to illustrate the gene function and its translational usage. These are being used in diverse crop improvement and protection programs to mitigate abiotic and biotic stresses. Due to several attractive features like efficiency, simplicity, high specificity, and amenability to multiplexing CRISPR technologies, genome editing technologies described here are transforming the way crop breeding is done and provide protection of crop against pests which had developed resistance against chemical control.

References

Arora R, Sandhu S (eds) (2017) Breeding insect resistant crops for sustainable agriculture. Springer Singapore

Beale MH, Birkett MA, Bruce TJ, Chamberlain K, Field LM, Huttly AK, Martin JL, Parker R, Phillips AL, Pickett JA, Prosser IM (2006) Aphid alarm pheromone produced by transgenic plants affects aphid and parasitoid behavior. PNAS 103(27):10509–10513

Bi HL, Xu J, Tan AJ, Huang YP (2016) CRISPR/Cas9-mediated targeted gene mutagenesis in Spodoptera litura. Insect Sci. 23(3):469–77.

Chen J, Luo J, Wang Y, Gurav AS, Li M, Akbari OS, Montell C (2021) Suppression of female fertility in *Aedes aegypti* with a CRISPR-targeted male-sterile mutation. PNAS 118(22)

Cong L, Ran FA, Cox D, Lin S, Barretto R, Habib N, Hsu PD, Wu X, Jiang W, Marraffini LA, Zhang F (2013) Multiplex genome engineering using CRISPR/Cas systems. Science 339(6121):819–823

Chen K, Stephanou A S, Roberts G A, White J H, Cooper LP, Houston P J, & Dryden DT (2016) The Type I restriction enzymes as barriers to horizontal gene transfer: determination of the DNA target sequences recognised by livestock-associated methicillin-resistant Staphylococcus aureus clonal complexes 133/ST771 and 398. In Biophysics of Infection (pp. 81-97). Springer, Cham.

Cho, S. W., Kim, S., Kim, J. M., & Kim, J. S. (2013). Targeted genome engineering in human cells with the Cas9 RNA-guided endonuclease. Nat. Biotechnol. 31(3):230–232.

Dhaliwal GS, Jindal V, Dhawan AK (2010) Insect pest problems and crop losses: changing trends. Ind J Ecol 37(1):1–7

Douris V, Steinbach D, Panteleri R, Livadaras I, Pickett JA, Van Leeuwen T, Nauen R, Vontas J (2016) Resistance mutation conserved between insects and mites unravels the benzoylurea insecticide mode of action on chitin biosynthesis. PNAS 113(51):14692–14697

Ferry N, Gatehouse AM (2010) Transgenic crop plants for resistance to biotic stress. In: Transgenic cop plants. Springer, Berlin/Heidelberg, pp 1–65

Gaj T, Gersbach C A, & Barbas III C F (2013) ZFN, TALEN, and CRISPR/Cas-based methods for genome engineering. Trends Biotechnol. 31(7):397-405.

Galizi R, Hammond A, Kyrou K, Taxiarchi C, Bernardini F, O'Loughlin SM, Papathanos PA, Nolan T, Windbichler N, Crisanti A (2016) A CRISPR-Cas9 sex-ratio distortion system for genetic control. Sci Rep 6(1):1–5

Gantz VM, Jasinskiene N, Tatarenkova O, Fazekas A, Macias VM, Bier E, James AA (2015) Highly efficient Cas9-mediated gene drive for population modification of the malaria vector mosquito Anopheles stephensi. Proc Natl Acad Sci 112(49):E6736–E6743

Gilles AF, Schinko JB, Averof M (2015) Efficient CRISPR-mediated gene targeting and transgene replacement in the beetle *Tribolium castaneum*. Development 142(16):2832–2839

Hafeez M, Liu S, Jan S, Shi L, Fernández-Grandon GM, Gulzar A, Ali B, Rehman M, Wang M (2019) Knock-down of gossypol-inducing cytochrome P450 genes reduced deltamethrin sensitivity in *Spodoptera exigua* (Hübner). Int J Mol Sci 20(9):2248

Hall AB, Basu S, Jiang X, Qi Y, Timoshevskiy VA, Biedler JK, Sharakhova MV, Elahi R, Anderson MA, Chen XG, Sharakhov IV (2015) A male-determining factor in the mosquito *Aedes aegypti*. Science 348(6240):1268–1270

Hammond A, Galizi R, Kyrou K, Simoni A, Siniscalchi C, Katsanos D, Gribble M, Baker D, Marois E, Russell S, Burt AA (2016) CRISPR-Cas9 gene drive system targeting female reproduction in the malaria mosquito vector *Anopheles gambiae*. Nature 34(1):78–83

Heu CC, McCullough FM, Luan J, Rasgon JL (2020) CRISPR-Cas9-based genome editing in the silverleaf whitefly (Bemisia tabaci). CRISPR J 3(2):89–96

Hsu PD, Scott DA, Weinstein JA, Ran FA, Konermann S, Agarwala V, Li Y, Fine EJ, Wu X, Shalem O, Cradick TJ (2013) DNA targeting specificity of RNA-guided Cas9 nucleases. Nature 31(9):827–832

Hsu, P. D., Scott, D. A., Weinstein, J. A., Ran, F. A., Konermann, S., Agarwala, V., & Zhang, F. (2013). DNA targeting specificity of RNA-guided Cas9 nucleases. Nature biotechnology, 31(9):827-832.

Huang Y, Chen Y, Zeng B, Wang Y, James AA, Gurr GM, Yang G, Lin X, Huang Y, You M (2016) CRISPR/Cas9 mediated knockout of the abdominal-A homeotic gene in the global pest, diamondback moth (Plutella xylostella). Insect Biochem Mol Biol 75:98–106. https://doi.org/10.1016/j.ibmb.2016.06.004

Itokawa K, Komagata O, Kasai S, Ogawa K, Tomita T (2016) Testing the causality between CYP9M10 and pyrethroid resistance using the TALEN and CRISPR/Cas9 technologies. Sci Rep 20;6(1):1–0

Jackson SA, Iwata A, Lee SH, Schmutz J, Shoemaker R (2011) Sequencing crop genomes: approaches and applications. New Phytol 191(4):915–925

Kim YG, Cha J, Chandrasegaran S (1996) Hybrid restriction enzymes: zinc finger fusions to Fok I cleavage domain. Proc Natl Acad Sci 93(3):1156–1160

Koutroumpa FA, Monsempes C, François MC, De Cian A, Royer C, Concordet JP, Jacquin-Joly E (2016) Heritable genome editing with CRISPR/Cas9 induces anosmia in a crop pest moth. Sci Rep 6(1):1–9

Kumar V, & Jain M (2015). The CRISPR–Cas system for plant genome editing: advances and opportunities. J. Exp. Bot., 66(1):47–57.

Larsson MC, Domingos AI, Jones WD, Chiappe ME, Amrein H, Vosshall LB (2004) Or83b encodes a broadly expressed odorant receptor essential for Drosophila olfaction. Neuron 43(5):703–714

Li Y, Zhang J, Chen D, Yang P, Jiang F, Wang X, Kang L (2016) CRISPR/Cas9 in locusts: successful establishment of an olfactory deficiency line by targeting the mutagenesis of an odorant receptor co-receptor (Orco). Insect Biochem Mol Biol 79:27–35

Lowder L, Malzahn A, Qi Y (2016) Rapid evolution of manifold CRISPR systems for plant genome editing. Front Plant Sci 2016 Nov 14;7:1683.

Lu HP, Luo T, Fu HW, Wang L, Tan YY, Huang JZ, Wang Q, Ye GY, Gatehouse AM, Lou YG, Shu QY (2018) Resistance of rice to insect pests mediated by suppression of serotonin biosynthesis. Nature 4(6):338–344

Mao, Y., Zhang, H., Xu, N., Zhang, B., Gou, F., & Zhu, J. K. (2013). Application of the CRISPR–Cas system for efficient genome engineering in plants. Mol Plant. 6(6):2008–2011.

Malone LA, Barraclough EI, Lin-Wang K, Stevenson DE, Allan AC (2009) Effects of red-leaved transgenic tobacco expressing a MYB transcription factor on two herbivorous insects, *Spodoptera litura* and *Helicoverpa armigera*. Entomol Exp Appl 133(2):117–127

Mamta B, Rajam MV (2018) RNA interference: a promising approach for crop improvement. In: Biotechnologies of crop improvement, vol 2. Springer, Cham, pp 41–65

McFarlane GR, Whitelaw CBA, Lillico SG (2018) CRISPR-based gene drives for pest control. Trends Biotechnol 36(2):130–133

Nemudryi AA, Valetdinova KR, Medvedev SP, Zakian SM (2014) TALEN and CRISPR/Cas genome editing systems: tools of discovery. Acta Naturae (англоязычная версия) (3:(22)

Noman A, Aqeel M, & He S (2016) CRISPR-Cas9: tool for qualitative and quantitative plant genome editing. Front. Plant Sci. 7:1740.

Petolino JF, Davies JP (2013) Designed transcriptional regulators for trait development. Plant Sci 201:128–136

Rathinam M, Mishra P, Mahato AK, Singh NK, Rao U, Sreevathsa R (2019) Comparative transcriptome analyses provide novel insights into the differential response of Pigeonpea (*Cajanus cajan* L.) and its wild relative (*Cajanus platycarpus* (Benth.) Maesen) to herbivory by *Helicoverpa armigera* (Hübner). Plant Mol Biol 101(1):163–182

Razzaq A, Saleem F, Kanwal M, Mustafa G, Yousaf S, Imran Arshad HM, Hameed MK, Khan MS, Joyia FA (2019) Modern trends in plant genome editing: an inclusive review of the CRISPR/Cas9 toolbox. Int J Mol Sci 20(16):4045

Sauer NJ., Mozoruk J, Miller RB, Warburg Z J, Walker KA, Beetham PR, & Gocal G F (2016). Oligonucleotide-directed mutagenesis for precision gene editing. Plant Biotechnol. J. 14(2):496:502.

Scott M J, Gould F, Lorenzen M, Grubbs N, Edwards O, & O'Brochta D (2018) Agricultural production: assessment of the potential use of Cas9-mediated gene drive systems for agricultural pest control. J. Responsible Innov. 5(sup1): S98-S120.

Singer MF (1979) In: Setlow JK, Hollaender A (eds) Introduction and historical background, in genetic engineering, vol 1. Plenum, New York, pp 1–13

Singh S, Kumar NR, Maniraj R, Lakshmikanth R, Rao KY, Muralimohan N, Arulprakash T, Karthik K, Shashibhushan NB, Vinutha T, Pattanayak D (2018) Expression of Cry2Aa, a *Bacillus thuringiensis* insecticidal protein in transgenic pigeon pea confers resistance to gram pod borer, *Helicoverpa armigera*. Sci Rep 8(1):1–2

Sun D, Guo Z, Liu Y, Zhang Y (2017) Progress and prospects of CRISPR/Cas systems in insects and other arthropods. Front Physiol 8:608

Tyagi S, Kesiraju K, Saakre M, Rathinam M, Raman V, Pattanayak D, Sreevathsa R (2020) Genome editing for resistance to insect pests: an emerging tool for crop improvement. ACS omega 5(33):20674–20683

Vats S, Kumawat S, Kumar V, Patil GB, Joshi T, Sonah H, Sharma TR, Deshmukh R (2019) Genome editing in plants: exploration of technological advancements and challenges. Cell 11:1386

Wang J, Zhang H, Wang H, Zhao S, Zuo Y, Yang Y, Wu Y (2016) Functional validation of cadherin as a receptor of Bt toxin Cry1Ac in *Helicoverpa armigera* utilizing the CRISPR/Cas9 system. Insect Biochem Mol Biol 76:11–17

Wang J, Ma H, Zuo Y, Yang Y, Wu Y (2020) CRISPR mediated gene knockout reveals nicotinic acetylcholine receptor (nAChR) subunit a6 as a target of spinosyns in *Helicoverpa armigera*. Pest Manag Sci 76. https://doi.org/10.1002/ps.5889

Wang H, Shi Y, Wang, L Liu S, Wu S, Yang, Y, & Wu, Y. (2018). CYP6AE gene cluster knockout in Helicoverpa armigera reveals role in detoxification of phytochemicals and insecticides. Nat. Commun. 9(1):1–8

Weeks DP, Spalding MH, Yang B (2016) Use of designer nucleases for targeted gene and genome editing in plants. Plant Biotechnol J 14(2):483–495

Zhang F, Wen Y, Guo X (2014) CRISPR/Cas9 for genome editing: progress, implications and challenges. Hum Mol Genet 23(R1):R40–R46

Zhu GH, Chereddy SC, Howell JL, Palli SR (2020) Genome editing in the fall armyworm, *Spodoptera frugiperda*: multiple sgRNA/Cas9 method for identification of knockouts in one generation. Insect Biochem Mol Biol 122:103373

Zhang F, Wen Y, Guo X (2014) CRISPR/Cas9 for genome editing: progress, implications and challenges. Hum. Mol. Genet. 23(R1):R40-6.

Zimmer CT, Garrood WT, Puinean AM, Eckel-Zimmer M, Williamson MS, Davies TE, Bass C (2016) A CRISPR/Cas9 mediated point mutation in the alpha 6 subunit of the nicotinic acetylcholine receptor confers resistance to spinosad in *Drosophila melanogaster*. Insect Biochem Mol Biol 73:62–69

Zuo Y, Wang H, Xu Y, Huang J, Wu S, Wu Y, Yang Y (2017) CRISPR/Cas9 mediated G4946E substitution in the ryanodine receptor of Spodoptera exigua confers high levels of resistance to diamide insecticides. Insect Biochem Mol Biol 89:79–85

Zuo YY, Ma HH, Lu WJ, Wang XL, Wu SW, Nauen R, Wu YD, Yang YH (2020) Identification of the ryanodine receptor mutation I4743M and its contribution to diamide insecticide resistance in *Spodoptera exigua* (Lepidoptera: Noctuidae). Insect Sci 4:791–800

Applications of Gene Drive for Weeds and Pest Management Using CRISPR/Cas9 System in Plants

Srividhya Venkataraman and Kathleen Hefferon

Abstract Solutions to world hunger include addressing the challenge of agricultural losses due to weeds and pests. New crop varieties which possess herbicide tolerance and pest resistance are required. Genome editing has proven to be an emerging technology which could successfully address these challenges in dozens of crop varieties. Gene drives, which enable new crop traits to be passed on to entire populations in the field, are also being explored for their ability to increase yield and combat food insecurity. In principle, gene drives offer great potential for controlling agricultural pests, insect vectors of disease, and weed management. This chapter thus describes the control of weeds and insect pests using gene drive technologies.

Keywords CRISPR-Cas9 · Gene drive · Weed containment · Agricultural pests · Vector-borne disease

1 Introduction

Losses in agricultural productivity due to damage by pests and weed management augment the goal of eradicating world poverty and hunger. New crop varieties which are pest resistant and tolerant to weeds could be a sustainably and environmentally friendly approach. Often crops lack the gene pool necessary to support these properties, and as a result, farmers must rely on costly chemical inputs such as insecticides or herbicides to maintain high crop yields. New breeding technologies, such as genome editing using CRISPR-Cas9, could play a role in improving the gene pool of crops to address biotic stresses. In this system, a guide RNA (gRNA)

S. Venkataraman · K. Hefferon (✉)
Cell and Systems Biology, University of Toronto, Toronto, ON, Canada
e-mail: klh22@cornell.edu

© The Author(s), under exclusive license to Springer Nature Switzerland AG 2022
S. H. Wani, G. Hensel (eds.), *Genome Editing*,
https://doi.org/10.1007/978-3-031-08072-2_14

251

that is independently expressed guides the Cas9 endonuclease to its target site. The sequence specified by the gRNA can be changed to determine a particular crop trait.

Genome editing has been used as a tool in dozens of plants, including essential crops like wheat, rice, and maize. The potential of genome editing has enhanced the potential for generating crops with new agronomic traits that can assist in their resistance to pests and tolerance to herbicides.

Gene drive entails the basic concepts of genome editing so that new traits can be passed on to an entire population during reproduction. In principle, individual organisms with the requisite genotype could be engineered at the laboratory level and subsequently released into the field to disseminate this genotype among wild populations. The approach functions through distortion of segregation due to which the resulting heterozygotes spread a desired allele at greater than 50% frequency compared to that provided by Mendelian inheritance (Li et al. 2020). As an alternative, the genomes of insects have been modified to create gene drives or to counteract resistance to various insecticides (Bisht et al. 2019). This chapter details the management of weeds and insect pests through the use of gene drive technologies.

2   Weed Management Through the Use of Gene Drives

Weed containment is a complex and emerging problem concerned with the interaction of natural-human systems in which it is challenging to apply simple fixes to contain the growth of invasive agricultural weeds. One rapidly advancing and highly critical issue is the widespread appearance of herbicide resistance (Gould et al. 2018). Another constraining issue is associated with the increasing growth of alien-invasive weeds that are difficult to contain as they often emerge in unstable, obscure, and rugged areas where traditional management practices are not easy to implement (Epanchin-Niell et al. 2010).

The basic tenet of gene drives functions through suppression or sensitization of populations (Neve 2018). Suppression drives are practiced through targeting key traits such as fecundity, formulation, and persistence of the weeds. This results in minimal-input suppression of intrusive, extraneous weed species in agricultural lands and natural habitats (Webber et al. 2015).

As opposed to the suppression drives, sensitizing gene drives can be used to make a given weed population target to become susceptible to particular intervention measures in order to achieve successful weed management. Weeds have widely acquired resistance to herbicides (Kreiner et al. 2018). A gene drive derived through endonucleases can mutate alleles coding for herbicide resistance and replace them with their wild-type equivalents that are herbicide susceptible, thus containing the weed population (Neve 2018).

Gene drive mechanisms applied to manage weed populations can preclude or eliminate many invasive weed species that challenge biodiversity or cause changes in pathogens responsible for destroying crops and spreading diseases. These gene drive systems can introduce new traits into existing populations and could protect species of endangered plants or could even revive those that are extinct. Conventionally, small and declining species are genetically rescued by transplanting a few of the genetically distinct immigrant plants into limited populations. The main function of such a strategy is to introduce favorable genetic traits into endangered populations from a limited number of immigrant plants that are adapted more efficiently to the prevailing conditions. This leads to the augmentation of adaptive potential, while at the local level, this adaptive genetic variation is not overloaded (Whiteley et al. 2015; Tallmon et al. 2004).

In the USA, gene drives inhibiting the spread of the spotted knapweed *Centaurea maculosa* that is non-indigenous have been devised in order to enable the protection of the biodiversity of native plant species. The knapweed damages naturally occurring ecosystems and leads to soil erosion. The unbridled dissemination of knapweed can be contained by means of gene drives formulated to suppress genes that determine the sex of this weed species in a methodology similar to the gene drives devised to control mosquito populations directed toward protecting humans from diseases transmitted by mosquitoes. Such a strategy could skew the sex ratio of the weed plants which can result in a population crash (Languin 2014). Notwithstanding, when compared with mosquitoes, the growth of the knapweed is slow and it is not yet apparent how factors such as the distance and frequency of pollen spread in the wild would affect the gene drive procedures (National Academies Press 2016). Yet another function afforded by gene drives in plant populations is the eradication of pigweed from arable fields. Pigweed has a rapid reproduction cycle and has established resistance against the extensively used herbicide, glyphosate. This resistant state can be reversed back to sensitivity using the gene drive technology. Alternatively, a suppression gene drive can be used so as to bias pigweed sex ratio leading to the collapse of this weed population (National Academies Press 2016).

Recently, genome editing technology has become popular which when used in conjunction with gene drives can potentially eliminate weed species. Nevertheless, the major caveat in practicing gene drives toward containing or eliminating invasive weed species is that they necessarily require that the target plants reproduce sexually and have a rapid rate of reproduction. Moreover, gene drive measures would have to be employed repeatedly over a given course of time as the target plant populations undergo natural selection, thereby consequently losing the trait that was genetically introduced (Callaway 2017). Also, there could emerge a few plants in the target plant population that generate resistance to the gene drive resulting in the re-emergence of plants refractory to its future use.

Table 1 enlists some instances of resistance mechanisms detected in weeds against a few of the widely used herbicide classes. While the extensive use of herbicides has diminished crop losses by and large, the problem of widespread herbicide resistance has increasingly emerged (Lucas et al. 2015; Ranson et al. 2011; Powles and Yu 2010). This is compounded by concerns regarding the impacts of herbicide applications on human health as well as off-target effects on the environment. Hence, this situation necessitates the adoption of alternative technologies for pest control. This calls for control of pest species using genome manipulation strategies to reduce their genetic fitness.

Table 1 Examples of mechanisms of resistance observed in weed populations for some of the highest used classes of herbicides

Herbicide group	Herbicide resistance: genetic and mechanistic causes	Determinant molecules	Biochemical/physiological outcomes	References
	TSR, acetyl-CoA carboxylase amino acid substitutions: I1781L/V/I/A/T; W1999C/L/S; W2027C; I2041N/V/T; D2078G; C2088 R; G2096A/S	Acetyl-CoA carboxylase	Reduced herbicide sensitivity of ACCase	Kaundun, 2014, Guo et al. (2017)
ALS inhibitors	TSR, ALS amino acid substitutions: G654D/E; S653I/N/T; W574L/M/G/R; A205F/V; D376E; R377H; P197W/T/K/M/Y/E/N/I/A/S/Q/L/R/H A122N/T/S/Y/V	Acetolactate synthase	Reduced herbicide sensitivity of ALS	Powles and Yu (2010), Yu and Powles (2014), Larran et al. (2017), Panozzo et al. (2017) and Brosnan et al. (2016)
Photosystem II inhibitors	Target site resistance, D1 polypeptide substitutions: N266T; S264T/G; F255V/I; A251V; V219I; L218V	D1 polypeptide	Diminished D1 protein sensitivity to herbicide	Masabni and Zandstra (1999), Li et al. (2014), Perez-Jones et al. (2009), Mechant et al. (2008), Park and Mallory-Smith (2006), Thiel and Varrelmann (2014) and Dumont et al. (2016)
Inhibitors of protoporphyrinogen oxidase	Target site resistance, protoporphyrinogen oxidase codon G210 deletion	Protoporphyrinogen oxidase	Diminished protoporphyrinogen oxidase sensitivity to herbicide	Patzoldt et al. (2006) and Salas et al. (2016)
	Target site resistance, protoporphyrinogen oxidase substitution mutations: R128M/G; S149I; G114E; G399A	Protoporphyrinogen oxidase	Decreased sensitivity of protoporphyrinogen oxidase to herbicide	Rangani et al. (2019), Giacomini et al. (2017) and Rousonelos et al. (2012)
Engineered auxins	Target site resistance, indole acetic acid 16 substitution mutation G127N	AUX/IAA Co-receptor of auxin/indole acetic acid	Modified signaling of auxin	LeClere et al. (2018)

Herbicide group	Herbicide resistance: genetic and mechanistic causes	Determinant molecules	Biochemical/physiological outcomes	References
5-enolpyruvylshikimate 3′-phosphate synthase inhibitors	Target site resistance, 5-enolpyruvylshikimate 3′-phosphate synthase substitution mutations: P106L/T/A/S; T120S	5-enolpyruvylshikimate 3′-phosphate synthase	Decreased sensitivity of 5-enolpyruvylshikimate 3′-phosphate synthase to herbicide	Sammons and Gaines (2014) and Li et al. (2018)

Adapted from Perotti et al. (2020)

3 Natural Gene Drives

3.1 Gene Drives That Occur at the Pre-gametic Phase

Gene drives taking place at the meiotic or pre-gametic phase operate by disrupting transmission ratios at the time of meiosis wherein gametes that harbor the drive allele possess enhanced likelihood of being generated (Lindholm et al. 2016). In plant systems, the meiotic drives defined thus far have developed molecular machinery biasing chromosome segregation to the plant cell which eventually becomes the female gamete. This is called "female-based meiotic drive." In almost all of the flowering plants, only a single product resulting from meiosis develops into a valid megaspore that will form the female gametocyte consisting of a fertile egg cell.

Contrarily, the remaining three meiosis products go through programmed cell death (Yang et al. 2010). Such genetic asymmetry at the time of meiosis enables the development of selfish genes as it indemnifies conflict between homologous chromosomes for survival in the remaining single egg cell. In maize (*Zea mays*), there exists the Ab10 chromosome which is the selfish equivalent of the normal chromosome (N10). The Ab10 chromosome is responsible for the abnormal (Ab10) meiotic gene drive mechanism in maize and this has been well-studied. The Ab10 chromosome contains large sections of repetitive DNA called knobs, and these are present on the long arm of this chromosome. While mechanisms enabling the transmission bias are multifaceted (Dawe et al. 2018), these Ab10 knob structures distort transmission by helping in the active movement of Ab10 to the fertile egg cell via meiotic spindles much before the N10 chromosomes. This leads to biased segregation of chromosome in which the Ab10 is conveyed to a majority (60–80%) of the progeny (Higgins et al. 2018).

In another example of natural gene drives, the centromere-associated distorter (D) locus present in *Mimulus guttatus* (monkey flower) has been well-studied (Fishman and Saunders 2008; Fishman and Willis 2005). This was initially detected by means of studies on non-Mendelian inheritance in hybrid plants generated from crosses between *M. guttatus* and its close relative *M. nasutus*. This gene locus D is equivalent to the Ab10 chromosome wherein it leads to the preferred transmittance of chromosomes containing the drive genetic element into the ovum in the course of female meiosis. Such skewed transmission generates progeny from heterospecific crosses, 98% of which contain the D locus. On the other hand, conspecific crosses result in 58% of the progeny carrying the D locus, a variation that could be attributed to the existence of suppressor genetic elements in *M. guttatus* (Fishman and Kelly 2015).

Among the selfish genetic material occurring in plant systems, the B chromosomes happen to be the most well-characterized. The B chromosomes are extra genetic elements that accrue in a non-Mendelian mode. They invoke detrimental outcomes on their host plants and have been shown to occur in more than a thousand flowering plant species (Burt and Trivers 2009; Jones 1995). The gene drive caused by the B chromosomes is enabled by several irregular meiotic and mitotic events which distort transmission ratios, and the B chromosomes aggregate selfishly in the germline (Jones 1995, 2018). Also, the B chromosomes often achieve the drive by creating genetic asymmetries in the course of gametogenesis (significantly pollen mitosis) in a manner similar to the meiotic drives described above. The dynamics of B chromosome populations could be explained by the opposing factors of bias in segregation and deleterious consequences on the host plants. This is proven by the fact that when occurring at low copy numbers, the B chromosomes have only a few harmful effects but when accrued in plants, they reduce plant growth and fertility (Jones 1995). These cumulative deleterious effects could reduce the accretion of B chromosomes and maintain their frequencies in plant populations (Werren 2011). While the precise mechanisms accounting for losses of fitness are not yet determined, from a physiological angle, these harmful outcomes could simply reflect losses caused by replication and subsistence of nonessential DNA concerned with the aggregation of B chromosomes (Lynch and Marinov 2015).

In general, premeiotic gene drives are maintained at medium frequencies in most plant systems. Nevertheless, despite the advantage in transmission, there exists loss of fitness related with the transport of the drive element that could explain the sustenance of polymorphisms and the deduction as to why these drives do not progress to fixation. In theory, the maize Ab10 system could attain a rate of transmission of 83% in heterozygotes, while the observed empirical transmission rates exist at between 60% and 80% (Higgins et al. 2018). Notwithstanding, in populations of wild teosintes, the Ab10 is present only at low to medium frequencies (Kanizay et al. 2013) the reason for which is likely to be diminished viabilities of the pollen and the seeds in addition to the seed sizes in Ab10 homozygotes (Higgins et al. 2018). In a similar manner, the D locus of *Mimulus canattain* exhibits nearly 100% transmission rate arising from interspecific crosses, whereas in natural populations it segregates at 30–40% frequencies (Fishman and Kelly 2015), demonstrating significant loss (20%) in fertility of pollen (Fishman and Saunders 2008) as well as reduced seed size (Fishman and Kelly 2015).

3.2   Gene Drives Occurring at the Post-gametic Phase

When distortion in chromosome segregation causing gamete inviability occurs after meiosis completion, this is called post-gametic gene drive. To accomplish the post-gametic drive, the drive allele must be able to distinguish between competing (self- versus non-self) meiotic products. Also, the drive allele must lead to inviability of the competing meiotic product (Bravo Núñez et al. 2018).

A majority of post-gametic drives in plants have been recognized by identification of hybrid sterility subsequent to crosses among crops and their equivalent wild counterparts. Successive investigations of traits determining sterility have identified gene loci that skew allelic transmittance by operating as "gamete killers" (Endo 2015). Several post-gametic gene drives described hitherto operate through a wide array of molecular processes. Several of them segregate in the form of a singular locus possessing many genes that are tightly linked that in turn cause direct obstruction of gametes transporting non-self-alleles or execute the generation of a toxic gene element and its antidote (Sweigart et al. 2019). The *Oryza sativa* qHMS7 locus establishes gene drive by means of a poison-antidote mechanism that functions on male gametes (Yu et al. 2018). In the backcrossed hybrids of sativa/meridionalis, the pollen inheriting the meridionalis qHMS7 haplotype becomes infertile. Herein, the qHMS7 locus of sativa has the closely linked ORF2 gene coding for a pollen-destroying toxin and the ORF3 gene that codes for an antidote molecule to the toxin encoded by ORF2. However, the qHMS7 locus of meridionalis codes for an ORF2-encoded molecule that is nontoxic, and additionally this locus does not possess ORF3.

The kinetics of gamete destroyers among populations of plant species as yet remains largely obscure. In all of the systems elucidated thus far, post-gametic gene drive loci have been identified among species where the plant phenotype is revealed only through interspecific matings. Such an observation could simply reflect the capability of these loci to drive through plant species at high efficacy. This robust advantage in transmission caused by the destroying competing gametes could mean that theoretically they could happen to be stabilized in plant populations at a rapid rate.

However, this transmission advantage could be offset by losses, the most significant of which is decreased fecundity which has potent ability to alter the post-gametic drive population dynamics that destroy pollen (Knight et al. 2005). How these drawbacks can be surmounted using synthetic or constructed gene drives warrants further investigations concerning the function of selfish element progression in driving their development (Sweigart et al. 2019).

4   Weed Control Using Indirect Genetic Engineering Programs

Agricultural practices such as selection of cultivars, breeding of plant, or crop germplasm genetic engineering have been variously applied to bring about agricultural pest control. In this context, many crops and their wild populations code for genetic variations toward resistance to plant pathogens (Rudd et al. 2001) as well as insect pests (Panda and Khush 1995). Varieties of crops could show differences in their weed resistance (Lemerle et al. 1996) and allelopathic potential (Belz 2007). Genetic traits that confer augmented resilience or even resistance against pathogens and pest species have been identified in modern crops, and these genetic traits can be re-instated via marker-enabled breeding as long as they do not result in negative agricultural and nutritional impacts. Moreover, modern agricultural practices have been directed by the emergence of transgenic crop technology that expresses inimitable herbicide as well as pest resistance gene traits in commercial crops (James 2010). Thus, transgenic crops encoding herbicide resistance have revolutionized the management of weed species by facilitating the use of non-selective, wide spectrum multitudes of herbicides toward control of weed populations (Duke 2015). Notwithstanding, despite some successful developments in this area, these crops do face significant hindrances in proliferating the respective engineered genes through widespread natural populations.

5 Controlling Agricultural Pests Using Direct Genetic Engineering Programs

Gene drives contain selfish genetic elements that are effectually spread between plant species that reproduce sexually despite their capability to lower their fitness levels (Hammond et al. 2016; Esvelt et al. 2014). Hence, gene drives can convey genetic traits with greater efficacy than that predicted through inheritance in the conventional Mendelian fashion (Dance 2015; Esvelt et al. 2014). Gene drives to achieve pest control have been under active investigation for over 10 years thus far (Bax and Thresher 2009; Burt 2003; Curtis 1994; Burt and Koufopanou 2004; Spielman 1994; Davis et al. 2000, 2001; Schliekelman et al. 2005; Sinkins and Gould 2006). Nevertheless, in practice there exist several challenges in their implementation. The recent advent of the CRISPR/Cas9 genetic engineering technology has revolutionized that scenario.

6 Weed Control Achieved Through the CRISPR/Cas9 Gene Drives

Containing the growth of agricultural weed species through the CRISPR/Cas9 technology is greatly promising even though overcoming the huge seed production capacity and self-fertilizing life cycle of many weed species makes it highly challenging. As such, gene drive technologies are still in their inception but would most likely develop into productive species-specific, reliable tools to control pests subject to containment in the geographical perspective (Marshall and Akbari 2018). A schematic representation of the CRISPR/Cas9 mechanism as applied toward weed management is detailed in Fig. 1.

The precept of weed population suppression employing the CRISPR/Cas9 technology rests with the presumption that these gene drives can be utilized to release and disperse a fitness load capable of circumscribing the installation, plenitude, dissemination, resilience, and effect of the weed populations. This could be achieved using genetic engineering technologies capable of targeting distinct genetic traits of the weed plants that determine their persistence, competence, seed dormancy, phenology, and morphology (Neve 2018).

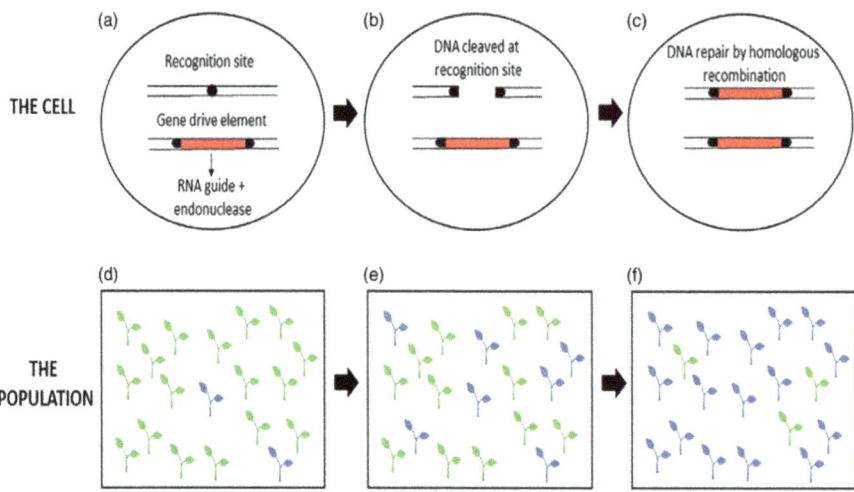

Fig. 1 Control of agricultural weeds using the CRISPR/Cas9 gene drive system. After a successful mating between wild-type weeds and the introduced genetically engineered weed species, plants containing the gene drive adaptation would be produced in their heterozygous state (**a**). Thereupon, the Cas9 nuclease led by the guide RNA would cleave the DNA present in the wild-type chromosome at the precise recognition site (**b**). Consequently, this drive chromosome would act as template to repair the cleavage through homologous recombination. This would result in plants homozygous for the trait encoded by the gene drive (**c**). At the level of the plant population, weeds containing the engineered genetic drive (blue plants) would be released into the wild population of the weed community (green plants) (**d**) and this drive would be propagated across several generations (**e, f**) till the phase at which weeds containing the gene drive allele would be predominant in this weed population. (Adapted from Neve 2018)

Nevertheless, this is constrained by our inadequate comprehension of the molecular biology of the genetic traits in weeds. In spite of this, for instance, the Rht-1 dwarfing homologues practiced in wheat breeding (Thomas 2017) could be recognized in the analogous grass weed plants. If these genes can be disseminated across weed populations, they could diminish the competency of the weeds. In a similar manner, research conducted in weedy rice show the molecular causes of the developed genetic traits including seed shattering and seed dormancy that are involved with the endurance and dispersal of wild rice, thereby conveying one more molecular target for measures to diminish weed fitness in agroecosystems (Nadir et al. 2017). Additionally, meiotic gene drive strategies which distort the sex ratio of the weeds that are dioecious like the *Amaranthus* species could be utilized. Other molecular targets to diminish the reproduction and fecundity of plants could concentrate on genetic engineering techniques obstructing gametogenesis to inhibit the production of ovule or pollen, thereby resulting in the prevention and/or biases in gamete formation.

Weed genomics, an evolving genetic engineering field, has elucidated the importance of research into the molecular genetics of the "weediness" traits to comprehend the emergence of those traits, and this could function as a likely unique source of genetic variability directed toward enhancement of crops (Ravet et al. 2018). With the promise of curtailment of the weed populations through direct genetic intervention, these efforts convey more incentives if these gene drive schemes could be used to change or knock out weedy genetic traits among wild populations.

Considering gene drives that are sensitizing, suppression of the weed population does not depend on the drive element as such but instead is dependent on the ensuing administration, the enforcement of which could be coordinated in space and time (Neve 2018). Due to the abovementioned advantage, sensitizing gene drives can be formulated to transmit engineered gene elements that do not have any intrinsic negative impacts on the fitness of the plants, thus allowing their dispersal but more significantly ensuring suppression of the population is under the control of distinct management measures. The most apparent purpose of these sensitizing gene drives toward weed management would be to restore populations that are herbicide-resistant back to susceptibility against the respective herbicide. Significantly, a majority of the herbicide targets are well-characterized wherein many mutations expressing resistance have been clearly defined (Powles and Yu 2010), because of which this trait reversal is technically achievable.

Steering these susceptibility traits across populations of weeds to the level where efficacious suppression of formerly resistant populations can be attainable would nevertheless require many generations in the course of which the target herbicide cannot be utilized as it will destroy the plant species containing the very drive elements executing this reversal. Importantly, the major caveat is that any resistant weed plants despite occurring at lower frequencies in the weed populations consequent to an efficacious drive event could cause rapid re-emergence of the resistance upon recommencing the use of the herbicide.

Carefully formulated resistance management techniques have to be adopted in order to manage the newly sensitive weed populations. Notwithstanding, this approach is problematic due to the fact that mechanisms driving the herbicide resistance are determined by complicated, possibly polygenic systems wherein their genetic architecture is not yet completely elucidated (Delye et al. 2013).

By means of a combination of synthetic and structural biology along with synthetic chemistry, sensitizing drives could be employed to modify essential, greatly conserved plant genes, thus sensitizing the plants to distinct herbicide categories (Esvelt et al. 2014). These edited plant genes would necessarily have to retain their regular enzymatic function to disseminate across populations. The above measures would enable design of particularly targeted chemical modifications while circumscribing any undesired non-target as well as environmental impacts.

Lately, technologies executing weed control by targeting and destroying mature weed seeds thus circumscribing seed bank replenishments are being used (Neve 2018). Such approaches are dependent on the seed retention in plants at stage of maturity during crop harvest times in order to enable their segregation and destruction. In the case of the major annual grass weed *Alopecurus myosuroides*, prevalent in Northwest Europe, early shattering of the seeds limits the prospects of the abovementioned strategies. Gene drives designed to target the seed-shattering gene loci could be employed to facilitate seed retention on the parent weeds until the time of

crop harvest. Integration of this approach along with the harvest weed control measures could be performed as a new control methodology for this as well as other
fractious weed species. Such a strategy is conceptually feasible since single-
nucleotide polymorphisms occurring within targeted genes have caused the disap-
pearance of the seed shattering trait during the course of domestication of rice
(Konishi et al. 2006).

7 Weed Gene Drives

There are several routes for agricultural weed management, the most significant of
which are the sensitizing gene drives. Nevertheless, there are many constraints in
the production of transformation systems toward weed control, in particular, those
compatible with genome editing CRISPR/Cas9 technology as well as the recognition of ideal molecular targets. In this light, gene drives for weed control must be
able to surmount barriers in the weed molecular genetics in addition to those exhibited by ecological parameters and biology of populations. Therefore, the fruitful
formulation and performance of gene drives is compellingly dependent on collabo-
rations between weed ecologists, molecular geneticists, and evolutionary biologists.

To achieve success in gene drives designed for weed management, the edited
genes must spread through the weed populations at rates high enough to facilitate
adequate population control within the constraints of a reasonable time frame. As
predicted by modeling studies, nearly 10–20 generation times are necessary to disseminate the edited genes so as to promote the fixation of a given trait in the natural
weed populations, subject to repercussions in the fitness of the edited genes (i.e.,
selection coefficient) and conversion effectiveness of the CRISPR/Cas9-based targeting and gene editing reactions (Unckless et al. 2017). In the light of the above,
there is a compelling requirement to perform additional modeling investigations to
recognize the full capability of gene drives as applied to agricultural weed
containment.

Even as genome engineering has been by and large used as a molecular device
for genetic enhancement of culturally and economically significant crop species,
synthetic molecular tools for gene regulation of wild populations of plants have
received relatively less attention (Barrett et al. 2019; Hay et al. 2010; Gould 2008;
Neve 2018).

Gene drives contain selfish genes that can skew the segregation ratios during
gamete development or meiosis (Burt and Trivers 2009). Hence, they can spread
across populations while imposing high costs of fitness on their plant hosts
(Lindholm et al. 2016). In theory, gene drives could be constructed to cause favorable genetic modifications in wild plant populations. The great promise of the use of
gene drives was foreseen previously; however, the genetic engineering technologies
needed to implement such drives did not exist before recent times.

Currently, novel insights into the CRISPR/Cas9 genome editing technologies
have paved the way for genetically modifying plants (Webber et al. 2015; Neve

2018). For the successful execution of a given gene drive, there has to exist synergy between intervention schemes developing at the genetic, molecular, cellular, organismic, and population levels. This is especially important while applying these gene drives to manage wild populations of plants, bearing in mind that plants exhibit a wide variety of DNA repair mechanisms, cytogenetic arrangements, and life histories.

Quickly driven novel, adaptive genetic variations in plants could improve fitness of plant populations by augmenting resistance to invasive pathogens and fluctuations in the environment (Novak et al. 2018). Such adaptive strategies would augment the frequency of dispersal of the beneficial genes and will ensure the future viability of these populations by activating the adaptive genetic traits even before the appearance of the concerned threat. Yet, the main caveat of this approach is the obstacle in identifying genes coding for these adaptive genetic traits.

8   Genetic Modification of Polyploids

Synthetic or artificial gene drives can moreover be formulated to work at reduced genomic levels. There are great challenges in the engineering of polyploid plant species because of the incidence of genome duplication as well as the occurrence of several homologous alleles. Hence, even while genome editing driven by the CRISPR/Cas9 mechanism has been engineered in various polyploid crop plants such as potato, strawberries, oilseed rape, and wheat, the outcome of such genetic manipulations is greatly variable ranging from the editing of only a single allele (Zhang et al. 2016; Andersson et al. 2017) through different assortments of homologous edited alleles (Wilson et al. 2019) to editing of all of the entire alleles in a genome (Braatz et al. 2017; Andersson et al. 2017; Zhang et al. 2016; Wilson et al. 2019). This poses a great challenge as gene editing across all of the gene alleles could be necessary to engineer a desirable altered genetic trait in plant species that are polyploid. In this context, gene drives derived through endonuclease technologies could be very much useful to engineer polyploid species as long as they are constructed to drive the respective gene edits through all of the copies of the genes and homologous alleles in their genomes.

Modern synthetic gene drive measures employ the CRISPR/Cas9 technology or the synthetic toxin-antidote approaches (e.g., Webber et al. 2015; Neve 2018; Novak et al. 2018). Still, there exist many selfish gene elements appearing naturally with capability to distort the segregation ratios described in plants (Lindholm et al. 2016; Bravo Núñez et al. 2018). Understanding the mechanisms enabling the drive of such elements and determinants affecting the drive frequencies within populations can possibly bring to light ideal routes to design and execute these synthetic drive technologies. A comprehensive approach to mechanistically regulate various varieties of the drive elements (Bravo Núñez et al. 2018) is to recognize whether they disrupt the ratios of transmission prior to or following gametogenesis.

In order to be effectual, gene drives should be conducted only in those populations undergoing sexual reproduction and having shortened generation times. Moreover, gene drive efficacy is determined by the target weed breeding structure, frequency of gene flow, and geographical distribution of the weeds (NASEM 2016). This implies that not all the weed species are suitable for population control by means of the gene drive technology. Additionally, there are several risks involved in the implementation of the gene drives.

9 Risks of Weed Gene Drives

Gene drives have the potential to permanently change plant communities within a short period of time, and this could have beneficial or detrimental effects. On the negative side, this could lead to deterioration of current ecosystems which in turn can result in irreversibly harmful outcomes. Therefore, gene drives have to be practiced bearing in mind that all the concerned stakeholders worldwide are committed to mandatory social contracts.

Before contemplating the application of pest control gene drives, many important issues have to be addressed (Medina 2018) such as the following: What are the ecological outcomes of the extinction of the respective pest species? How many of the pest species can we eliminate until we encounter severe negative outcomes? Who can determine which pest species deserve to become extinct? What moral issues are involved in doing away with the survival of a given pest species? Thus, risks encountered in practicing these gene drives globally would be very high and from this perspective could preclude their deployment.

10 The Potential of Gene Drives to Reduce Vector-Borne Diseases (VBD)

Almost one fifth of all infectious diseases are primarily spread by vectors such as mosquitoes and ticks (de la Fuente et al. 2017). Dengue fever, Lyme disease, and Zika have routinely been found in warm, tropical climates; however, more recently they have been isolated in temperate regions such as Europe and North America (Tjaden et al. 2018). The reason for this shift in geography is in part due to international trade, climate change, and globalization; together these are responsible for the emerging new threats (United Nations 2018).

Today, transmission of infectious diseases carried by mosquitoes can be augmented by the use of protective clothing, improved sanitation, bed nets, improved water management, and insecticide sprays. However, insecticides can be costly and can create insect resistance, while protective clothing and bed nets are difficult to widely distribute (Rakotoson et al. 2017). Global initiatives such as the World

Mosquito Program make use of *Wolbachia*, an endosymbiont bacterium that can disrupt disease transmission in biting insects (Curtis and Sinkins 1998). These natural gene drives could potentially control mosquito and tick populations in the future.

Modern gene drive technology makes it possible to limit the spread of diseases such as dengue, Zika, and malaria by controlling mosquito populations. Oxitec, a biotech company, has a mobile laboratory in Brazil and develops genetically engineered mosquitoes with a gene added that prevents their offspring from reaching maturity. GM mosquitoes released into regions that include nearby urban areas can thus reduce diseases that are prevalent, such as chikungunya and dengue. The male mosquitoes contain a self-limiting gene that prevents their survival before they are mature enough to reproduce. The construct includes a reporter gene with a promoter under the control of tetracycline. When this antibiotic is included in the mosquitoes' water, the self-limiting gene is inactivated. Within the lab, the mosquitoes can reach maturity and be released into the environment. Lab-grown males can mate with wild females, and since the offspring lack access to tetracycline, the self-limiting gene is activated, resulting in an early death. In 2016, Oxitec used this technology to reduce mosquito populations in the Cayman Islands by 60%.

As another example, Target Malaria uses a CRISPR-Cas9 gene drive strategy for specific species of mosquitoes that carry malaria. When gene drive males are released and mate with wild females, they will reduce the number of viable offspring. After multiple generations, most of the modified mosquitoes will be eliminated from the population. The gene drive is so highly specific that it will not impact other mosquito species found within that ecosystem (McFarling 2017). Much work in this field continues to be explored (Carballar-Lejarazú et al. 2020; Simoni et al. 2020; Williams et al. 2020).

11   Using Gene Drive to Combat Lyme Disease

Tick-transmitted Lyme disease is created by *Borrelia* bacteria and can cause neurological problems to arthritis. Mice are the main reservoir of *Borrelia*. This microbe has a complex genome including multiple plasmids that create unique manifestations of the disease (Casjens et al. 2018). Gene drive mice that harbor immunity to *Borrelia* through production of an antibody is one way to block Lyme disease (Hammond et al. 2016). Another, broader approach would be to immunize mice against a protein found in tick saliva. Unlike conventional vaccination, the immunity could be passed on from one generation of mice to the next via gene drive (Enzmann 2018). By releasing gene drive mice into the wild in island communities such as Nantucket and Martha's Island, where Lyme disease is highly prevalent, the short reproductive cycle of mice could spread immunity to this disease within a short time period (Bouchard 2017). Such a study is currently ongoing (Buchthal et al. 2019).

12   Other Examples of Gene Drives for Pest Management

To date, many other examples of gene drive for pest management have been studied. These have been implemented by insertion of gene drive technology into the insect pests themselves rather than the plants. Some of these include alterations in reproduction, pupae development, and flightlessness traits in *Plutella xylostella*, *Cochliomyia hominivorax*, *Lucilia cuprina*, *Spodoptera litura*, *Bactrocera dorsalis*, and *Drosophila suzukii* (Chen et al. 2019; Paulo et al. 2019; Bi et al. 2019; Zhao et al. 2019; Peng et al. 2015; Huang et al. 2017; Ahmed et al. 2019; Zheng et al. 2019).

Gene drives that focus on plant resistance to insect pests are also under development. For example, Block et al. (2019) have shown that by altering maize terpenoid biosynthesis, it would be possible to increase insect resistance. This research is currently underway.

13   Conclusions and Future Prospects

Gene drives undoubtedly present promise for use in controlling insect vectors of diseases and agricultural pests and for ecosystem conservation yet also come with potential risks such as unknown downstream effects or accidental or nefarious spread. A code of ethics may be one step toward the development and regulation of gene drives. This could be used to ensure that a balanced analysis of risks and benefits is taken into consideration for each gene drive application. This has been accomplished with the work of Target Malaria, and this case could serve as an example of an effective introduction of gene drive technology to society. It is important to remember that events such as the current COVID-19 pandemic can negatively impact public trust. For example, a high number of vaccine-hesitant people have been fed misinformation and conspiracy theories. The role of science communication for the use of gene drive technology to combat some of our most pressing challenges in agriculture today will be paramount.

References

Ahmed HMM, Hildebrand L, Wimmer EA (2019) Improvement and use of CRISPR/Cas9 to engineer a sperm-marking strain for the invasive fruit pest Drosophila suzukii. BMC Biotechnol 19(1):85. https://doi.org/10.1186/s12896-019-0588-5. PMID: 31805916; PMCID: PMC6896403

Andersson M, Turesson H, Nicolia A, Fält A-S, Samuelsson M, Hofvander P (2017) Efficient targeted multiallelic mutagenesis in tetraploid potato (Solanum tuberosum) by transient CRISPR–Cas9 expression in protoplasts. Plant Cell Rep 36:117–128. https://doi.org/10.1007/s00299-016-20

Annas GJ, Beisel CL, Clement K, Crisanti A, Francis S, Galardini M, Galizi R, Grünewald J, Immobile G, Khalil AS, Müller R, Pattanayak V, Petri K, Paul L, Pinello L, Simoni A, Taxiarchi

C, Joung JK (2021) A code of ethics for gene drive research. CRISPR J 4(1):19–24. https://doi.org/10.1089/crispr.2020.0096. Epub 2021 Feb 10. PMID: 33571044; PMCID: PMC7898401

Barrett LG, Legros M, Kumaran N, Glassop D, Raghu S, Gardiner DM (2019) Gene drives in plants: opportunities and challenges for weed control and engineered resilience. Proc Biol Sci 286(1911):20191515. https://doi.org/10.1098/rspb.2019.1515

Bax NJ, Thresher RE (2009) Ecological, behavioral, and genetic factors influencing the recombinant control of invasive pests. Ecol Appl 19:873–888

Belz RG (2007) Allelopathy in crop/weed interactions – an update. Pest Manag Sci 63:308–326

Bi HL, Xu J, He L, Zhang Y, Li K, Huang YP (2019) CRISPR/Cas9-mediated ebony knockout results in puparium melanism in Spodoptera litura. Insect Sci 26(6):1011–1019. https://doi.org/10.1111/1744-7917.12663. Epub 2019 Mar 5

Bisht DS, Bhatia V, Bhattacharya R (2019) Improving plant-resistance to insect-pests and pathogens: the new opportunities through targeted genome editing. Semin Cell Dev Biol 96:65–76. https://doi.org/10.1016/j.semcdb.2019.04.008. Epub 2019 May 8. PMID: 31039395.62-3

Block AK, Vaughan MM, Schmelz EA, Christensen SA (2019) Biosynthesis and function of terpenoid defense compounds in maize (Zea mays). Planta 249(1):21–30. https://doi.org/10.1007/s00425-018-2999-2. Epub 2018 Sep 6

Bouchard S (2017) Gene 'editing' on mice tested in war on ticks. News of Maine's Coast and Islands. http://www.islandinstitute.org/working-waterfront/%E2%80%8Bgene-%E2%80%98editing%E2%80%99-mice-tested-war-ticks

Braatz J, Harloff H-J, Mascher M, Stein N, Himmelbach A, Jung C (2017) CRISPR-Cas9 targeted mutagenesis leads to simultaneous modification of different homoeologous gene copies in polyploid oilseed rape (Brassica napus). Plant Physiol 174:935–942. https://doi.org/10.1104/pp.17.00426

Bravo Núñez MA, Nuckolls NL, Zanders SE (2018) Genetic villains: killer meiotic drivers. Trends Genet 34:424–433. https://doi.org/10.1016/j.tig.2018.02.003

Brosnan JT, Vargas JJ, Breeden GK, Grier L, Aponte RA, Tresch S, Laforest M (2016) A new amino acid substitution (Ala-205-Phe) in acetolactate synthase (ALS) confers broad spectrum resistance to ALS-inhibiting herbicides. Planta 243:149–159

Buchman A, Marchall JM, Ostrovski D, Yang T, Akbari OS (2018) Synthetically engineered Medea gene drive system in the worldwide crop pest Drosophila suzukii. PNAS 115(18):4725–4730

Buchthal J, Evans SW, Lunshof J, Telford SR 3rd, Esvelt KM (2019) Mice against ticks: an experimental community-guided effort to prevent tick-borne disease by altering the shared environment. Philos Trans R Soc Lond Ser B Biol Sci 374(1772):20180105. https://doi.org/10.1098/rstb.2018.0105. PMID: 30905296; PMCID: PMC6452264

Burt A (2003) Site-specific selfish genes as tools for the control and genetic engineering of natural populations. Proc R Soc B Biol Sci 270:921–928

Burt A, Koufopanou V (2004) Homing endonuclease genes: the rise and fall and rise again of a selfish element. Curr Opin Genet Dev 14:609–615

Burt A, Trivers R (2009) Genes in conflict: the biology of selfish genetic elements. Harvard University Press, Cambridge, MA

Callaway E (2017) Gene drives thwarted by emergence of resistant organisms. Nature News, January 31

Carballar-Lejarazú R, Ogaugwu C, Tushar T, Kelsey A, Pham TB, Murphy J, Schmidt H, Lee Y, Lanzaro GC, James AA (2020) Next-generation gene drive for population modification of the malaria vector mosquito, Anopheles gambiae. Proc Natl Acad Sci U S A 117(37):22805–22814. https://doi.org/10.1073/pnas.2010214117. Epub 2020 Aug 24. PMID: 32839345; PMCID: PMC7502704

Casjens SR, Di L, Akther S, Mongodin EF, Luft BJ, Schutzer SE, Fraser CM, Qiu WG (2018) Primordial origin and diversifcation of plasmids in Lyme disease agent bacteria. BMC Genomics 19(1):218

Chen W, Yang F, Xu X, Kumar U, He W, You M (2019) Genetic control of Plutella xylostella in omics era. Arch Insect Biochem Physiol 102(3):e21621. https://doi.org/10.1002/arch.21621. Epub 2019 Sep 20

Curtis CF (1994) The case for malaria control by genetic manipulation of its vectors. Parasitol Today 10:371–374

Curtis CF, Sinkins SP (1998) Wolbachia as a possible means of driving genes into populations. Parasitology 116(Suppl):S111–S115

Dance A (2015) Core concept: CRISPR gene editing. Proc Natl Acad Sci 112:6245–6246

Davis SA, Catchpole EA, Fulford GR (2000) Periodic triggering of an inducible gene for control of a wild population. Theor Popul Biol 58:95–106

Davis S, Bax N, Grewe P (2001) Engineered underdominance allows efficient and economical introgression of traits into pest populations. J Theor Biol 212:83–98

Dawe RK et al (2018) A kinesin-14 motor activates neocentromeres to promote meiotic drive in maize. Cell 173:839–850.e18. https://doi.org/10.1016/j.cell.2018.03.009

de la Fuente J, Antunes S, Bonnet S, Cabezas-Cruz A, Domingos AG, Estrada-Peña A, Johnson N, Kocan KM, Mansfeld KL, Nijhof AM, Papa A, Rudenko N, Villar M, Alberdi P, Torina A, Ayllón N, Vancova M, Golovchenko M, Grubhoffer L, Caracappa S, Fooks AR, Gortazar C, Rego RT (2017) Tick-pathogen interactions and vector competence: identification of molecular drivers for tick-borne diseases. Front Cell Infect Microbiol 7:114. https://doi.org/10.3389/fcimb.2017.0011

Délye C, Jasieniuk M, Le Corre V (2013) Deciphering the evolution of herbicide resistance in weeds. Trends Genet 29(11):649–658. https://doi.org/10.1016/j.tig.2013.06.001

Duke SO (2015) Perspectives on transgenic, herbicide-resistant crops in the United States almost 20 years after introduction. Pest Manag Sci 71:652–657

Dumont M, Letarte J, Tardif FJ (2016) Identification of a psbA mutation (valine 219 to isoleucine) in Powell amaranth (Amaranthus powellii) conferring resistance to linuron. Weed Sci 64:6–11

Endo TR (2015) Gametocidal genes. In: Molnár-Láng M, Ceoloni C, Doležel J (eds) Alien introgression in wheat: cytogenetics, molecular biology, and genomics. Springer, Cham, pp 121–131

Enzmann B (2018) CRISPR and the community are teaming up to combat lyme disease. https://www.synthego.com/blog/crispr-lyme-disease

Epanchin-Niell RS, Hufford MB, Aslan CE, Sexton JP, Port JD, Waring TM (2010) Controlling invasive species in complex social landscapes. Front Ecol Environ 8:210–216. https://doi.org/10.1890/090029

Esvelt KM, Smidler AL, Catteruccia F, Church GM (2014) Concerning RNA-guided gene drives for the alteration of wild populations. eLife J Responsible Innov 5:S81–S97. https://doi.org/10.7554/eLife.03401

Evans BR, Kotsakiozi P, Costa-da-Silva AL, Ioshino RS, Garziera L, Pedrosa MC, Powell JR (2019) Transgenic Aedes aegypti mosquitoes transfer genes into a natural population. Sci Rep 9(1):1–6

Fishman L, Kelly JK (2015) Centromere-associated meiotic drive and female fitness variation in Mimulus: female fitness costs of meiotic drive. Evolution 69:1208–1218. https://doi.org/10.1111/evo.12661

Fishman L, Saunders A (2008) Centromere-associated female meiotic drive entails male fitness costs in monkey flowers. Science 322:1559–1562. https://doi.org/10.1126/science.1161406

Fishman L, Willis JH (2005) A novel meiotic drive locus almost completely distorts segregation in Mimulus (monkey flower) hybrids. Genetics 169:347–353. https://doi.org/10.1534/genetics.104.032789

Giacomini DA, Umphres AM, Nie H, Mueller TC, Steckel LE, Young BG, Scott RC, Tranel PJ (2017) Two new PPX2 mutations associated with resistance to PPO-inhibiting herbicides in Amaranthus palmeri. Pest Manag Sci 73:1559–1563

Gould F (2008) Broadening the application of evolutionarily based genetic pest management. Evolution 62:500–510. https://doi.org/10.1111/j.1558-5646.2007.00298.x

Gould F, Brown ZS, Kuzma J (2018) Wicked evolution: can we address the sociobiological dilemma of pesticide resistance? Science 360:728–732. https://doi.org/10.1126/science.aar3780

Guo W, Zhang L, Wang H, Li Q, Liu W, Wang J (2017) A rare Ile-2041-Thr mutation in the ACCase gene confers resistance to ACCase-inhibiting herbicides in shortawn foxtail (Alopecurus aequalis). Weed Sci 65:239–246

Hammond A, Galizi R, Kyrou K, Simoni A, Siniscalchi C, Katsanos D, Gribble M et al (2016) A CRISPR-Cas9 gene drive system-targeting female reproduction in the malaria mosquito vector Anopheles Gambiae. Nat Biotechnol 34:78–83

Hay BA, Chen C-H, Ward CM, Huang H, Su JT, Guo M (2010) Engineering the genomes of wild insect populations: challenges, and opportunities provided by synthetic Medea selfish genetic elements. J Insect Physiol 56:1402–1413. https://doi.org/10.1016/j.jinsphys.2010.05.022

Higgins DM, Lowry EG, Kanizay LB, Becraft PW, Hall DW, Dawe RK (2018) Fitness costs and variation in transmission distortion associated with the abnormal chromosome 10 meiotic drive system in maize. Genetics 208:297–305. https://doi.org/10.1534/genetics.117.300060

Huang Y, Wang Y, Zeng B, Liu Z, Xu X, Meng Q, Huang Y, Yang G, Vasseur L, Gurr GM, You M (2017) Functional characterization of Pol III U6 promoters for gene knockdown and knockout in Plutella xylostella. Insect Biochem Mol Biol 89:71–78. https://doi.org/10.1016/j.ibmb.2017.08.009. Epub 2017 Sep 7

James C (2010) A global overview of biotech (GM) crops: adoption, impact and future prospects. GM Crops 1:8–12

Jones RN (1995) B chromosomes in plants. New Phytol 131:411–434. https://doi.org/10.1111/j.1469-8137.1995.tb03079.x

Jones RN (2018) Transmission and drive involving parasitic B chromosomes. Gene 9:388. https://doi.org/10.3390/genes9080388

Kanizay LB, Pyhäjärvi T, Lowry EG, Hufford MB, Peterson DG, Ross-Ibarra J, Dawe RK (2013) Diversity and abundance of the abnormal chromosome 10 meiotic drive complex in Zea mays. Heredity 110:570–577. https://doi.org/10.1038/hdy.2013.2

Kaundun SS (2014) Resistance to acetyl-CoA carboxylase-inhibiting herbicides. Pest Manag Sci 70:1405–1417

Knight TM et al (2005) Pollen limitation of plant reproduction: pattern and process. Annu Rev Ecol Evol Syst 36:467–497. https://doi.org/10.1146/annurev.ecolsys.36.102403.115320

Konishi S, Izawa T, Lin SY, Ebana K, Fukuta Y, Sasaki T et al (2006) An SNP caused loss of seed shattering during rice domestication. Science 312:1392–1396

Kreiner JM, Stinchcombe JR, Wright SI (2018) Population genomics of herbicide resistance: adaptation via evolutionary rescue. Annu Rev Plant Biol 69:611–635. https://doi.org/10.1146/annurev-arplant-042817-040038

Languin K (2014) Genetic engineering to the rescue against invasive species? National Geographic, July 18

Larran AS, Palmieri VE, Perotti VE, Lieber L, Tuesca D, Permingeat HR (2017) Target-site resistance to acetolactate synthase (ALS)-inhibiting herbicides in Amaranthus palmeri from Argentina. Pest Manag Sci 73:2578–2584

LeClere S, Wu C, Westra P, Sammons RD (2018) Cross-resistance to dicamba, 2, 4-D, and fluroxypyr in Kochia scoparia is endowed by a mutation in an AUX/IAA gene. Proc Natl Acad Sci U S A 115:E2911–E2920

Lemerle D, Verbeek B, Cousens RD, Coombes NE (1996) The potential for selecting wheat varieties strongly competitive against weeds. Weed Res 36:505–513

Li Z, Boyd N, McLean N, Rutherford K (2014) Hexazinone resistance in red sorrel (Rumex acetosella). Weed Sci 62:532–537

Li J, Peng Q, Han H, Nyporko A, Kulynych T, Yu Q, Powles S (2018) Glyphosate resistance in Tridax procumbens via a novel EPSPS Thr-102-Ser substitution. J Agric Food Chem 66:7880–7888

Li J, Aidlin Harari O, Doss AL, Walling LL, Atkinson PW, Morin S, Tabashnik BE (2020) Can CRISPR gene drive work in pest and beneficial haplodiploid species? Evol Appl

13(9):2392–2403. https://doi.org/10.1111/eva.13032. PMID: 33005229; PMCID: PMC7513724

Lindholm AK, Dyer KA, Firman RC, Fishman L, Forstmeier W, Holman L, Johannesson H, Knief U, Kokko H, Larracuente AM, Manser A, Montchamp-Moreau C, Petrosyan VG, Pomiankowski A, Presgraves DC, Safronova LD, Sutter A, Unckless RL, Verspoor RL, Wedell N, Wilkinson GS, Price TAR (2016) The ecology and evolutionary dynamics of meiotic drives. Trends Ecol Evol 31:315–326

Lucas JA, Hawkins NJ, Fraaije BA (2015) The evolution of fungicide resistance. Adv Appl Microbiol 90:29–92

Lynch M, Marinov GK (2015) The bioenergetic costs of a gene. Proc Natl Acad Sci USA 112:15 690–15 696. https://doi.org/10.1073/pnas.1514974112

Marshall J, Akbari O (2018) Can CRISPR-based gene drive be confined in the wild? A question for molecular and population biology. ACS Chem Biol 13:424–430

Masabni JG, Zandstra BH (1999) A serine-to-threonine mutation in linuron-resistant Portulaca oleracea. Weed Sci 47:393–400

McFarling UL (2017) Could this zoo of mutant mosquitoes lead the way to eradicating Zika? Stat News. https://www.statnews.com/2017/12/13/gene-drive-mosquitoes-darpa/

Mechant E, De Marez T, Hermann O, Olsson R, Bulcke R (2008) Target site resistance to metamitron in Chenopodium album L. J Plant Dis Prot Spec 21:37–40

Medina RF (2018) Gene drives and the management of agricultural pests. J Responsible Innov 5(S1):S255–S262

Nadir S, Xiong HB, Zhu Q, Zhang XL, Xu HY, Li J et al (2017) Weedy rice in sustainable rice production. A review. Agron Sustain Dev 37:46

NASEM (National Academies of Sciences, Engineering, and Medicine) (2016) Gene drives on the horizon: advancing science, navigating uncertainty, and aligning research with public values. The National Academies Press, Washington, DC. https://doi.org/10.17226/23405

National Academies Press (2016) Gene drives on the horizon: advancing science, navigating uncertainty, and aligning research with public values. Available at https://www.nap.edu/catalog/23405/gene-drives-on-the-horizon-advancing-science-navigating-uncertainty-and. Aligning Research with Public Values, viewed on May 3, 2017

Neve P (2018) Gene drive systems: do they have a place in agricultural weed management? Pest Manag Sci 74:2671–2679. https://doi.org/10.1002/ps.5137

Novak BJ, Maloney T, Phelan R (2018) Advancing a new toolkit for conservation: from science to policy. CRISPR J 1:11–15. https://doi.org/10.1089/crispr.2017.0019

Panda N, Khush GA (1995) Host plant resistance to insects. CAB International, Wallingford

Panozzo S, Scarabel L, Rosan V, Sattin M (2017) A new Ala-122-Asn amino acid change confers decreased fitness to ALS-resistant Echinochloa crus-galli. Front Plant Sci 8:2042

Park KW, Mallory-Smith CA (2006) psbA mutation (Asn266 to Thr) in Senecio vulgaris L. confers resistance to several PS II-inhibiting herbicides. Pest Manag Sci 62:880–885

Patzoldt WL, Hager AG, McCormick JS, Tranel PJ (2006) A codon deletion confers resistance to herbicides inhibiting protoporphyrinogen oxidase. Proc Natl Acad Sci U S A 103:12329–12334

Paulo DF, Williamson ME, Arp AP, Li F, Sagel A, Skoda SR, Sanchez-Gallego J, Vasquez M, Quintero G, Pérez de León AA, Belikoff EJ, Azeredo-Espin AML, McMillan WO, Concha C, Scott MJ (2019) Specific Gene Disruption in the Major Livestock Pests Cochliomyia hominivorax and Lucilia cuprina Using CRISPR/Cas9. G3 (Bethesda) 9(9):3045–3055. https://doi.org/10.1534/g3.119.400544. PMID: 31340950; PMCID: PMC6723136

Peng W, Zheng W, Handler AM, Zhang H (2015) The role of the transformer gene in sex determination and reproduction in the tephritid fruit fly, Bactrocera dorsalis (Hendel). Genetica 143(6):717–727. https://doi.org/10.1007/s10709-015-9869-7

Perez-Jones A, Intanon S, Mallory-Smith C (2009) psbA mutation (Phe 255 to Ile) in Capsella bursa-pastoris confers resistance to triazinone herbicides. Weed Sci 57:574–578

Perotti VE, Larran AS, Palmieri VE, Martinatto AK, Permingeat HR (2020) Herbicide resistant weeds: a call to integrate conventional agricultural practices, molecular biology knowledge and new technologies. Plant Sci 290:110255

Powles SB, Yu Q (2010) Evolution in action: plants resistant to herbicides. Annu Rev Plant Biol 61:317–347

Rakotoson JD, Fornadel CM, Belemvire A, Norris LC, George K, Caranci A, Lucas B, Dengela D (2017) Insecticide resistance status of three malaria vectors, Anopheles gambiae (s.l.), An. funestus and An. mascarensis, from the South, Central and East coasts of Madagascar. Parasit Vectors 10(1):396. https://doi.org/10.1186/s13071-017-2336-9

Rangani G, Salas-Pérez RA, Aponte RA, Knapp M, Craig IR, Meitzner T, Langaro AC, Noguera MM, Porri A, Roma-Burgos N (2019) A novel single-site mutation in the catalytic domain of Protoporphyrinogen oxidase IX (PPO) confers resistance to PPO-inhibiting herbicides. Front Plant Sci 10:568

Ranson H, N'Guessan R, Lines J, Moiroux N, Nkuni Z, Corbel V (2011) Pyrethroid resistance in African anopheline mosquitoes: what are the implications for malaria control? Trends Parasitol 27:91–98

Ravet K, Patterson EL, Krähmer H, Hamouzová K, Fan L, Jasieniuk M et al (2018) The power and potential of genomics in weed biology and management. Pest Manag Sci 74:2216–2225. https://doi.org/10.1002/ps.5048

Rousonelos SL, Lee RM, Moreira MS, VanGessel MJ, Tranel PJ (2012) Characterization of a common ragweed (Ambrosia artemisiifolia) population resistant to ALS- and PPO-inhibiting herbicides. Weed Sci 60:335–344

Rudd JC, Horsley RD, McKendry AL, Elias EM (2001) Host plant resistance genes for Fusarium head blight. Crop Sci 41:620–627

Salas RA, Burgos NR, Tranel PJ, Singh S, Glasgow L, Scott RC, Nichols RL (2016) Resistance to PPO-inhibiting herbicide in Palmer amaranth from Arkansas. Pest Manag Sci 72:864–869

Sammons RD, Gaines TA (2014) Glyphosate resistance: state of knowledge. Pest Manag Sci 70:1367–1377

Schliekelman P, Ellner S, Gould F (2005) Pest control by genetic manipulation of sex ratio. J Econ Entomol 98:18–34

Schorderet-Weber S, Noack S, Selzer PM, Kaminsky R (2017) Blocking transmission of vector borne diseases. Int J Parasitol Drugs Drug Resist 7(1):90–109

Simoni A, Hammond AM, Beaghton AK, Galizi R, Taxiarchi C, Kyrou K, Meacci D, Gribble M, Morselli G, Burt A, Nolan T, Crisanti A (2020) A male-biased sex-distorter gene drive for the human malaria vector Anopheles gambiae. Nat Biotechnol 38(9):1054–1060. https://doi.org/10.1038/s41587-020-0508-1. Epub 2020 May 11. Erratum in: Nat Biotechnol. 2020 Aug 6;: PMID: 32393821; PMCID: PMC7473848

Sinkins SP, Gould F (2006) Gene drive systems for insect disease vectors. Nat Rev Genet 7:427–435

Spielman A (1994) Why entomological antimalaria research should not focus on transgenic mosquitoes. Parasitol Today 10:374–376

Sweigart AL, Brandvain Y, Fishman L (2019) Making a murderer: the evolutionary framing of hybrid gamete-killers. Trends Genet 35:245–252. https://doi.org/10.1016/j.tig.2019.01.004

Tallmon DA, Luikart G, Waples RS (2004) The alluring simplicity and complex reality of genetic rescue. Trends Ecol Evol 19:489–496. https://doi.org/10.1016/j.tree.2004.07.003

Thiel H, Varrelmann M (2014) Identification of a new PSII target site psbA mutation leading to D1 amino acid leu218val exchange in the Chenopodium album D1 protein and comparison to cross-resistance profiles of known modifications at positions 251 and 264. Pest Manag Sci 70:278–285

Thomas SG (2017) Novel Rht-1 dwarfing genes: tools for wheat breeding and dissecting the function of DELLA proteins. J Exp Bot 68:354–358

Tjaden NB, Caminade C, Beierkuhnlein C, Thomas SM (2018) Mosquito-borne diseases: advances in modelling climate-change impacts. Trends Parasitol 34(3):227–245. https://doi.org/10.1016/j.pt.2017.11.006

Unckless RL, Clark AG, Messer PW (2017) Evolution of resistance against CRISPR/Cas9 gene drive. Genetics 205:827–841

United Nations Department of Economic and Social Affairs (2018). https://www.un.org/development/desa/en/news/population/2018-revision-of-world-urbanization-prospects.html

Webber BL, Raghu S, Edwards OR (2015) Opinion: is CRISPR-based gene drive a biocontrol silver bullet or global conservation threat? Proc Natl Acad Sci USA 112:10 565–10 567. https://doi.org/10.1073/pnas.1514258112

Werren JH (2011) Selfish genetic elements, genetic conflict, and evolutionary innovation. Proc Natl Acad Sci USA 108:10 863–10 870. https://doi.org/10.1073/pnas.1102343108

Whiteley AR, Fitzpatrick SW, Funk WC, Tallmon DA (2015) Genetic rescue to the rescue. Trends Ecol Evol 30:42–49. https://doi.org/10.1016/j.tree.2014.10.009

Williams AE, Franz AWE, Reid WR, Olson KE (2020) Antiviral effectors and gene drive strategies for mosquito population suppression or replacement to mitigate arbovirus transmission by *Aedes aegypti*. Insects 11(1):52. https://doi.org/10.3390/insects11010052. PMID: 31940960; PMCID: PMC7023000

Wilson FM, Harrison K, Armitage AD, Simkin AJ, Harrison RJ (2019) CRISPR/Cas9-mediated mutagenesis of phytoene desaturase in diploid and octoploid strawberry. Plant Methods 15:45. https://doi.org/10.1186/s13007-019-0428-6

Yang W-C, Shi D-Q, Chen Y-H (2010) Female gametophyte development in flowering plants. Annu Rev Plant Biol 61:89–108. https://doi.org/10.1146/annurev-arplant-042809-112203

Yu Q, Powles SB (2014) Resistance to AHAS inhibitor herbicides: current understanding. Pest Manag Sci 70:1340–1350

Yu X et al (2018) A selfish genetic element confers non-Mendelian inheritance in rice. Science 360:1130–1132. https://doi.org/10.1126/science.aar4279

Zhang Y, Liang Z, Zong Y, Wang Y, Liu J, Chen K, Qiu J-L, Gao C (2016) Efficient and transgene-free genome editing in wheat through transient expression of CRISPR/Cas9 DNA or RNA. Nat Commun 7:12617. https://doi.org/10.1038/ncomms12617

Zhao S, Xing Z, Liu Z, Liu Y, Liu X, Chen Z, Li J, Yan R (2019) Efficient somatic and germline genome engineering of Bactrocera dorsalis by the CRISPR/Cas9 system. Pest Manag Sci 75(7):1921–1932. https://doi.org/10.1002/ps.5305. Epub 2019 Jan 30. PMID: 30565410

Zheng W, Li Q, Sun H, Ali MW, Zhang H (2019) Clustered regularly interspaced short palindromic repeats (CRISPR)/CRISPR-associated 9-mediated mutagenesis of the multiple edematous wings gene induces muscle weakness and flightlessness in Bactrocera dorsalis (Diptera: Tephritidae). Insect Mol Biol 28(2):222–234. https://doi.org/10.1111/imb.12540. Epub 2018 Oct 26. PMID: 30260055

Recent Trends in Targeting Genome Editing of Tomato for Abiotic and Biotic Stress Tolerance

S. Anil Kumar, Suman Kumar Kottam, M. Laxmi Narasu, and P. Hima Kumari

Abstract Crop genetic modification is in high demand to address the environmental difficulties for agricultural sustainability, particularly in light of the current condition of global climate change and a drop in the world food production/population rate ratio, which are both on the rise. Most of the times, breeding tomatoes for salt, drought, and fungal stresses is hampered due to polygenic inheritance, linkage drag, availability of limited information on resistance screening, and markers linked to quantitative trait loci (QTL). A revolutionary approach to crop improvement has recently been made possible by the development of clustered regularly interspaced short palindromic repeats (CRISPR) and clustered regularly interspaced short palindromic repeats-associated (Cas) protein. The CRISPR-Cas9 technology is just precise, but also very efficient.

Keywords CRISPR/Cas9 · Tomato · Abiotic stress · Biotic stress · Genome editing · Plant breeding

1 Introduction

Tomatoes originally originated from South America and were considered as food in Mexico for the first time. In the sixteenth century, tomatoes were introduced in Europe and spread throughout the world after Spanish colonization. It belongs to the family Solanaceae, which includes more than 3000 species (Weese and Bohs

S. A. Kumar · S. K. Kottam
Vignan's Foundation for Science, Technology & Research (Deemed to be University), Guntur, Andhra Pradesh, India

M. L. Narasu · P. H. Kumari (✉)
Centre for Biotechnology, Jawaharlal Nehru Technological University, Kukatpally, Hyderabad, Telangana, India

© The Author(s), under exclusive license to Springer Nature Switzerland AG 2022
S. H. Wani, G. Hensel (eds.), *Genome Editing*,
https://doi.org/10.1007/978-3-031-08072-2_15

2007). An overview of the tomato history from cultivation to biopharming has been explained by Bergougnoux (2013). The term "tomato" originates from the Spanish word *tomate*. Tomatoes are also called "love apples" or "poor man's orange." Tomato has become a model organism for research purposes due to its ability to grow in a wide range of climatic conditions, short life span, photoperiod insensitivity, relatively smaller genome size of 950 Mb, absence of gene redundancy, ease of development of haploids, high self-fertilization and homozygosity, availability of wide array of mutants, and capacity for vegetative propagation by grafting (McCormick et al. 1986; Menda et al. 2004). Tomatoes were the first genetically modified crop to be used for commercialization in the USA (Bruening and Lyons 2000).

2   Productivity and Total Acreage

After potatoes, tomato is the second most abundant and consumed vegetable. After maize, rice, wheat, potato, soybean, and cassava, it is the seventh most economic crop in the world. The cultivated land and its production are increasing day by day. Tomatoes are produced on a global scale in a quantity of around 186 million tons, and its cultivation is reported in 144 countries (FAOSTAT 2020). China is the primary producer, accounting for around one fourth of global supply, followed by India and the USA. While China ranks first with 50 million tons, India ranks second with 17.5 million tons, followed by the USA with 13.2 million tons, and then Turkey, Egypt, Iran, Italy, Brazil, and Spain. The estimated area used for tomato cultivation in India is about 8.7 lakh hectares.

3   Diversification of Tomato

Tomatoes display extraordinary diversity in terms of physical characteristics and spatial location. It normally grows to a height of 1–3 m and has a delicate stem that frequently tumbles over the ground and creeps over adjacent plants. Branches are often sub-opposite to the inflorescence of the main cyme. In its natural environment, it grows as a perennial but as an annual in temperate regions (Müller 1940). Tomatoes are largely self-pollinating species; however, controlled hybridization is allowed. Tomatoes range in size from 5 mm for tom berries to 1–2 cm for cherry tomatoes to 10 cm or over for wild beefsteak tomatoes. Commercial tomatoes with a diameter of 5–6 cm are the most frequently cultivated. Tomatoes are applied in several ways, including beefsteak sandwiches, sauces and pastes, and salads. The most common tomatoes used in food industry and processing are "slicing" or "globe" tomatoes. While most varieties produce red fruits, a variety of cultivars also produce green, yellow, pink, orange, purple, black, and white fruits. The striped and variegated fruits are particularly remarkable. Tomato fruits appear in a range of forms, including spherical, oblong-shaped, cylindrical, torpedo, and bell-shaped. The size and morphology of tomatoes cultivated in the field are associated with more than ten quantitative trait loci (QTLs) (Tanksley 2004). Tomatoes cultivated for processing sauces are typically elongated, measuring 7–9 cm in length and 4–5 cm in diameter and are referred as plum tomatoes having less moisture. Tomato domestication began in Mexico in the nineteenth century and expanded globally. Domestication and breeding resulted in morphological and physiological changes with less genetic diversity due to artificial selection.

4   Tomato Nutritional Value and Applications

Tomato has high impact on nutrition with low-calorific value and bears no cholesterol. Nonetheless, they are an incredible source of various vital nutrients, β-carotene, γ-carotene, lycopene, ascorbic acid, and supplements rich in antioxidants such as folic acid and selenium. Tomato fruit contains 23 types of major and minor elements like carbon, oxygen, hydrogen, phosphorous, potassium, nitrogen, magnesium, calcium, sulfur, copper, boron, iron, chlorine, molybdenum, manganese, and zinc (Table 1). Consumption of tomatoes and their related products reduces the risk of gastrointestinal and prostate cancer (Franceschi et al. 1994; Giovannucci et al. 1995; Ali et al. 2021). Intake of lycopene suppresses cell proliferation and growth of cancer cells (Levy et al. 1995; Clinton 1998). Tomato consumptions reduce LDL cholesterol, homocysteine, platelet aggregation and protect against cardiovascular diseases (Willcox et al. 2003). Additionally, tomato inhibits breast, colorectal, cervical, ovarian, pancreatic, and lung cancer. It is also used to keep glaucoma, arthritis, sugar levels, and bronchitis at bay. Tomato has a total ORAC (oxygen radical absorbance capacity) of 367 μM TE/100gm. Tomatoes were often alluded to as "poor man's oranges" owing to their high concentration of essential nutrients and their ability to treat a wide range of ailments.

5   Factors Limiting Tomato Production

Plants need optimal environmental conditions for their growth and development. Due to their sedentary nature, plants are always confronted with various biotic and abiotic stresses during different developmental phases. The stress factors are interrelated and cause a cascade of metabolic changes resulting in plant mortality (Rodriguez et al. 2005). These stresses cause osmotic changes in the cell, leading to the metabolic derailment of lower cellular activity, growth, and yield (Zhou et al. 2011). Annually, biotic and abiotic stress factors reduce agricultural output up to 50% (Wang et al. 2003; Oerke 2006). Plants sense and respond to biotic and abiotic stress factors in an intricate and coherent way (Atkinson and Urwin 2012). As a result, plants develop a variety of cascading interactions (Jones and Dangl 2006).

Table 1 Nutritional value of tomato (per 100 grams)

Type of nutrient	Range	References
Carbohydrates	4.5–6.0 g	Ali et al. (2021), Elbadrawy and Sello (2016), Abdullahi et al.
Protein	11–25 g	(2016), Navarro-Gonzalez et al. (2011), Quinet et al. (2019),
Lipid	3.6–5.0	Canene-Adams et al. (2005) and Salunkhe et al. (1974)
Vitamin B9	15 μg	
Niacin	9.7 mg	
Vitamin B6	1.3–1.7 mg	
Thiamin	0.04-0-9 mg	
Vitamin A	833 IU	
Vitamin C	11–85 IU	
Vitamin E	15 mg	
Vitamin K	98 μg	
Sodium	57 mg	
Potassium	18–1000 mg	
Calcium	49–162 mg	
Iron	1.5–6 mg	
Zinc	0.17 mg	
Phosphorus	174–380 mg	
Carotene-β	449 μg	
Carotene-α	101 μg	
Lutein-zeaxanthin	123 μg	
Lycopene	2573 μg	

Upon exposure to abiotic stress, several genes and gene products are induced, which help in ion homeostasis, stimulation of detoxifying enzymes, and formation of late embryogenesis abundant proteins. Numerous intracellular defense mechanisms are initiated in response to biotic stressors, culminating in the production of bactericidal and pathogenesis-related (PR) proteins (Yun et al. 1997; Veronese et al. 2003).

6   Impact of Abiotic and Biotic Stress Factors

Abiotic stress is defined as the damage done to plants by nonliving factors in a specific environment. In many agricultural plants, abiotic stresses are the principal restrictions determining crop productivity and nutrient content (Salava et al. 2021). Abiotic stress includes salinity, drought, heat and cold, high light intensities including UV, flooding, oxidative, metal, and nutrients (Kaplan et al. 2004; Havlin et al. 2005; Chinnusamy et al. 2007; Parent et al. 2008; Sanchez et al. 2008; Gill and Tuteja 2010). When plants are subjected to excessive high salinity and dehydration, osmotic stress takes place (Hasegawa et al. 2000; Yamaguchi-Shinozaki and Shinozaki 2006). Salinity and drought are the two most significant abiotic stressors, impacting 6% and 64% of the world's cultivated area respectively (Cramer et al. 2011). Salt and drought have a detrimental effect on plant growth, reproduction, phenotypic expression, water-nutrient ratios, photosynthesis, absorption, and respiration. During salt and drought stresses, chemical or hormonal signals are sent from roots to leaves, leading to reduced leaf growth (Westgate et al. 1996).

The term "biotic stress" refers to the damage inflicted to plants by other living beings. Biotic stresses include bacterial, fungal, viral, insect, wounding, and herbivory (Selitrennikoff 2001; Kessler and Baldwin 2002; Poupard et al. 2003; Anssour and Baldwin 2010). Biotic stress in tomato leads to significant loss of production (Bai and Lindhout 2007). Tomatoes are the focus of almost 200 different diseases, including bacterial, fungal, viral, nematodes, and mycoplasma-like organisms. In addition to these diseases, insect pests cause damage to crop by voracious feeding or by transmitting diseases. Insect pests include aphids, flea beetles, leaf miners, spider mites, fruit worm, budworm, stink bugs, and green house whiteflies. In most instances, tomato plants are invaded by a multitude of fungal pathogens, viz., *Fusarium oxysporum* and *Alternaria solani*, resulting in reduced yield and fruit quality.

7 Plant Genome Editing

The ability to create genetic variation in crops is crucial for sustainable agriculture. Reverse genetics is enabled via genome editing using site-specific nucleases. Genome editing and experiments involving targeted gene insertion must be performed efficiently and accurately. Targeted DNA double-strand breaks (DSBs) are introduced into the DNA strand using a synthetic nuclease, enabling the cellular DNA repair pathways to become activated (Bortesi and Fischer 2015). Non-homologous end joining (NHEJ) and homologous recombination (HR) methods can be used to edit the genome. Random insertions and deletions (indels) can cause a gene to be knocked out by NHEJ in many cases. Strategies like upregulation of HR-related proteins or the introduction of negative selection markers from outside homologous regions of the inserted cassette to avoid random integration events from enduring can result in modest increases in gene targeting efficacy (Puchta and Fauser 2013). Genome editing accelerates plant breeding by allowing precise and predictable mutations to be introduced directly into an elite background, and the CRISPR/Cas9 method is particularly useful since many characteristics can be tweaked concurrently (Gao 2019). The volume of information about the CRISPR/Cas9 system's properties is currently acquired from research in mammalian species, and while it appears that most of the observations can be extrapolated, recent studies in plants are still essential to maintain that system properties are generalizable to various organisms.

Until 2013, zinc finger nucleases (ZFNs) and transcription activator-like effector nucleases (TALENS) were the predominant genome editing technologies. Both are fusion proteins composed of a customized DNA-binding domain linked to the restriction enzyme FokI nonspecific nuclease domain, and they have been effectively employed in a variety of species, including plants (Palpant and Dudzinski 2013; Jankele and Svoboda 2014). The most cutting-edge technique for genome editing is centered on RNA-guided designed nucleases, which have already demonstrated significant promise due to their accessibility, efficacy, and adaptability. The most extensively utilized system is *Streptococcus pyogenes* type II clustered regularly interspaced short palindromic repeat (CRISPR)/Cas9 system (Jinek et al. 2012). CRISPR/Cas systems are components of bacteria's and archaea's adaptive immune systems, defending them from intruding nucleic acids such as viruses by cutting the foreign DNA in a sequence-dependent mechanism. Immunity is acquired by the incorporation of short segments of intruding DNA known as spacers between two adjoining repeats at the CRISPR locus' distal end. The presence of a conserved protospacer-adjacent motif (PAM) downstream of the target sequence is a prerequisite for fragmentation, which is typically 5′-NGG-3′ but can also be NAG (Hsu et al. 2003; Gasiunas et al. 2012; Jinek et al. 2012). Specificity is enabled from the "seed sequence" which must match between the RNA and target DNA around 12 nucleotides upstream of the PAM. Using data from mammalian systems, an advanced in silico approach of the nuclear genome-wide analysis of eight noteworthy plant species (*Arabidopsis*, *Oryza*, *Zea mays*, *Medicago truncatula*, *tomato*, *soybean*, and *Brachypodium distachyon*) was performed to anticipate specific Grna spacers with the minimal possible of off-target cleavage (Xie et al. 2014).

8   Genome Editing Tools for Biotic and Abiotic Stress in Tomato

ZFN, CRISPR/Cas, and cytidine base editor (CBE), a kind of DNA base editor, have all been applied to tomato (Fig. 1) for core genome editing technologies (Brooks et al. 2014; Hilioti et al. 2016; Shimatani et al. 2017; Xia et al. 2021). lZF3 is a C2H2 zinc-finger protein transcription factor. Plant stress response/tolerance is mediated by zinc finger TFs.

Salt stress induces SlZF3 expression. SlZF3 RNAi silenced lines are sensitive to salt stress, but SlZF3 overexpression lines are not (Li et al. 2018). Multiplexed editing of SlHyPRP1 with CRISPR/Cas9 led to targeted deletions of its regulatory motif (s) and resulted in salt stress-tolerant events in grown tomato (Bao et al. 2015; Tran et al. 2021). Silenced GABA pathway genes such as SlGAD and SlGABA-T in tomato using a virus-induced gene silencing (VIGS) method resulted in enhanced salinity sensitivity, whereas SlSSADH demonstrated decreased salinity sensitivity. Osmotic stress causes oxidative damage and forms excessive reactive oxygen species (ROS), causing cell damage and tissue death. Osmotic stress activates the SNF1-related protein kinase2 (SnRK2) family of proteins, which are required for stress signaling. Plants with SnRK2.1 and SnRK2.2 overexpression

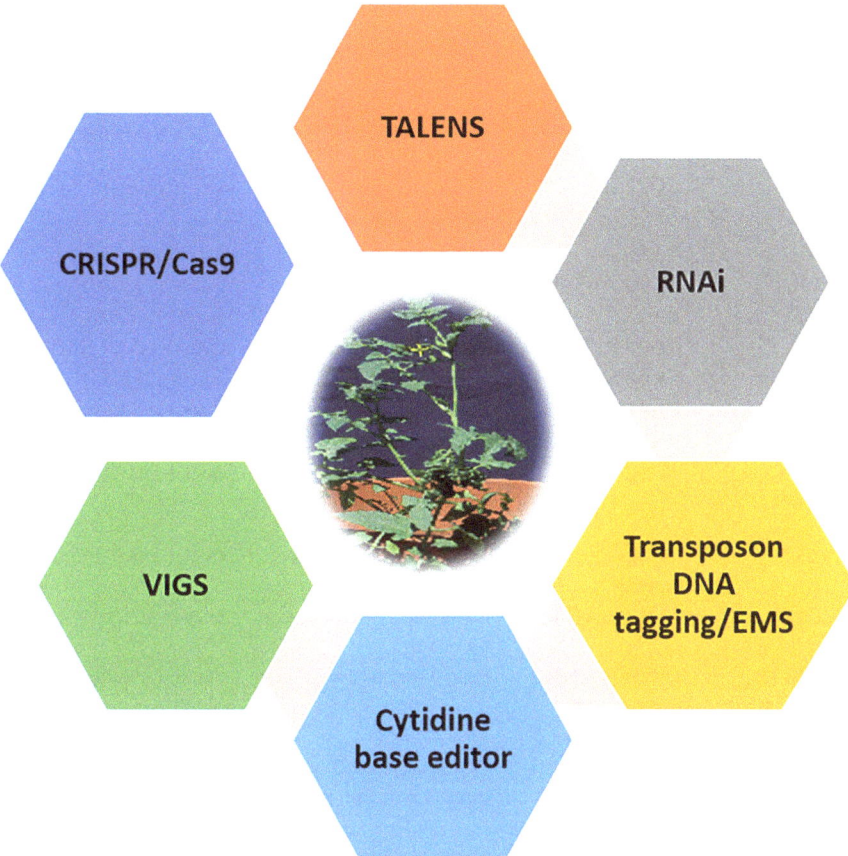

Fig. 1 Core genome editing methods used to study tomato plant

had lower tolerance to salt stress, whereas RNAi silenced lines had higher resistance to salt stress (Yang et al. 2015). The negative GA signaling regulator PROCERA (PRO) was effectively mutated in tomato using TALENs controlled by an estrogen-inducible promoter (Lor et al. 2014). Veillet et al. (2019) utilized *Agrobacterium*-mediated CRISPR/Cas9 gene transfer to efficiently change the cytidine bases in the acetolactate synthase (ALS) gene and generated chlorsulfuron-resistant tomato plants with 71% efficacy. Various phenotypic changes occurred in tomato plants due to the latest genome editing methods available currently are mentioned in Table 2.

Table 2 List of genome editing methods for stress and phenotypic variations in tomato

Targeted gene	Genome editing tool used	Type of stress/mutation developed in tomato	References
MYB12	CRISPR/Cas9	Fruit color change	Deng et al. (2018), Yang et al. (2017) and Zhu et al. (2018)
SlNL33	RNAi	Oxidative stress and bacterial resistance	Ye et al. (2019)
Hog1	RNAi	Fungal resistance	Pareek and Rajam (2017)
SlLBD40	CRISPR/Cas9	Dehydration	Liu et al. (2020)
SlHZ24	RNAi	Oxidative stress	Hu et al. (2016)
JAZ2	CRISPR/Cas9	Viral resistance	Ortigosa et al. (2019)
PROCERA	CRISPR/Cas9	Heritable mutation	Lor et al. (2014)
LeERF2	RNAi	Chilling stress	Zhang and Huang (2010)
LeCOLD1	RNAi	Chilling stress	Zhang et al. (2020)
SELF-PRUNING 5G (SP5G)	CRISPR/Cas9	Early flowering and rapid yield with stunted growth of plants	Soyk et al. (2017)
SlARS1	Transposon DNA tagging	Salt and drought stress	Campos et al. (2016)
SlAGL6	EMS mutagenesis	Parthenocarpy due to early flowering	Klap et al. (2017)
SLPMR4	CRISPR/Cas9	Fungal resistance	Santillan Martinez et al. (2020)
eIF4E1	CRISPR/Cas9	Viral resistance	Yoon et al. (2020)
ALS	CRISPR/Cas9	Herbicide resistance	Danilo et al. (2019)
Ripening inhibitor (RIN)	CRISPR/Cas9	Induction of fruit ripening	Ito et al. (2020)
ALC	CRISPR/Cas9	Long shelf life	Yu et al. (2017)
CLV3	CRISPR/Cas9	Increased fruit size and yield	Rodriguez-Leal et al. (2017)
SlDDB1, SlDET1, and SlCYC-B	Cytosine base editor	Enhanced pigments content	Hunziker et al. (2020)
SlSR1	Viral-induced gene silencing (VIGS)	Bacterial resistance and drought stress	Liu et al. (2014)
SlAMI	RNAi	Salt and oxidative stress	AbuQamar et al. (2009)

9   Conclusion

Stress-resistant agricultural crops are needed to feed a growing global population and diminishing natural resources. In traditional breeding, crosses are frequently made between superior cultivar lines or between superior cultivar lines and wild species or near relatives. Based on the compatibility and complexity of the parents, it takes 5–7 years to generate a new cultivar from an established cultivar and therefore more than a decade from a wild species. Introducing new genes with genome editing techniques can generate transgenic crops that are resistant to diverse environmental challenges.

References

Abdullahi II, Abdullahi N, Abdu AM, Ibrahim AS (2016) Proximate, mineral and vitamin analysis of fresh and canned tomato. Biosci Biotechnol Res Asia 13:1163–1169

AbuQamar S, Luo H, Laluk K, Mickelbart MV, Mengiste T (2009) Crosstalk between biotic and abiotic stress responses in tomato is mediated by the AIM1 transcription factor. Plant J 58:347–360. https://doi.org/10.1111/j.1365-313x.2008.03783.x

Ali MY, Sina AAI, Khandker SS, Neesa L, Tanvir EM, Kabir A, Khalil I, Gan SH (2021) Nutritional composition and bioactive compounds in tomatoes and their impact on human health and disease: a review. Food 10:45. https://doi.org/10.3390/foods10010045

Anssour S, Baldwin IT (2010) Variation in antiherbivore defense responses in synthetic *Nicotiana* allopolyploids correlates with changes in uniparental patterns of gene expression. Plant Physiol 153:1907–1918. https://doi.org/10.1104/pp.110.156786

Atkinson NJ, Urwin PE (2012) The interaction of plant biotic and abiotic stresses: from genes to the field. J Exp Bot 63:3523–3543. https://doi.org/10.1093/jxb/ers100

Bai Y, Lindhout P (2007) Domestication and breeding of tomatoes: what have we gained and what can we gain in the future? Ann Bot 100:1085–1094

Bao H, Chen X, Lv S, Jiang P, Feng J, Fan P, Nie L, Li Y (2015) Virus-induced gene silencing reveals control of reactive oxygen species accumulation and salt tolerance in tomato by γ-aminobutyric acid metabolic pathway. Plant Cell Environ 38:600–613. https://doi.org/10.1111/pce.12419

Bergougnoux V (2013) The history of tomato: From domestication to biopharming. Biotechnol Adv 32:170–189. https://doi.org/10.1016/j.biotechadv.2013.11.003

Bortesi L, Fischer R (2015) The CRISPR/Cas9 system for plant genome editing and beyond. Biotechnol Adv 33:41–52. https://doi.org/10.1016/j.biotechadv.2014.12.006

Brooks C, Nekrasov V, Lippman ZB, Van Eck J (2014) Efficient gene editing in tomato in the first generation using the clustered regularly interspaced short palindromic repeats/CRISPR-associated9 system. Plant Physiol 166:1292–1297. https://doi.org/10.1104/pp.114.247577

Bruening G, Lyons JM (2000) The case of the FLAVR SAVR tomato. Calif Agric 54:6–7. https://doi.org/10.3733/ca.v054n04p6

Campos JF, Cara B, Pérez-Martín F, Pineda B, Egea I, Flores FB, Fernandez-Garcia N, Capel J, Moreno V, Angosto T (2016) The tomato mutant ars1 (alteed response to salt stress 1) identifies an R1-type MYB transcription factor involved in stomatal closure under salt acclimation. Plant Biotechnol J 14:1345–1356. https://doi.org/10.1111/pbi.12498

Canene-Adams K, Campbell JK, Zaripheh S, Jeffery EH, Erdman JJW (2005) The tomato as a functional food. J Nutr 135:1226–1230. https://doi.org/10.1093/jn/135.5.1226

Chinnusamy V, Zhu J, Zhu JK (2007) Cold stress regulation of gene expression in plants. Trends Plant Sci 12:444–451. https://doi.org/10.1016/j.tplants.2007.07.002

Clinton SK (1998) Lycopene: chemistry, biology and implications for human health and disease. Nutr Rev 56:35–51. https://doi.org/10.1111/j.1753-4887.1998.tb01691.x

Cramer GR, Urano K, Delrot S, Pezzotti M, Shinozaki K (2011) Effects of abiotic stress on plants: a systems biology perspective. BMC Plant Biol 11:163. https://doi.org/10.1186/1471-2229-11-163

Danilo B, Perrot L, Mara K, Botton E, Nogue F, Mazier M (2019) Efficient and transgene-free gene targeting using agrobacterium-mediated delivery of the CRISPR/Cas9 system in tomato. Plant Cell Rep 38:459–462. https://doi.org/10.1007/s00299-019-02373-6

Deng L, Wang H, Sun CL, Li Q, Jiang HL, Du MM, Li CB, Li CY (2018) Efficient generation of pink-fruited tomatoes using CRISPR/Cas9 system. J Genet Genomics 45:51–54. https://doi.org/10.1016/j.jgg.2017.10.002

Elbadrawy E, Sello A (2016) Evaluation of nutritional value and antioxidant activity of tomato peel extracts. Arab J Chem 9:S1010–S1018. https://doi.org/10.1016/j.arabjc.2011.11.011

FAOSTAT (2020) https://www.fao.org/faostat/en/#compare

Franceschi S, Bidoli E, Vecchia CL, Talamini R, D'Avanzo B, Negri E (1994) Tomatoes and risk of digestive-tract cancers. Int J Cancer 59:181–184. https://doi.org/10.1002/ijc.2910590207

Gao C (2019) Precision plant breeding using genome editing technologies. Transgenic Res 28:53–55. https://doi.org/10.1007/s11248-019-00132-7

Gasiunas G, Barrangou R, Horvath P, Siksnys V (2012) Cas9-crRNA ribonucleoprotein complex mediates specific DNA cleavage for adaptive immunity in bacteria. Proc Natl Acad Sci U S A 109:E2579–E2586. https://doi.org/10.3410/f.718148789.793504799

Gill SS, Tuteja N (2010) Reactive oxygen species and antioxidant machinery in abiotic stress tolerance in crop plants. Plant Physiol Biochem 48:909–930. https://doi.org/10.1016/j.plaphy.2010.08.016

Giovannucci E, Ascherio A, Rimm EB, Stampfer MJ, Colditz GA, Willett WC (1995) Intake of carotenoids and retino in relation to risk of prostate cancer. J Natl Cancer Inst 87:1767–1776. https://doi.org/10.1016/j.plaphy.2010.08.016

Hasegawa PM, Bressan RA, Zhu JK, Bohnert HJ (2000) Plant cellular and molecular responses to high salinity. Annu Rev Plant Physiol Plant Mol Biol 51:463–499. https://doi.org/10.1146/annurev.arplant.51.1.463

Havlin JL, Beaton JD, Tisdale SL, Nelson WL (2005) Soil fertility and nutrient management: an introduction to nutrient management, 7th edn. Pearson/Prentice Hall, Upper Saddle River

Hilioti Z, Ganopoulos I, Ajith S, Bossis I, Tsaftaris A (2016) A novel arrangement of zinc finger nuclease system for in vivo targeted genome engineering: the tomato LEC1-LIKE4 gene case. Plant Cell Rep 35:2241–2255. https://doi.org/10.1007/s00299-016-2031-x

Hsu PD, Scott DA, Weinstein JA, Ran FA, Konermann S, Agarwala V, Li Y, Fine EJ, Wu X, Shalem O, Cradick TJ (2003) DNA targeting specificity of RNA-guided Cas9 nucleases. Nat Biotechnol 31:827–832. https://doi.org/10.1038/nbt.2647

Hu T, Ye J, Tao P, Li H, Zhang J, Zhang Y, Ye Z (2016) The tomato HD-zip transcription factor Sl HZ 24 modulates ascorbate accumulation through positive regulation of the d-mannose/l--galactose pathway. Plant J 85:16–29. https://doi.org/10.1111/tpj.13085

Hunziker J, Nishida K, Kondo A, Kishimoto S, Ariizumi T, Ezura H (2020) Multiple gene sub-stitution by target-AID base-editing technology i tomato. Sci Rep 10:20471. https://doi.org/10.1111/tpj.13085

Ito Y, Sekiyama Y, Nakayama H, Nishizawa-Yokoi A, Endo M, Shima Y, Nakamura N, Kotake-Nara E, Kawasaki S, Hirose S (2020) Allelic mutations in the ripening-inhibitor locus gener-ate extensive variation in tomato ripening. Plant Physiol 183:80–95. https://doi.org/10.1104/pp.20.00020

Jankele R, Svoboda P (2014) TAL effectors: tools for DNA targeting. Brief Funct Genomics 13:409–419. https://doi.org/10.1093/bfgp/elu013

Jinek M, Chylinski K, Fonfara I, Hauer M, Doudna JA, Charpentier E (2012) A programmable dual-RNA-guided DNA endonuclease in adaptive bacterial immunity. Science 337:816–821. https://doi.org/10.1126/science.1225829

Jones JD, Dangl JL (2006) The plant immune system. Nature 444:323–329. https://doi.org/10.1038/nature05286

Kaplan F, Kopka J, Haskell DW, Zhao W, Schiller KC, Gatzke N, Sung DY, Guy CL (2004) Exploring the temperature stress metabolome of *Arabidopsis*. Plant Physiol 136:4159–4168. https://doi.org/10.1104/pp.104.052142

Kessler A, Baldwin IT (2002) Plant responses to insect herbivory: the emerging molecular analysis. Annu Rev Plant Biol 53:299–328. https://doi.org/10.1146/annurev.arplant.53.100301.135207

Klap C, Yeshayahou E, Bolger AM, Arazi T, Gupta SK, Shabtai S, Usadel B, Salts Y, Barg R (2017) Tomato facultative parthenocarpy results from SlAGAMOUS-LIKE 6 loss of function. Plant Biotechnol J 15:634–647. https://doi.org/10.1111/2Fpbi.12662

Levy J, Bosin E, Feldman B, Giat Y, Miinster A, Danilenko M, Sharoni Y (1995) Lycopene is a more potent inhibitor of human cancer cell proliferation than either alpha-carotene or beta-carotene. Nutr Cancer 24:257–266. https://doi.org/10.1080/01635589509514415

Li Y, Chu Z, Luo J, Zhou Y, Cai Y, Lu Y, Xia J, Kuang H, Ye Z, Ouyang B (2018) The C2H2 zinc-finger protein Sl ZF 3 regulates as a synthesis and salt tolerance by interacting with CSN5B. Plant Biotechnol J 16:1201–1213. https://doi.org/10.1111/pbi.12863

Liu B, Ouyang Z, Zhang Y, Li X, Hong Y, Huang L, Liu S, Zhang H, Li D, Song F (2014) Tomato NAC transcription factor SlSRN1 positively regulates defense response against biotic stress but negatively regulates abiotic stress response. PLoS One 9:e102067. https://doi.org/10.1371/journal.pone.0102067

Liu L, Zhang J, Xu J, Li Y, Guo L, Wang Z, Zhang X, Zhao B, Guo Y-D, Zhang N (2020) CRISPR/Cas9 targeted mutagenesis of SlLBD40, a lateral organ boundaries domain transcription factor, enhances drought tolerance in tomato. Plant Sci 301:110683. https://doi.org/10.1016/j.plantsci.2020.110683

Lor VS, Starker CG, Voytas DF, Weiss D, Olszewski NE (2014) Targeted mutagenesis of the tomato PROCERA gene using transcription activator-like effector nucleases. Plant Physiol 166:1288–1291. https://doi.org/10.1104/pp.114.247593

McCormick S, Niedermeyer J, Fry J, Barnason A, Horsch R, Fraley R (1986) Leaf disc transformation of cultivated tomato (L. esculentum) using agrobacterium tumefaciens. Plant Cell Rep 5:81–84. https://doi.org/10.1007/BF00269239

Menda N, Semel Y, Peled D, Eshed Y, Zamir D (2004) In silico screening of a saturated mutation library of tomato. Plant J 38:861–872. https://doi.org/10.1111/j.1365-313x.2004.02088.x

Müller C (1940) A revision of the genus *solanum*. Misc Pub-USDA 382:1–28

Navarro-Gonzalez I, García-Valverde V, García-Alonso J, Periago-Castón MJ (2011) Chemical profile, functional and antioxidant properties of tomato peel fiber. Food Res Int 44:1528–1535. https://doi.org/10.1016/j.foodres.2011.04.005

Oerke EC (2006) Crop losses to pests. J Agric Sci 144:31–43. https://doi.org/10.1017/S0021859605005708

Ortigosa A , Gimenez-Ibanez S, Leonhardt N, Solano R (2019) Design of a bacterial speck resistant tomato by CRISPR/Cas9-mediated editing of SlJAZ2. Plant Biotechnol J 17:665-673. https://doi.org/10.1111/pbi.13006

Palpant NJ, Dudzinski D (2013) Zinc finger nucleases: looking toward translation. Gene Ther 20:121–127. https://doi.org/10.1038/gt.2012.2

Pareek M, Rajam MV (2017) RNAi-mediated silencing of MAP kinase signalling genes (Fmk1, Hog1, and Pbs2) in *fusarium oxysporum* reduces pathogenesis on tomato plants. Fungal Biol 121:775–784. https://doi.org/10.1016/j.funbio.2017.05.005

Parent C, Berger A, Folzer H, Dat J, Crevecoeur M, Badot PM, Capelli N (2008) A novel non symbiotic hemoglobin from oak: Cellular and tissue specificity of gene expression. New Phytol 177:142–154. https://doi.org/10.1111/j.1469-8137.2007.02250.x

Poupard P, Parisi L, Campion C, Ziadi S, Simo-neau P (2003) A wound and ethephon inducible PR-10 gene subclass from apple is differentially expressed during infection with a compatible and incompatible race of *Venturia inaequalis*. Physiol Mol Plant Pathol 62:3–12. https://doi.org/10.1016/s0885-5765(03)00008-0

Puchta H, Fauser F (2013) Gene targeting in plants: 25 years later. Int J Dev Biol 57:629–637. https://doi.org/10.1387/ijdb.130194hp

Quinet M, Angosto T, Yuste-Lisbona FJ, Blanchard-Gros R, Bigot S, Martinez J-P, Lutts S (2019) Tomato fruit development and metabolism. Front Plant Sci 10:1554. https://doi.org/10.3389/fpls.2019.01554

Rodriguez M, Canales E, Borras-Hidalgo O (2005) Molecular aspects of abiotic stress in plants. Biotecnol Apl 22:1–10

Rodriguez-Leal D, Lemmon ZH, Man J, Bartlett ME, Lippman ZB (2017) Engineering quantitative trait variation for crop improvement by genome editing. Cell 171:470–480. https://doi.org/10.1016/j.cell.2017.08.030

Salava H, Thula S, Mohan V, Kumar R, Maghuly F (2021) Application of genome editing in tomato breeding: mechanisms, advances, and prospects. Int J Mol Sci 22:682. https://doi.org/10.3390/ijms22020682

Salunkhe DK, Jadhav SJ, Yu MH (1974) Quality and nutritional composition of tomato fruit as influenced by certain biochemical and physiological changes. Plant Food Hum Nutr 24:85–113. https://doi.org/10.1007/BF01092727

Sanchez DH, Siahpoosh MR, Roessner U, Ud-vardi M, Kopka J (2008) Plant metabolomics reveals conserved and divergent metabolic responses to salinity. Physiol Plant 132:209–219. https://doi.org/10.1111/j.1399-3054.2007.00993.x

Santillan Martinez MI, Bracuto V, Koseoglou E, Appiano M, Jacobsen E, Visser RGF, Wolters AA, Bai Y (2020) CRISPR/Cas9- targeted mutagenesis of the tomato susceptibility gene PMR4 for resistance against powdery mildew. BMC Plant Biol 20:284. https://doi.org/10.1186/s12870-020-02497-y

Selitrennikoff C (2001) Antifungal proteins. Appl Environ Microbiol 67:2883–2894. https://doi.org/10.1128/AEM.67.7.2883-2894.2001

Shimatani Z, Kashojiya S, Takayama M, Terada R, Arazoe T, Ishii H, Teramura H, Yamamoto T, Komatsu H, Miura K, Ezura H, Nishida K, Ariizumi T, Kondo A (2017) Targeted base editing in rice and tomato using a CRISPR-Cas9 cytidine deaminase fusion. Nat Biotechnol 35:441–443. https://doi.org/10.1038/nbt.3833

Soyk S, Müller NA, Park SJ, Schmalenbach I, Jiang K, Hayama R, Zhang L, Van Eck J, Jimenez-Gomez JM, Lippman ZB (2017) Variation in the flowering gene SELF PRUNING 5G promotes day-neutrality and early yield in tomato. Nat Genet 49:162–168. https://doi.org/10.1038/ng.3733

Tanksley SD (2004) The genetic, developmental and molecular bases of fruit size and shape variation in tomato. Plant Cell 16:181–189. https://doi.org/10.1105/tpc.018119

Tran MT, Doan DTH, Kim J, Song YJ, Sung YW, Das S, Kim EJ, Son GH, Kim SH, Van Vu T, Kim JY (2021) CRISPR/Cas9-based precise excision of SlHyPRP1 domain(s) to obtain salt stress-tolerant tomato. Plant Cell Rep 40:999–1011. https://doi.org/10.1007/s00299-020-02622-z

Veillet F, Perrot L, Chauvin L, Kermarrec MP, Guyon-Debast A, Chauvin JE, Nogué F, Mazier M (2019) Transgene-free genome editing in tomato and potato plants using Agrobacterium-mediated delivery of a CRISPR/Cas9 cytidine base editor. Int J Mol Sci 20:402. https://doi.org/10.3390/ijms20020402

Veronese P, Ruiz MT, Coca MA, Hernandez-Lopez A, Lee H, Ibeas JI, Damsz B, Pardo JM, Hasegawa PM, Bressan RA, Narasimhan ML (2003) In defense against pathogens. Both plant sentinels and foot soldiers need to know the enemy. Plant Physiol 131:1580–1590. https://doi.org/10.1104/pp.102.013417

Wang W, Vinocur B, Altman A (2003) Plant responses to drought, salinity and extreme temperatures: towards genetic engineering for stress tolerance. Planta 218:1–14. https://doi.org/10.1007/s00425-003-1105-5

Weese TL, Bohs L (2007) A three gene phylogeny of the genus solanum (Solanaceae). Syst Bot 32:445–463. https://doi.org/10.1600/036364407781179671

Westgate ME, Passioura JB, Munns R (1996) Water status and ABA content of floral organs in drought-stressed wheat. Aust J Plant Physiol 23:763–772. https://doi.org/10.1071/pp9960763

Willcox JK, Catignani GL, Lazarus S (2003) Tomatoes and cardiovascular health. Crit Rev Food Sci Nutr 43:1–18. https://doi.org/10.1080/10408690390826437

Xia X, Cheng X, Li R, Yao J, Li Z, Cheng Y (2021) Advances in application of genome editing in tomato and recent development of genome editing technology. Theor Appl Genet 134:2727–2747. https://doi.org/10.1007/s00122-021-03874-3

Xie K, Zhang J, Yang Y (2014) Genome-wide prediction of highly specific guide RNA spacers for CRISPR-Cas9-mediated genome editing in model plants and major crops. Mol Plant 7:923–926. https://doi.org/10.1093/mp/ssu009

Yamaguchi-Shinozaki K, Shinozaki K (2006) Transcriptional regulatory networks in cellular responses and tolerance to dehydration and cold stresses. Annu Rev Plant Biol 57:781–803. https://doi.org/10.1146/annurev.arplant.57.032905.105444

Yang Y, Tang N, Xian Z, Li Z (2015) Two SnRK2 protein kinases genes play a negative regulatory role in the osmotic stress response in tomato. Plant Cell Tissue Org Cult 122:421–434. https://doi.org/10.1007/s11240-015-0779-2

Yang L, Huang W, Xiong FJ, Xian ZQ, Su DD, Ren MZ, Li ZG (2017) Silencing of SlPL, which encodes a pectate lyase in tomato, confers enhanced fruit firmness, prolonged shelf-life and reduced susceptibility to grey mould. Plant Biotechnol J 15:1544–1555. https://doi.org/10.1111/pbi.12737

Ye J, Liu G, Chen W, Zhang F, Li H, Ye Z, Zhang Y (2019) Knockdown of SlNL33 accumulates ascorbate, enhances disease and oxidative stress tolerance in tomato (*Solanum lycopersicum*). Plant Growth Regul 89:49–58. https://doi.org/10.1007/s10725-019-00512-3

Yoon YJ, Venkatesh J, Lee JH, Kim J, Lee HE, Kim DS, Kang BC (2020) Genome editing of eIF4E1 in tomato confers resistance to pepper mottle virus. Front Plant Sci 11:1098. https://doi.org/10.3389/2Ffpls.2020.01098

Yu QH, Wang B, Li N, Tang Y, Yang S, Yang T, Xu J, Guo C, Yan P, Wang Q, Asmutola P (2017) CRISPR/Cas9-induced targeted mutagenesis and gene replacement to generate long-shelf life tomato lines. Sci Rep 7:11874. https://doi.org/10.1038/s41598-017-12262-1

Yun DJ, Bressan RA, Hasegawa PM (1997) Plant antifungal proteins. In: Janick J (ed) Plant breeding reviews, vol pp 14. Wiley, New York, pp 39–88

Zhang Z, Huang R (2010) Enhanced tolerance to freezing in tobacco and tomato overexpressing transcription factor TERF2/LeERF2 is modulated by ethylene biosynthesis. Plant Mol Biol 73:241–249. https://doi.org/10.1007/s11103-010-9609-4

Zhang L, Guo X, Qin Y, Feng B, Wu Y, He Y, Wang A, Zhu J (2020) The chilling tolerance divergence 1 protein confers cold stress tolerance in processing tomato. Plant Physiol Biochem 151:34–46. https://doi.org/10.1016/j.plaphy.2020.03.007

Zhou G, Yang LT, Li YR, Zou CL, Huang LP, Qiu LH, Huang X, Srivastava MK (2011) Proteomic analysis of osmotic stress-responsive proteins in sugarcane leaves. Plant Mol Biol Rep 30:349–359. https://doi.org/10.1007/s11105-011-0343-0

Zhu GT, Wang SC, Huang ZJ, Zhang SB, Liao QG, Zhang CZ, Lin T, Qin M, Peng M, Yang CK, Cao X, Han X, Wang XX, van der Knaap E, Zhang ZH, Cui X, Klee H, Fernie AR, Luo J, Huang SW (2018) Rewiring of the fruit metabolome in tomato breeding. Cell 172:249–261. https://doi.org/10.1016/j.cell.2017.12.019

Part IV
Legal Framework for CRISPR/Cas Technology

Biosafety Issue Related to Genome Editing in Plants Using CRISPR-Cas9

Ramesh Katam, Fatemeh Hasanvand, Vinson Teniyah, Jessi Noel, and Virginia Gottschalk

Abstract Genetic modification of plants was in practice long before establishing the genetic paradigm. In their quest to attain superior plants, humans continually modified plants through selection and mutation in plant breeding. Genome editing based on clustered, regularly interspaced short palindromic repeats and associated protein 9 (CRISPR-Cas9) has played a lead role in crop development. With the adaptation of technologies, CRISPR is now known as the most versatile genome editing tool, revolutionizing plant breeding with broad applications in genetic manipulations of crops with unparalleled simplicity, precision, and superior proficiency. While the system has potential for food products to accommodate an exponentially growing world population, there have been increased concerns of methodologies and biosafety of the engineered products for consumption for humans and other living organisms. Here, we provide a fully comprehensive analysis of the patent, ethics, and biosafety regulation concerning the application of CRISPR technology in plant systems. We focused on the precautions taken by the scientific communities to address the several biosafety and ethical issues raised by various governments across the globe. We also highlight this technology's patent, ethical issues, and regulatory frameworks founded by multiple nations to regulate

R. Katam (✉)
Department of Biological Sciences, Florida A&M University, Tallahassee, FL, USA

Department of Biology, University of Florida, Gainesville, FL, USA
e-mail: Ramesh.katam@famu.edu

F. Hasanvand
Department of Biological Sciences, Florida A&M University, Tallahassee, FL, USA

Department of Horticultural Science, Lorestan University, Khorramabad, Iran

V. Teniyah
Department of Biological Sciences, Florida A&M University, Tallahassee, FL, USA

Department of Epidemiology, New York University, New York City, NY, USA

J. Noel · V. Gottschalk
Department of Biological Sciences, Florida A&M University, Tallahassee, FL, USA

© The Author(s), under exclusive license to Springer Nature
Switzerland AG 2022
S. H. Wani, G. Hensel (eds.), *Genome Editing*,
https://doi.org/10.1007/978-3-031-08072-2_16

289

CRISPR-Cas-modified organisms/products. This review presents the current regulatory framework and ethical barriers that may impact technology implementation and recognition and additionally discussed in the viewpoint for an alternative regulatory system that matches the scientific progress.

Keywords CRISPR-Cas9 · Bioethics · Biosafety · Agriculture · Genome editing

1 Introduction

Decades of practicing traditional plant breeding attenuated food scarcity efforts. However, they created limited genetic diversity among agriculture crops. Due to an expanding population, the current agriculture productivity and yield potential model cannot meet the increased demand for food (Ray et al. 2013). Thus, improving desirable traits such as yield and tolerance to abiotic and biotic stresses in crops and horticultural plants must be implemented in modern biotechnological tools. In the past years, with the looming of a genomic sequence, plant breeding witnessed several advances in technologies contributing to exponential growth in agriculture research. Despite advances toward food security and sustainability, crops continue to face several yield losses provoked by the varying climatic conditions that have caused several abiotic stresses. Researchers have been using speed breeding, genome editing tools, and high-throughput phenotyping to increase crop quality and productivity (Raza et al. 2019). Recent developments in genome editing technology, using programmable nucleases, clustered regularly interspaced short palindromic patterns (CRISPR), and CRISPR-associated (Cas) proteins have opened an opportunity to a new era of breeding technology. Various genome editing techniques have been adopted to address the heightened demand for food in the future. However, these technologies have come with some challenges, such as their specificity and sensitivity, off-target effects, and technique expertise, thus raising the concern for biosafety to humans and other host systems.

2 Genome Editing and CRISPR-Cas9

The systematic genome editing methods use four endonucleases: (i) zinc finger nucleases, (ii) transcription activator-like effector nucleases (TALENS), (iii) the combination, CRISPR-Cas9, and (iv) orthologs for plant genome editing (Bao et al. 2019; El-Mounadi et al. 2020). TALENs and ZFNs are used less due to their design complexity, limitation on multiplexed mutations, and transfection inefficiency (Doudna and Charpentier 2014). The CRISPR-Cas system is currently the most prospective genome editing technology used worldwide, known for its efficiency and precision (Fig. 1).

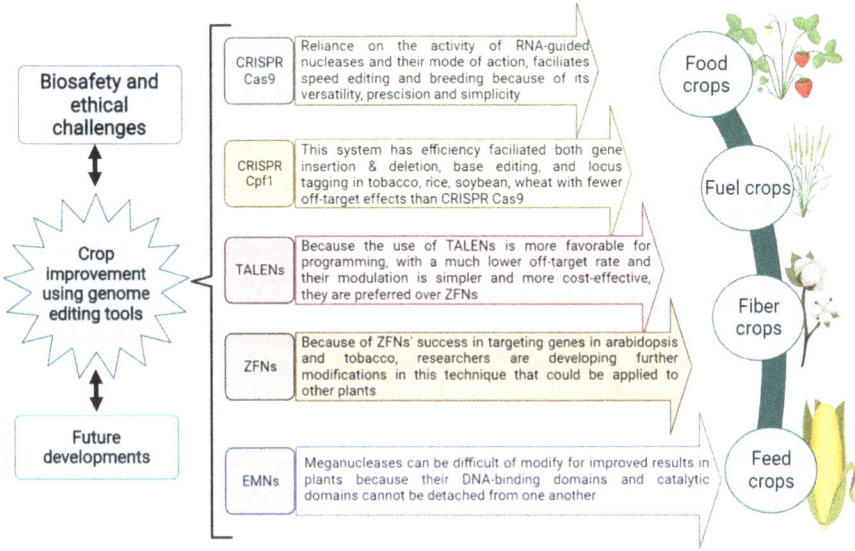

Fig. 1 CRISPR-Cas9 mode of action and the components involved in the genome editing

The CRISPR-Cas system, first discovered in prokaryotes, is widely used in archaea and in several prokaryotes to provide resistance to plasmids and bacterio-phages (Ayanoğlu et al. 2020). The CRISPR sequences are short DNA repetitions of viral or plasmid origin, defined as clustered regularly interspaced short palindromic repeats. Their purpose is to attain catalytic inactivation of the Cas9-associated "effector domain" by substituting the Cas9 protein. These effector complexes differ widely, and based on their complexity, six types of CRISPR-Cas systems have been discovered and grouped as Type II classes (Makarova et al. 2018).

Currently, the Type II CRISPR-Cas9 serves as a means for genetic engineering, in which specific Cas protein targets binding and cleavage. Their effector machinery is comparatively simple to adopt; hence, it has become an impactful tool for genome manipulation for application in prokaryotic and eukaryotic cells. A particular technique is used to precisely edit a single base pair in the genome without introducing double-strand breaks (DSB) using engineered Cas9-based editors. This system functions as an RNA-guided endonuclease enzyme that cleaves DNA and generates DSB (Fig. 2). The system is composed of Cas9, a single-guide RNA (sgRNA), and trans-activating CRISPR RNA (tracrRNA), which are RNA structures crucial for the assembly of the effector complex and DNA recognition (Globus and Qimron 2018).

CRISPR-Cas system aids in the shield and the attainment of invading nucleotides. The CRISPR array comprises conserved domains known as direct repeats and embedded variable sequences with the same length as "spacers." The CRISPR array helps to recall the invader specified line by detecting and severing foreign nucleotides (Wang et al. 2019). Cas9 needs a specific protospacer adjoining motif (PAM) localized on the non-target DNA strand downstream of the target DNA sequence. Further adaptation of the Cas9 enables manipulation of the methyl groups at specific positions on the DNA, thus allowing researchers to assess how these alterations

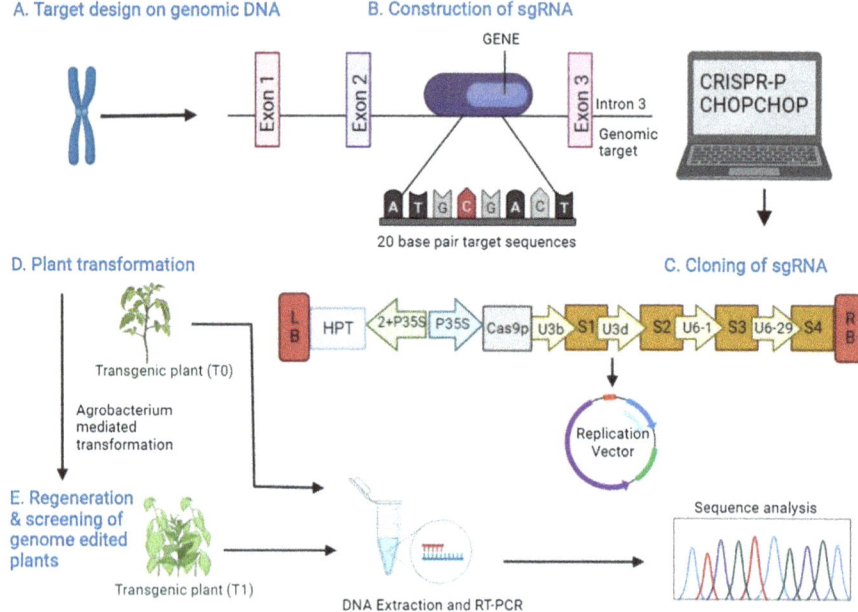

Fig. 2 General method of plant genome editing and their applications of crop improvement

affect gene expression (Cribbs and Perera 2017). More recently, CRISPR-Cas9 has been used to turn cells into programmable computers, where researchers have engineered molecular switches to control cell fate to enable conditional behaviors (Fu et al. 2014). The goal of deletion is frequently to eliminate or change the gene expression, providing the crop with added resistance to the contagious agent (Agapito-Tenfen et al. 2018). This example proves the adaptability of the CRISPR-Cas9 system in producing essential research tools in vitro.

3   CRISPR Technology for Crop Improvement

Successful genetic transformations have been achieved using CRISPR-Cas9 for enhanced nutritional value, biotic and abiotic tolerance, immunity metabolic pathways, resistance to biotic or abiotic stresses, and herbicide resistance in major crops (Fig. 3). Several crops are altered employing gene-editing techniques to eliminate the objective genes for either knocking out or modifying the gene functions. These crops thus have altered genetic composition without integrating transgene DNA into the genome of these plants (Ricroch et al. 2017). Various plant species and over a hundred genes have been subjected to successful editing with CRISPR/Cas9 creating many desirable traits insignificant crops, as summarized in the Table 1. For several crop plants, practices have been created to separate the protoplast to be transfected with cassettes carrying the CRISPR-Cas9 that serves genome editing (El-Mounadi et al. 2020).

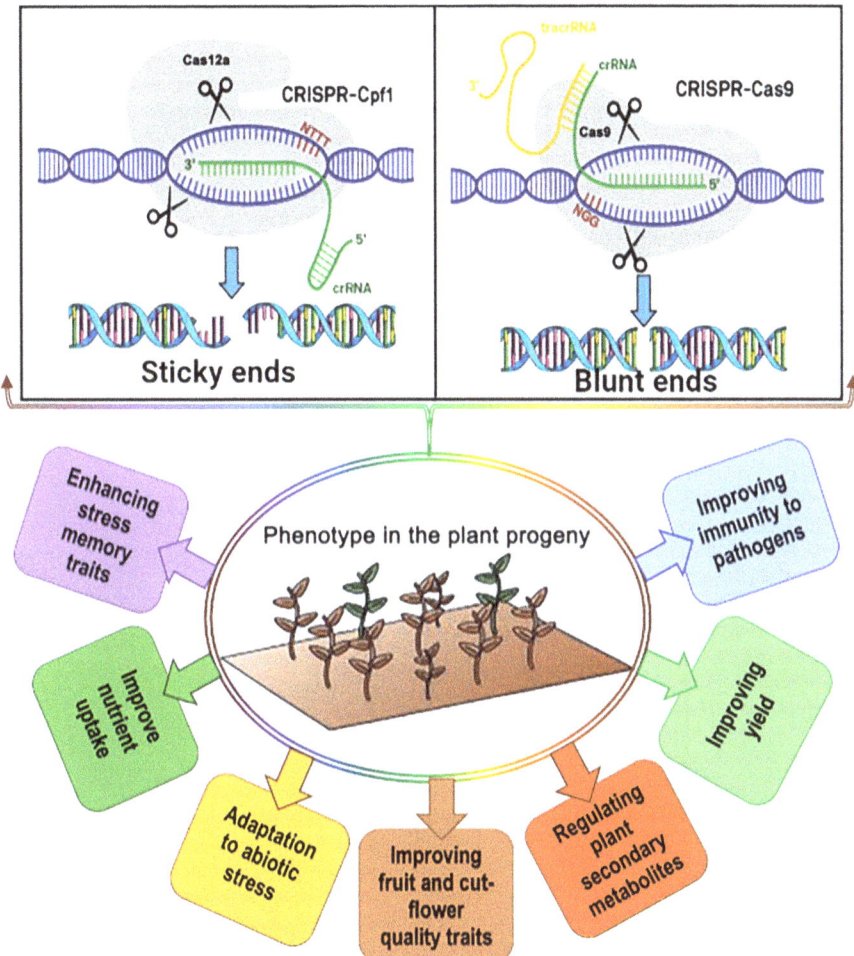

Fig. 3 CRISPR genome editing advancements in plant advances in agriculture and horticulture crop productivity using CRISPR genome editing technology

4 Biosafety Concerns to Humans

Although the CRISPR-Cas system allows manipulating genomic DNA effectively, this raises many ethical, political, and scientific concerns (Lyon 2017). It has become imperative to elucidate concepts and foster discussion among all involved stakeholders to articulate an optimal, ethical, and legal framework. An international panel of genetics groups has weighed in on some of the ethical and scientific implications facing the clinical practice of gene editing. Despite the technological advancement in gene editing, there are several challenges of CRISPR-Cas9 associated with germline gene therapy, particularly the validation of the technology and

Table 1 Summary of the successful examples of CRISPR-Cas9 genome editing in crop plants

Crop	Gene	Trait	Reference
Model organisms	*Arabidopsis thaliana*	GH3 genes	Able to conjugate indole-3-acetic acid (IAA) to amino acids
Guo et al. (2022)			
OsMORE1a, MORE1	Enhanced resistance against hemibiotrophic pathogens *M. oryzae* and *Xanthomonas oryzae* pv. *oryzae* but increased susceptibility to *Cochliobolus miyabeanus*	Kim et al. (2022)	TTG1
	Regulates cell fate determination, including trichome initiation and root hair formation, flavonoid biosynthesis, seed coat mucilage production	Wang et al. (2021a, b, c, d, e, f, g, h, i)	SVB
	Trichome formation	Hussain et al. (2021)	LFY
	Master transcription factor for flowering	Choi et al. (2021a, b)	AITR
	Play a role in regulating plant responses to ABA and abiotic stresses	Chen et al. (2021)	YKT61
	Essential for male and female gametogenesis	Ma et al. (2021a, b)	ADR1, SOC3/SOC3-L1, TN2, NRG1**
	ADR1: may participate in the activation of immune responses, enhanced resistance to *G. cichoracearum*. NLR & SOC3: chilling sensitivity	Wang et al. (2021a, b, c, d, e, f, g, h, i)	*Nicotiana benthamiana* CRISPR-Cas13
Nicotiana benthamiana	LbaCas13a, LbuCas13a	Reduced viral accumulation	Sharma et al. (2022)
	NLR	Resistance to soybean mosaic virus	Yin et al. (2021)

ADR1, SOC3/SOC3-L1, TN2, NRG1**	ADR1: may participate in the activation of immune responses, enhanced resistance to G. cichoracearum. NLR & SOC3: chilling sensitivity	Wang et al. (2021a, b, c, d, e, f, g, h, i)	CLCuV**
	Controls presence of cotton leaf curl disease	Mubarik et al. (2021)	SOC1**
	LsCYCD1-1, LsCYCD3-2, LsCYCD6-1, and LsCYCU4-1: cell division-related genes	Choi et al. (2021a, b)	OsAUX4**
	May function in the regulation of root development, supports primary root elongation	Ye et al. (2021)	Cereal crops
Hordeum vulgare		HvARE1	Nitrogen use efficiency improvement
Karunarathne et al. (2022)			
HvEIF4E	Resistance to barley yellow mosaic virus (BaYMV) and barley mild mosaic virus (BaMMV)	Hoffie et al. (2021)	HVP10
	Enhances salt tolerance via promoting Na+ translocation into root vacuoles	Fu et al. (2022)	Hordeum vulgare CRISPR/Cas9v
	APETALA2, HvAPETALA2, HvMADS29	HvAPETALA2: controls floret organ identity, floret boundaries, and maternal tissue differentiation and elimination during grain development HvMADS29: aids in maternal tissue differentiation	Shoesmith et al. (2021)
Oryza sativa	Bsr-d1, Pi21 and ERF922	Increased resistance to rice blast	Zhou et al. (2021)

(Continued)

Table 1 (Continued)

Crop	Gene	Trait	Reference
OsVDE	Negatively regulated the OsNECD2/4/5 expressions, ABA biosynthesis, and salt stress tolerance	Wang et al. (2021a, b, c, d, e, f, g, h, i)	OsGW2
	Improves aleurone layer and grain nutritional quality	Achary and Reddy (2021)	FAR1
	Structural divergence via frameshift and nonsense mutations that caused premature translation termination or specific domain truncations	Chae et al. (2021)	OsCly1
	Involved in cleistogamous flowering	Ohtsuki et al. (2021)	eIF4G
	Partial resistance to rice black-streaked dwarf virus	Wang et al. (2021a, b, c, d, e, f, g, h, i)	PDS, YSA
	PDS: visible phenotypes—variegated plants, dwarf albino plants, and normal-size albino plants. YSA: variegated leaves characterized by both white and green streaks or variegated plants with white leaf tips	Molina-Risco et al. (2021)	ZEP1
	Increased genetic recombination frequency	Liu et al. (2021a, b, c, d, e)	c2BGC, c4BGC
	Deletion of c2BGC, associated with phytocassane biosynthesis, but not c4BGC, momilactone biosynthesis leading to a lesion mimic phenotype	Li et al. (2022)	GSK2, OsBZR1/2, PPKL2
	Showed diversity of plant height, leaf angle, and grain morphology	Liu et al. (2021a, b, c, d, e)	OsHXK1
	Improved rice photosynthetic efficiency and yield	Zheng et al. (2021)	OsHOL1
	Major player in methyl iodide production, since its knockout abolished the process	Carlessi et al. (2021)	UBMERGENCE 1A-1
	Disruption of SUB1A-1 is enough to limit key traits associated with the quiescence strategy of submergence survival	Liang et al. (2021)	OsLHW, OSH6
	OsLHW: may be a repressor of drought-induced metaxylem plasticity. OSH6: showed larger shoots with deeper roots	Reeger et al. (2021)	OsCaM1, OsCCaMK
	Promote lateral root growth and IAA content, and enhance transcription of OsDREB2A, OsDREB2B, and OsNAC6 under salt stress	Yang et al. (2021)	
	FAD2	Controls linoleic acid formation and oleic acid accumulation	Bahariah et al. (2021)

OsANTH	Play a role in rice pollen germination	Lee et al. (2021a, b)	Grain Size 3
	Dictates grain length	Usman et al. (2021)	MSH2
	DNA mismatch repair gene	Karthika et al. (2021)	DT-A
	Inactivates EF2, leading to inhibition of protein synthesis	Nishizawa-Yokoi and Toki (2021)	Hygromycin phosphotransferase
	Hygromycin resistance	Majumder et al. (2021)	TALE, Tal2c
	Activates the expression of the rice 2-oxoglutarate-dependent dioxygenase gene OsF3H03g to promote infection in the highly virulent strain; virulence factor to target rice OsF3H$_{04g}$, a homologue of OsF3H$_{03g}$	Wu et al. (2021)	*Oryza sativa* CRISPR/Sc^{++} -mediated
OsGS1, OsMPK15	OsGS1: exhibits tolerance to osmotic and salinity stress at the seedling stage. OsMPK15: manages ROS and disease resistance	Ma et al. (2021a, b)	
Sorghum bicolor	BTx623	Short-stature, early maturing	Massel et al. (2022)
RTx430, SC187, BTx642, BTx623, Ramada	RTx430, formed more somatic embryos continuously. BTx642 tissues formed smaller numbers and sizes of somatic embryos. Most RTx430 tissue under IMZ selection had some healthy tissues that were not brown or necrotic	Aregawi et al. (2021)	LIGULELESS-1
	Alternates leaf inclination angle	Brant et al. (2021)	*Triticale* cv. Bogo cas9/guide RNA
	ABA8'OH1	Putative determinants of pre-harvest sprouting of grains	Michalski et al. (2021)
Triticum aestivum	TaMBF1c	Regulating the heat stress response via the translation process	Tian et al. (2022)

(Continued)

Table 1 (Continued)

Crop	Gene	Trait	Reference
TaBAK1-2, Ta-eIF4E, Ta-eIF(iso)4E		BAK1 is a regulator of plant immunity and development, Ta-eIF4E and Ta-eIF(iso)4E act as susceptibility factors required for plant viruses from the Potyviridae family to complete their life cycle	Hahn et al. (2021)
	ARE1	Showed enhanced tolerance to N starvation and delayed senescence and increased grain yield in field conditions	Zhang et al. (2021a, b, c, d, e)
	TaQsd1	Seed germination	Liu et al. (2021a, b, c, d, e)
	TaSBEIIa	Starch composition, structure, and properties	Li et al. (2021)
	Mlo**	Resistance to the powdery mildew pathogen *Blumeria graminis*	Smedley et al. (2021)
Triticum turgidum	Mlo**	Resistance to the powdery mildew pathogen *Blumeria graminis*	Smedley et al. (2021)
Zea mays CRISPR/Cas12a	Cas12a	Improved editing efficiency in more recalcitrant target sequence	Dong et al. (2021)
Zea mays	FFMM	Strong ability to produce embryogenic callus from immature embryos and maintain resemblance to FFMM-A in flowering time and stature	McCaw et al. (2021)
qLW4a, ZmNL4		qLW4a, a major quantitative trait locus controlling leaf width; ZmNL4 regulates cell division	Gao et al. (2021)
	Wus2, Bbm	Morphogenic genes controlled by Axig1 and PLTP promoters	Peterson et al. (2021)
	En-SFR/Cas9, Em-SFR/Cas9, Em/En- SFR/Cas9	Allowed the selection of seeds carrying the transgene of having segregated the transgene out	Yan et al. (2021)
	CLE	Meristem size	Liu et al. (2021a, b, c, d, e)
Zizania palustris	OsAUX4**	May function in the regulation of root development, supports primary root elongation	Ye et al. (2021)
Oil seeds and legumes; *Brassica napus*	BnaFtsH1: BnaC06.FtsH1, BnaA07.FtsH1	Biomass accumulation, photosynthesis related	

Species	Gene	Trait / function	References
Xu et al. (2022)			
	BnaSBE	Enzymes that catalyze starch biosynthesis	Wang et al. (2021a, b, c, d, e, f, g, h, i)
	BnQCR8	Crop resistance against *Sclerotinia sclerotiorum* and *Botrytis cinerea*	Zhang et al. (2021a, b, c, d, e)
	BnSHPI, BnSHP2	BnSHPI: controls lignin contents at dehiscence zone. BnSHP2: reduced lignified layer and separation layer adjacent to valves and replum	Zaman et al. (2021)
	BnaC04EPSPS	Glyphosate tolerance	Wang et al. (2021a, b, c, d, e, f, g, h, i)
	BnaEOD3	Could act maternally to promote cotyledon cell expansion and proliferation to regulate seed growth in rapeseed	Khan et al. (2021)
Camelina sativa	FAD2	Increased monosaturated fatty acid content	Lee et al. (2021a, b)
Cicer arietinum	4CL, RVE7	RVE7: gene that encodes the transcription factor involved in circadian rhythm and the opening of cotyledon mediated by phytochrome; 4CL: associated with drought	Badhan et al. (2021)
Elaeis guineensis	EgPDS, EgBRI1	Enables the production of transgenic oil palm shoots exhibiting chimeric albino phenotypes as a result of DNA insertions	Yeap et al. (2021)
Glycine max	AITR	Enhance salt tolerance	Wang et al. (2021a, b, c, d, e, 2021f, g, h, i)

(continued)

Table 1 (Continued)

Crop	Gene	Trait	Reference
GmHSP17.9	Efficiently prevented the thermal aggregation of MDH at 45 °C and chemically induced aggregation of insulin by DTT	Yang et al. (2022)	GUS, GFP
	Root regeneration	Cheng et al. (2021)	Avr1b-1
	Essential for avirulence of *P. sojae* on Rps1b soybean plants but not essential for full virulence, may be susceptible to transcriptional polymorphisms	Gu et al. (2021)	GmIPK1, GmIPK2
	Codes for enzymes from the phytic acid synthesis pathway	Carrijo et al. (2021)	Rpp1L, Rps1
	Play a role in effector-triggered immunity	Nagy et al. (2021)	NLR**
	Resistance to soybean mosaic virus	Yin et al. (2021)	Fruits and vegetables
	Allium cepa	GFP	Exhibits bright green fluorescence when exposed to UV light
Miller et al. (2021)			
Brassica oleracea	MYB28	A-GSL biosynthesis genes and accumulation of the methionine-derived glucosinolate, glucoraphanin, in leaves and florets of field-grown	Neequaye et al. (2021)
Citrullus lanatus	ClGRF4, ClGIF1	Successfully create diploid seedless watermelon	Feng et al. (2021)
Cucumis melo, Cucurbita moschata, Citrullus lanatus, Cucumis sativus	Homologs of the ERECTA family of receptor kinase genes	Regeneration, genes encode receptor kinases known to control plant internode	Xin et al. (2002)
Lactuca sativa	SOC1**	LsCYCD1-1, LsCYCD3-2, LsCYCD6-1, and LsCYCU4-1: cell division-related genes	Choi et al. (2021)
Solanum lycopersicum	cpk28	Phosphorylates ascorbate peroxidase and enhances thermotolerance	Hu et al. (2021)

ABCG44 ABCG45	Reduces strigolactones in the root exudate and abolishes germination of the root-parasitic weed *P. aegyptiaca*	Bari et al. (2021)	SlPelo, SlMlo1
	SlMlo1: showed complete resistance to powdery mildew fungus. SlPelo: suppressed accumulation of TYLCV and restricted the spread to non-inoculated plant parts	Pramanik et al. (2021)	SlPUB24
	Enhances resistance to *Xanthomonas euvesicatoria*	Liu et al. (2021a, b, c, d, e)	Carotenoid isomerase
	Key gene in the carotenoid biosynthesis pathway	Jayraj et al. (2021)	Rx4
	Rx4: encodes a nucleotide-binding leucine-rich repeat protein	Zhang et al. (2021a, b, c, d, e)	HyPRP1
	Negative regulator of salt stress responses	Tran et al. (2021)	Slald1, Slfmo1
	Play a critical role in drought tolerance	Wang et al. (2021a, b, c, d, e, f, g, h, i)	*Solanum tuberosum* CRISPR/dMac3-Cas9
	StSBE3	Predicted to encode a protein containing the transit peptide for translocation to amyloplasts	Takeuchi et al. (2021)
Vitis vinifera	VvbZIP36	Shown to play a role in drought stress	Wang et al. (2021a, b, c, d, e, f, g, h, i)
Others		*Brachypodium distachyon*	BdFAR4
Production of C20:0 and C22:0 fatty alcohols associated with root suberin	Wang et al. (2021a, b, c, d, e, f, g, h, i)		
Cannabis sativa	DMG278	Shoot regeneration	Zhang et al. (2021a, b, c, d, e)
Cichorium intybus	CiGAS	Bitter tasting compound producer	Cankar et al. (2021)
Gossypium hirsutum	GhImA	Associated with the non-fluffy fiber phenotype	Zhang et al. (2021a, b, c, d, e)

(continued)

Table 1 (Continued)

Crop	Gene	Trait	Reference
	112 plant development-related genes	Play a vital role in the response of male reproductive organs to high temperature stress	Ramadan et al. (2021)
CLCuV**		Controls presence of cotton leaf curl disease	Mubarik et al. (2021)
Humulus lupulus	Phytoene desaturase	Involved in carotenoid biosynthesis	Awasthi et al. (2021)
Nicotiana tabacum	UBIQUITIN10	Showed their capacity to direct high levels of transgene expression in transient and stable transformation assays	Kumar et al. (2022)
	ADR1, SOC3/SOC3-L1, TN2, NRG1**	ADR1: may participate in the activation of immune responses, enhanced resistance to *G. cichoracearum*. NLR & SOC3: chilling sensitivity	Wang et al. (2021a, b, c, d, e, f, g, h, i)
Petunia × *hybrida*	F3H	Exhibited a pale purplish pink flower color	Yu et al. (2021)
Populus tremula × *alba*	LBD12, LBD4, YUC4, PLT1, SHR	SHR: short root. YUC4: part of a pathway linking auxin biosynthesis and gynoecium development. LBD12: apical meristem size. LBD4: controls vascular amount; PLT1: embryonic pattern formation	Triozzi et al. (2021)
Saccharum spp. hybrid	ALS	Herbicide tolerance	Oz et al. (2021)
	MgCh	Severely reduced chlorophyll contents	Eid et al. (2021)
Setaria viridis	gfp	Displayed Mendelian segregation in the T1 generation	Basso et al. (2021)

ethical concerns (Liang et al. 2021; Ayanoglu et al., 2020). Although CRISPR genome editing has been commonly applied in several crops, the technology suffers largely from low stability. Concerns about the CRISPR-Cas9 technology are related to Cas9 protein, which induced immunogenic side effects when delivered by the adeno-associated virus in mice.

A fundamental feature is a distinction between genetically modified, transgenic, and genome-edited plants (Eckerstorfer et al. 2019a, b). Genome-edited plants are essentially not transgenic. If the transgene carrying the CRISPR-Cas9 cassette is removed by gene segregation, it will be classified as non-transgenic. Prior to the development of genome editing, humans were already exposed to Cas9 proteins in their food and environment (El-Mounadi et al. 2020). CRISPR-induced selection of mutant cells may also occur in other species, as the CRISPR edits often occur in genes related to cell cycle arrest and DNA repair (Agapito-Tenfen et al. 2018). Thus, to correct and deter the spread of misconceptions, teaching the public the principles of genome editing is vital to prevent the spread of falsehoods.

4.1 Delivery of CRISPR-Cas Technology and Biosafety Concerns

We project that the focal point of apprehensiveness of CRISPR-Cas is off-target editing. Genome editing is considered a complex methodology that occurs through the subsequent process of trait segregation by way of breeding selection, to obtain a desired plant genome edit that allows for detecting and evaluating off-target edits, which can inhibit unnecessary downstream phenotypic effects. Gene editing with ZFNs has demonstrated high frequencies of off-target edits and high toxicity due to their low affinity and low specificity (Zhao and Wolt 2017).

4.1.1 Off-Targets and Their Effects

Any alteration of the genetic architecture that is not specific could lead to unintended effects (Ladics et al. 2015). Different genome editing methods exhibit different levels of precision, hence exhibiting levels of unintended genetic modifications, generally referred to as off-target effects (Li et al. 2019). CRISPR/Cas9 technology claims high precision and none too low off-target activity (Feng et al. 2018). However, off-target action does occur in organisms and plants (Braatz et al. 2017; Anderson et al. 2018).

Off-target activity depends on the frequency of homologous sequences in the genome, type, and expression level of nuclease (Yee 2016). Off-target editing can cause chromosomal relocations and other mutations, such as integrating DNA mismatches into the PAM-distal position of the sgRNA sequence (Ghosh et al. 2019). There are some risks associated with downstream attributes of the phenotype derived from genome editing that may be ascribed to unintended changes in the genome: cas9 protein bind and cleave DNA at off-target sites. Unfortunately, CRISPR-Cas9 tolerates a certain degree of mismatches between the sgRNA and the target, leading to off-target editing. Some collateral effects of CRISPR-Cas 9 use can also be attributed to the delivery vectors (Globus and Qimron 2018).

4.1.2 Players to Reduce the Off-Target Activity

Off-target editing produces fewer genetic differences than radiation mutagenesis through *Agrobacterium*-mediated transformation (Anderson et al. 2016). Plant genome editing demonstrates that the prevalence of relatively low off-target mutation is due to CRISPR editing. Studies imply that off-target outcomes of the CRISPR-Cas system in plants are infrequent, with undetectable or few low-frequency off-target mutations (Feng et al. 2018). Some unexpected off-target editing by CRISPR-Cas9 was reported in *Arabidopsis thaliana* (Zhang et al. 2018) and a rare off-target change in cotton plants (Li et al. 2019); hence, an in-depth research is focused on reducing off-target editing, specifically in human genome editing (Manghwar et al. 2020). In the efforts to reduce off-target activity, gRNA design, including RNA to DNA nucleotide replacements (Yin et al. 2018), length and composition of gRNA binding domain (Cho et al. 2014), as well as mismatches between gRNA and target DNA (Fu et al. 2014), there are molecular diagnostic tools to trace an off-target cutting (Duensing et al. 2018); this plays a crucial role in mitigating the adverse effects of off-target activity (Agapito-Tenfen et al. 2018).

The PAM (protospacer adjacent motif) sequence and its immediate upstream and downstream nucleotides, GC content of the gRNA, and epigenetics and chromatin structure of the target all engage in prospective roles in off-target activity. One or two mismatches within the seed sequence of the sgRNA in the PAM proximal region will also produce higher numbers of off-target edits. Many sgRNA design tools assist with the identification of specific sgRNA sequences to improve targeting and reduce off-target effects. A modified Cas9 nuclease with less stringent requirements for matching a typical protospacer adjacent motif (PAM) unexpectedly displays a higher overall specificity (Hu et al. 2018). The use of truncated sgRNA of 17 or 18 nt with a shortened 5′-end was shown to decrease undesired mutagenic effects at off-target sites in mammalian cell systems without sacrificing on-target genome editing efficiencies. In plants, condensed sgRNA in a CRISPR/Cas9 system as a constitutive promoter demonstrated high on-target mutation rates with no off-target effects identified (Hajiahmadi et al. 2019). In addition, merging truncated sgRNAs with pairs of Cas9 nickase led to further reductions in off-target mutations (Zhao and Wolt 2017).

5 Role of Specificity of Genome Editing on Limiting the Off-Target Activity

Specificity is an integral part of efficient genome editing, mainly when used as a tool for GMO production, which merit strict regulation. There are concerns about the specificity of Cas9 and the targeted sites due to PAM requirements (El-Mounadi et al. 2020).

Restrict the off-target activity of SDN-mediated genome editing by:

1. Selecting suitable developers and applying high-level specificity
2. Choosing the exact genomic target sequence, e.g., determining target sequences that display a low homology to other genomic sequences

5.1 Preventing Off-Target Mutations

Targeted phenotypical assessment is used to determine the significance of the unintentional effects and modify the current principles for risk assessment founded for GMOs.

6 Identification of Potential Off-Target Activity

In vitro techniques are there to help detect positions of potential off-target activity in the genome:

Genome-wide, unbiased identification of DSBs enabled by sequencing (GUIDE-seq), high-throughput genomic translocation sequencing (HTGTS), breaks labeling, enrichments on streptavidin and next-generation sequencing (BLESS), and digested genome sequencing (Digenome-seq) provide unbiased whole-genome screens for such sites.

7 Less Risky Vector Applications

In higher plants, the traditional *Agrobacterium*-mediated plasmid delivery system successfully distributes genome editing reagents. It can control transgene copy numbers to moderate concentration of sgRNA-Cas9 complexes and, thus, reduce off-target activity. Dose-dependent effects of the reagent may alter ratios of on-target and off-target edits before breeding selection to eliminate the gene-editing reagent. The delivery of pre-assembled Cas9 protein-gRNA ribonucleoproteins (RNPs) in place of plasmids can control Cas9 protein concentration to inhibit genotoxicity and ease overexpression of sgRNA while reducing off-target effects. This expression, in turn, avoids transferring genome editing reagents to the next generation. The possibility of RNP delivery to plant cells has been shown with biolistic particle bombardment, where recombinase delivery will lead to site-specific, heritable edits (Zhao and Wolt 2017). Replacing extracellular DNA with pre-assembled ribonucleoproteins (Cas9 + sgRNA) aids in decreasing the CRISPR-Cas activity in cells since its components are not continuously being created while simultaneously helping to alleviate the collateral effects ascribed to the delivery vector crops. An alternative method is performing whole-genome sequencing, Digenome-seq, or other susceptible methods to exclude GMOs with off-target mutations. These sequencing techniques guarantee that only the desired changes are introduced into the GMO crops. A third method is selecting precise target sequences using BLAST or a database specific to crop to reduce off-targets. Higher fidelity Cas9 or other Cas9 orthologues with higher fidelity or shortening the gRNA crops can all be used to decrease the off-target activity significantly (Globus and Qimron 2018). In addition, vetting pre-selected possible off-target sequences can often risk overlooking mutations at further loci in the plant genome. Whole-genome sequencing in plants will allow the identification of off-target effects in a less restricted way. Using both viral and nonviral methods, CRISPR-Cas9 is immediately transmitted to cells in the body (Roh et al. 2018).

PGD is a reproductive option that is an effective and ethically acceptable mode of treatment for most diseases. Perhaps one may be able to justify the use of gene editing research for these genetic conditions as a way of developing research tools and conditions for more complex applications in the future (Cribbs and Perera 2017).

8   Develop a Reversal Mechanism to Revert in Cases of Unintended Effects

Transgene integration and the risk of off-target mutations by delivering in vitro pre-assembled CRISPR-Cas9 ribonucleoproteins can prevent transgene integration. Base editing is being modified to improve the specificity of base editors by limiting deaminase activity outside of Cas9 binding by using different deaminase effectors or rationally engineering the deaminase to decrease its DNA binding ability.

9   Develop Detection Mechanism to Ensure Safety Before Field Application

Protein engineering efforts have resulted in detecting mutations in Cas9 that modify its PAM recognition, heighten its reliability, and identify other motifs (El-Mounadi et al. 2020).

10   Monitoring Mechanism for Post-release Stability

Identification of DNA signatures by developers would help identify a clear signature in the DNA, for example, the same stretch of nucleotides being eliminated. Suppose the developer reveals that the signature, the same PCR technology used for detecting GMOs, can be applied to detect and monitor genome-edited products (Duensing et al. 2018). The combined abilities for stress resistance and accurate DNA repair mechanisms in plants contribute to the genetic stability of the resulting edit in crops.

There must be accurate DNA repair mechanisms in coding regions to maintain genome stability, fertility, and genetic diversity.

11   Safe Alternatives from This Technology

The development of CRISPR/Cas9 techniques with the additive benefits from greater specificity in sgRNA design, minimal off-target binding activity, and the absence of off-target mutations have been achieved in zebrafish, mice, chicken, stem cells, *Arabidopsis*, and rice. Advancement in targeting minimized CRISPR/Cas9 off-target effects while safeguarding the wild genetic pool (Zhao and Wolt 2017).

The new traits made by non-GM may impact the species-specific traits of altered plants and are, thus, needed to assess their overall risks associated. Introducing traits present in native populations or related species is one way in which genome editing can be used. Many traits developed with genome editing should be considered advanced in crop plants. Comparable traits are absent in stable, planted populations of the plant species at substantial levels (Ecckerstprfer et al. 2019). Safety considerations associated with genome editing applications must be based on the characteristics of the application. Due to their different approaches, the issues for risk assessment can be dissimilar.

It improved RNA guide-design strategies, ribonucleoprotein delivery, protein engineering, spatiotemporally controlled Cas9, and gRNAs through many chemicals, environmental inducers, or synthetic genetic circuits that modulate CRISPR function according to predefined logic (El-Mounadi et al. 2020). To overcome the regeneration of immature cells *in planta*, particle bombardment (PB) that targets mature plant tissue was introduced in wheat (Hamada et al. 2017). Cell-penetrating peptides (CPPs) are short, positively charged peptides, which will translocate across cellular membranes, capable of binding site-specific nucleases (Rádis-Baptista et al. 2017).

Because of widespread applications, manipulating CRISPR-Cas without integrating a transgene is a relatively minor risk compared to former GM techniques, causing it to be more acceptable (Bartkowski et al. 2018). GMOs have been produced employing a transgenic method by inserting an exogenous gene into the host genome. In CRISPR-Cas genome editing systems (NHEJ repair mechanism), genetic alteration is primarily constructed on small indels of nucleotides within the endogenous target gene, similar to the natural variations. Genome editing can be accomplished with high specificity and efficacy through chemical and physical mutagenesis (Globus and Qimron 2018). The altered expression cascades introduced via genome editing are removed from the progeny through genetic segregation. Thus, the genetic properties of genome-edited crops are different from traditional GMOs but are considerably similar to the crops produced naturally or by mutagenesis. Therefore, genome-edited crops are considered a low risk to human safety (Li et al. 2019).

11.1 Regulation Country-Wise

11.1.1 EU Regulatory Framework

The regulatory requirements for registering plant varieties in EU or national catalogs fail to provide for breadth and standard of risk assessment compared with the requirements according to the corresponding biosafety frameworks. All countries of the EU are drafting their national regulatory recommendations; therefore, there's a need to redefine the assessment type for GMOs and related risks assessment and protocols.

Australia The implementation of genome editing technologies, regulated as the first- and second-generation ones, requires biosafety clearance from the Office of the Gene Technology Regulator (OGTR). The Australian government's new regulatory developments inspired the utility of genome editing technologies producing transgene-free plants, having no risk to human health and the environment.

Brazil The regulatory framework was established by the Brazilian National Biosafety Technical Commission (CTNBio) in 2014. The Normative Resolution No. 16 (RN16) of CTNBio was approved during 2018 and stated that every product derived out of NBTs or through genome editing technologies, that is, CRISPR/Cas system, will be dealt the same as traditional or transgenic plant, animal, or microorganism on a case-by-case basis. The regulation also highlighted that the next generations having transgene-free either developed through classical, induced mutagenesis by chemical/physical mutagens or produced naturally will be evaluated on a case-by-case basis and will be dealt with as a customary product (Eriksson 2018).

Canada In Canada, this aspect is crucial for the denomination of products that are subject to oversight for biosafety, e.g., according to the "plants with novel traits (PNT)" regulations. Canada regulates PNTs generated by conventional plant breeding approaches for biosafety in regulatory frameworks based on novelty as a product-oriented regulatory trigger (Eckerstorfer et al. 2019a, b).

China In 2001, China's state council issued the "Regulation on Administration of Agricultural Genetically Modified Organisms Safety." This regulation determines that GMOs are plants, animals, microorganisms, and derived products with genetic architecture modified through genome manipulation technologies utilized during agricultural production and development.

Moreover, during 2016, a working group was established within the National Biosafety Committee (NBC) for delivering technical expertise on risk assessment of novel technologies, including genome manipulation tools (Gao et al. 2018). The enforced regulation on genome-edited crops is under profound observation. In contrast, the plant breeders urged the regulatory bodies to deal with transgene-free plants as traditionally bred plants. These regulations will aid in reducing the efforts made during the evaluation process while also nurturing scientific research (Xiaoyu 2019).

Japan The Japanese authorities are still debating the regulatory framework dealing with genome-edited plants, animals, microorganisms, and derived products. In August 2018, the advisory session on GMOs from Japan's Ministry of Environment reconsidered the observations forward by the experts' committee on regulations of genome editing techniques. The committee recommended that any living entity with any foreign DNA/gene, detectable/non-detectable, be dealt with through GMOs regulations resulting in SDN-1 editing escape from the strict rules of GMOs (Zannoni 2019).

New Zealand The 1996 Act for Hazardous Substances and New Organisms (HSNO) deals with the process-based GMOs. According to HSNO, GMO is the outcome of any manipulated gene/genome through in vitro technologies. Therefore, many genome editing techniques are regulated through the HSNO regulatory framework (Fritsche et al. 2018). Section 26 of the HSNO Act explains that organism/product modified through genome editing techniques with no foreign DNA/gene, similar to chemically induced mutagenesis, will be excluded from GMO regulations. However, the sustainability council challenged the outcome, and the court passed the ruling that the HSNO regulation 1998 contains a closed list of techniques employed for genome manipulation. However, the addition of any technology to an already existing inventory will be a political judgment rather than an administrative decision (Kershen 2015). Therefore, the derived products from any genome editing technology can be regulated through the GMOs framework (Fritsche et al. 2018).

The US regulatory framework is presently challenged in its ability to appropriately weigh and analyze novel breeding approaches while over-regulating transgenic technologies with a clear record of safety (Camacho et al. 2014). This limitation affects the product developer's need for greater certainty in the regulatory process and the public's desire for appropriate governance of new technologies (Wolt et al., 2016).

11.2   Community Awareness

https://crispr-gene-editing-regs-tracker.geneticliteracyproject.org

12   Conclusions

Adopting new technologies has always presented challenges, including the various perceptions and public awareness. The public must be engaged in a series of dialogues on genome-edited crops, as this is the sole solution for meeting the needs of the growing world population. Government regulations are a significant bottleneck for the large-scale adoption of genome-edited crops all over the world. CRISPR-mediated technologies cannot be discounted, as this method provides sustainable agriculture. To make these crops acceptable for consumer usage, it is essential to demark between gene-edited plants with foreign DNA and gene-edited plants without any foreign DNA. Risk assessment tools must be designed case-by-case for non-genetically modified plants that exhibit a level of uncertainty. A thorough molecular classification to identify unintended sequence modifications introduced, changed, and validated the absence of unwarranted transgenic sequences, off-target improvements, and other genetic changes that might result in negative phenotype traits. These efforts will be meaningful from a safety standpoint and a superior option to entirely exempting nGM applications from biosafety assessments. Despite the ability of biotechnology to solve future problems of the human population, there is often reluctance among the public to accept and support biotechnological products in medicine, industry, or agriculture. Although agriculture is economically viable, ethical considerations are crucial. The public has rejected genetic modifications in plants and animals and GM food products, the central issue. (FAO 2003). GM foods have become the target of public concern due to unknown and unseen fears of their effects on the ecosystem and risks to human health. GM foods are also not labeled, leading to distrust and anxiety in consumers, regarding food safety. The central idea in agricultural ethics is that social responsibility overlaps with scientific responsibility (Smith 1990). Due to the rising importance of agriculture, a minimum of 14 universities in the USA have offered a course in agricultural ethics (Ruehr 2015). Hopefully, these efforts can contribute to public's ease of advancing and integrating genetic modification in society.

Acknowledgments KR, VT, JN, and VG gratefully acknowledge the support from National Science Foundation Research Experience for Undergraduate students (#15600349).

References

Achary VMM, Reddy MK (2021) CRISPR-Cas9 mediated mutation in GRAIN WIDTH and WEIGHT2 (GW2) locus improves aleurone layer and grain nutritional quality in rice. Sci Rep 11(1):21941

Agapito-Tenfen SZ, Okoli AS, Bernstein MJ, Wikmark OG, Myhr AI (2018) Revisiting risk governance of GM plants: the need to consider new and emerging gene-editing techniques. Front Plant Sci 9:1874

Anderson KR, Haeussler M, Watanabe C, Janakiraman V, Lund J, Modrusan et al (2018) CRISPR off-target analysis in genetically engineered rats and mice. Nat Methods 15(7):512–514

Anderson KR, Haeussler M, Watanabe C, Janakiraman V, Lund J, Modrusan Z, Stinson J, Bei Q, Buechler A, Yu C, Thamminana SR, Tam L, Sowick MA, Alcantar T, O'Neil N, Li J, Ta L, Lima L, Roose-Girma M, Rairdan X, Durinck S, Warming S. CRISPR off-target analysis in genetically engineered rats and mice. Nat Methods. 2018 Jul;15(7):512-514. doi: 10.1038/s41592-018-0011-5. Epub 2018 May 21. PMID: 29786090; PMCID: PMC6558654.

Aregawi K, Shen J, Pierroz G, Sharma MK, Dalhberg J, Owiti J et al (2021) Morphogene-assisted transformation of *Sorghum bicolor* allows more efficient genome editing. Plant Biotechnol J 20:748–760

Awasthi P, Kocábek T, Mishra AK, Nath VS, Shrestha A, Matoušek J (2021) Establishment of CRISPR/Cas9 Mediated targeted mutagenesis in hop (*Humulus lupulus*). Plant Physiol Biochem 160:1–7

Ayanoğlu FB, Elçin AE, Elçin YM (2020) Bioethical issues in genome editing by CRISPR-Cas9 technology. Turk J Biol 44(2):110–120

Ayanoğlu, F. B., Elçin, A. E., & Elçin, Y. M. (2020). Bioethical issues in genome editing by CRISPR-Cas9 technology. Turkish journal of biology Turk biyoloji dergisi, 44(2), 110–120.

Badhan S, Ball AS, Mantri N (2021) First report of CRISPR/Cas9 mediated DNA-free editing of *4CL* and *RVE7* genes in chickpea protoplasts. Int J Mol Sci 22(1):396

Bahariah B, Masani MYA, Rasid OA, Parveez GKA (2021) Multiplex CRISPR/Cas9-mediated genome editing of the FAD2 gene in rice: a model genome editing system for oil palm. J Genet Eng Biotechnol 19(1):86

Bao A, Burritt DJ, Chen H, Zhou X, Cao D, Tran LP (2019) The CRISPR/Cas9 system and its applications in crop genome editing. Crit Rev Biotechnol 39(3):321–336

Bari VK, Nassar JA, Meir A, Aly R (2021) Targeted mutagenesis of two homologous ATP-Binding cassette subfamily G (ABCG) genes in tomato confers resistance to parasitic weed *Phelipanche aegyptiaca*. J Plant Res 134(3):585–597

Bartkowski B, Theesfeld I, Pirscher F, Timaeus J (2018) Snipping around for food: economic, ethical and policy implications of CRISPR/Cas genome editing. Geoforum 96:172–180

Basso MF, Duarte KE, Santiago TR, de Souza WR, de Oliveira GB, Brito da Cunha B et al (2021) Efficient genome editing and gene knockout in *Setaria viridis* with CRISPR/Cas9 directed gene editing by the non-homologous end-joining pathway. Plant Biotechnol (Tokyo) 38(2):227–238

Braatz J, Harloff HJ, Mascher M, Stein N, Himmelbach A, Jung C (2017) CRISPR-Cas9 targeted mutagenesis leads to simultaneous modification of different homoeologous gene copies in polyploid oilseed rape (*Brassica napus*). Plant Physiol 174(2):935–942

Brant EJ, Baloglu MC, Parikh A, Altpeter F (2021) CRISPR/Cas9 mediated targeted mutagenesis of LIGULELESS-1 in *Sorghum* provides a rapidly scorable phenotype by altering leaf inclination angle. Biotechnol J 16(11):e2100237

Camacho A, Van Deynze A, Chi-Ham C, Bennett AB (2014) genetically engineered crops that fly under the US regulatory radar. Nat Biotechnol 32(11):1087–1091

Cankar K, Bundock P, Sevenier R, Häkkinen ST, Hakkert JC, Beekwilder J et al (2021) Inactivation of the germacrene a synthase gene by CRISPR/Cas9 eliminates the biosynthesis of sesquiterpene lactones in *Cichorium intybus L*. Plant Biotechnol J 19(12):2442–2453

Carlessi M, Mariotti L, Giaume F, Fornara F, Perata P, Gonzali S (2021) Targeted knockout of the gene OsHOL1 removes methyl iodide emissions from rice plants. Sci Rep 11(1):17010

Carrijo J, Illa-Berenguer E, LaFayette P, Torres N, Aragão FJL, Parrott W et al (2021) Two efficient CRISPR/Cas9 systems for gene editing in soybean. Transgenic Res 30(3):239–249

Chen S, Zhang N, Zhou G, Hussain S, Ahmed S, Tian H et al (2021) Knockout of the entire family of AITR genes in *Arabidopsis* leads to enhanced drought and salinity tolerance without fitness costs. BMC Plant Biol 21(1):137

Cheng Y, Wang X, Cao L, Ji J, Liu T, Duan K (2021) Highly efficient agrobacterium rhizogenes-mediated hairy root transformation for gene functional and gene editing analysis in soybean. Plant Methods 17(1):73

Cho SW, Kim S, Kim Y, Kweon J, Kim H, Bae S (2014) Analysis of off-target effects of CRISPR/Cas-derived RNA-guided endonucleases and nickases. Genome Res 24(1):132–141

Choi M, Yun JY, Kim JH, Kim JS, Kim ST (2021a) The efficacy of CRISPR-mediated cytosine base editing with the RPS5a promoter in *Arabidopsis thaliana*. Sci Rep 11(1):8087

Choi SH, Lee MH, Jin DM, Ji JS, Ahn WS, Jie EY et al (2021b) TSA promotes CRISPR/Cas9 editing efficiency and expression of cell division-related genes from plant protoplasts. Int J Mol Sci 22(15):7817

Cribbs AP, Perera SMW (2017) Science and bioethics of CRISPR-Cas9 gene editing: an analysis towards separating facts and fiction. Yale J Biol Med 90(4):625–634

Dong S, Qin YL, Vakulskas CA, Colingwood MA, Marand M, Rigoulot S et al (2021) Efficient targeted mutagenesis mediated by CRISPR-Cas12a ribonucleoprotein complexes in maize. Front Genome Ed 3:670529

Doudna JA, Charpentier E (2014) Genome editing. The new frontier of genome engineering with CRISPR-Cas9. Science 346(6213):1258096

Duensing N, Sprink T, Parrott WA, Fedorova M, Lema MA, Wolt JD et al (2018) Novel features and considerations for ERA and regulation of crops produced by genome editing. Front Bioeng Biotechnol 6:79

Eckerstorfer MF, Dolezel M, Heissenberger A, Miklau M, Reichenbecher W, Steinbrecher RA et al (2019a) An EU perspective on biosafety considerations for plants developed by genome editing and other new genetic modification techniques (nGMs). Front Bioeng Biotechnol 7:31

Eckerstorfer MF, Engelhard M, Heissenberger A, Simon S, Teichmann H (2019b) Plants developed by new genetic modification techniques-comparison of existing regulatory frameworks in the EU and Non-EU countries. Front Bioeng Biotechnol 7:26

Eid A, Mohan C, Sanchez S, Wang D, Altpeter F (2021) Multiallelic, targeted mutagenesis of magnesium chelatase with CRISPR/Cas9 provides a rapidly scorable phenotype in highly polyploid sugarcane. Front Genome Ed 3:654996

El-Mounadi K, Morales-Floriano ML, Garcia-Ruiz H (2020) Principles, applications, and biosafety of plant genome editing using CRISPR-Cas9. Front Plant Sci 11:56

Eriksson D (2018) The Swedish policy approach to directed mutagenesis in a european context. Physiol Plant 164(4):385–395

FAO. Regulating GMOs in developing and transition countries. Summary Document to Conference 9 of the FAO Biotechnology Forum 28 April to 1 June 2003

Feng C, Su H, Bai H, Wang R, Liu Y, Guo X et al (2018) High-efficiency genome editing using a Dmc1 promoter-controlled CRISPR/Cas9 system in maize. Plant Biotechnol J 16(11):1848–1857

Feng Q, Xiao L, He Y, Liu M, Wang J, Tian S et al (2021) Highly efficient, genotype-independent transformation and gene editing in watermelon (*Citrullus lanatus*) using a chimeric ClGRF4-GIF1 gene. J Integr Plant Biol 63(12):2038–2042

Fritsche S, Poovaiah C, MacRae E, Thorlby G (2018) A New Zealand perspective on the application and regulation of gene editing. Front Plant Sci 9:1323

Fu Y, Sander JD, Reyon D, Cascio VM, Joung JK (2014) Improving CRISPR-Cas nuclease specificity using truncated guide RNAs. Nat Biotechnol 32(3):279–284

Fu L, Wu D, Zhang X, Xu Y, Kuang L, Cai S et al (2022) Vacuolar H+-pyrophosphatase HVP10 enhances salt tolerance via promoting Na+ Translocation into Root Vacuoles. Plant Physiol 188(2):1248–1263

Gao W, Xu W, Huang K, Guo M, Luo Y (2018) Risk Analysis for genome editing-derived food safety in China. Food Control 84:128–137

Gao L, Yang G, Li Y, Sun Y, Xu R, Chen Y et al (2021) A kelch-repeat superfamily gene, ZmNL4, controls leaf width in maize (*Zea mays L.*). Plant J 107(3):817–830

Ghosh D, Venkataramani P, Nandi S, Bhattacharjee S (2019) CRISPR-Cas9 A boon or bane: the bumpy road ahead to cancer therapeutics. Cancer Cell Int 19:12

Globus R, Qimron U (2018) A Technological and regulatory outlook on CRISPR crop editing. J Cell Biochem 119(2):1291–1298

Gu B, Shao G, Gao W, Miao J, Wang Q, Liu X (2021) Transcriptional variability associated with CRISPR-Mediated gene replacements at the *Phytophthora sojae Avr1b-1* Locus. Front Microbiol 12:645331

Guo R, Hu Y, Aoi Y, Hira H, Ge C, Dai X et al (2022) Local conjugation of auxin by the GH3 amido synthetases is required for normal development of roots and flowers in *Arabidopsis*. Biochem Biophys Res Commun 589:16–22

Hahn F, Sanjurjo Loures L, Sparks CA, Kanyuka K, Nekrasov V (2021) Efficient CRISPR/Cas-mediated targeted mutagenesis in spring and winter wheat varieties. Plants (Basel) 10(7):1481

Hajiahmadi Z, Movahedi A, Wei H, Li D, Orooji Y, Ruan H et al (2019) Strategies to increase on-target and reduce off-target effects of the CRISPR/Cas9 system in plants. Int J Mol Sci 20(15):3719

Hamada H, Linghu Q, Nagira Y, Miki R, Taoka N, Imai R (2017) An in planta biolistic method for stable wheat transformation. Sci Rep 7(1):11443

Hoffie RE, Otto I, Perovic D, Budhagatapalli N, Habekuß A et al (2021) Targeted knockout of eukaryotic translation initiation factor 4E confers bymovirus resistance in winter barley. Front Genome Ed 3:784233

Hu JH, Miller SM, Geurts MH, Tang W, Chen L, Sun N, Zeina CM, Gao X, Rees HA, Lin Z, Liu DR. Evolved Cas9 variants with broad PAM compatibility and high DNA specificity. Nature. 2018 Apr 5;556(7699):57-63. doi: 10.1038/nature26155.

Hu Z, Li J, Ding S, Cheng F, Li X, Jiang Y et al (2021) The protein kinase CPK28 phosphorylates ascorbate peroxidase and enhances thermotolerance in tomato. Plant Physiol 186(2):1302–1317

Hussain S, Zhang N, Wang W, Ahmed S, Cheng Y, Chen S et al (2021) Involvement of ABA responsive *SVB* genes in the regulation of trichome formation in *arabidopsis*. Int J Mol Sci 22(13):6790

Karthika V, Chandrashekar BK, Kiranmai K, Ag S, Makarla U, Ramu VS (2021) Disruption in the DNA mismatch repair gene *MSH2* by CRISPR-*Cas9* in indica rice can create genetic variability. J Agric Food Chem 69(14):4144–4152

Karunarathne SD, Han Y, Zhang XQ, Li C (2022) CRISPR/Cas9 gene editing and natural variation analysis demonstrate the potential for HvARE1 in improvement of nitrogen use efficiency in barley. J Integr Plant Biol 64(3):756–770

Kershen DL (2015) Sustainability Council of New Zealand Trust v. The environmental protection authority: gene editing technologies and the law. GM Crops Food 6(4):216–222

Khan MHU, Hu L, Zhu M et al (2021) Targeted mutagenesis of EOD3 gene in *Brassica napus L.* regulates seed production. J Cell Physiol 236:1996–2007

Kim CY, Park JY, Choi G, Kim S, Vo KTX, Jeon J et al (2022) A rice gene encoding glycosyl hydrolase plays contrasting roles in immunity depending on the type of pathogens. Mol Plant Pathol 23(3):400–416

Kumar M, Ayzenshtat D, Marko A, Bocobza S (2022) Optimization of T-DNA configuration with UBIQUITIN10 promoters and tRNA-sgRNA complexes promotes highly efficient genome editing in allotetraploid tobacco. Plant Cell Rep 41(1):175–194

Ladics GS, Bartholomaeus A, Bregitzer P, Doerrer NG, Gray A, Holzhauser T et al (2015) Genetic basis, and detection of unintended effects in genetically modified crop plants. Transgenic Res 24(4):587–603

Lee KR, Jeon I, Yu H, Kim S, Kim H, Ahn S et al (2021a) Increasing monounsaturated fatty acid contents in hexaploid *Camelina sativa* seed oil by *FAD2* gene knockout using CRISPR-Cas9. Front Plant Sci 12:702930

Lee SK, Hong WJ, Silva J, Kim E, Park SK, Jung K et al (2021b) Global identification of *ANTH* genes involved in rice pollen germination and functional characterization of a key member, OsANTH3. Front Plant Sci 12:609473

Li J, Manghwar H, Sun L, Wang P, Wang G, Sheng H et al (2019) Whole genome sequencing reveals rare off-target mutations and considerable inherent genetic or/and somaclonal variations in CRISPR/Cas9-edited cotton plants. Plant Biotechnol J 17(5):858–868

Li J, Jiao G, Sun Y, Chen J, Zhong Y, Yan L et al (2021) Modification of starch composition, structure, and properties through editing of TaSBEIIa in both winter and spring wheat varieties by CRISPR/Cas9. Plant Biotechnol J 19(5):937–951

Li R, Zhang J, Li Z, Peters RJ, Yang B (2022) Dissecting the labdane-related diterpenoid biosynthetic gene clusters in rice reveals directional cross-cluster phytotoxicity. New Phytol 233(2):878–889

Liang Y, Biswas S, Kim B, Bailey-Serres J, Septiningsih EM (2021) Improved transformation and regeneration of indica rice: disruption of *SUB1A* as a test case via CRISPR-Cas9. Int J Mol Sci 22(13):6989

Liu C, Cao Y, Hua Y, Du G, Liu Q, Wei X et al (2021a) Concurrent disruption of genetic interference and increase of genetic recombination frequency in hybrid rice using CRISPR/Cas9. Front Plant Sci 12:757152

Liu D, Yu Z, Zhang G, Yin W, Li L, Niu M et al (2021b) Diversification of plant agronomic traits by genome editing of brassinosteroid signaling family genes in rice. Plant Physiol 187(4):2563–2576

Liu L, Gallagher J, Arevalo ED, Chen R, Skopelitis T, Wu Q et al (2021c) Enhancing grain-yield-related traits by CRISPR-Cas9 promoter editing of maize CLE genes. Nat Plants 7(3):287–294

Liu X, Meng G, Wang M, Qian Z, Zhang Y, Yang W (2021d) Tomato SlPUB24 enhances resistance to *Xanthomonas euvesicatoria* pv. *perforans* race T3. Hortic Res 8(1):30

Liu Y, Luo W, Linghu Q, Abe F, Hisano H, Sato K et al (2021e) *In planta* genome editing in commercial wheat varieties. Front Plant Sci 12:648841

Lyon J. (2017). Bioethics Panels Open Door Slightly to Germline Gene Editing. JAMA, 318(17), 1639–1640.

Ma G, Kuang Y, Lu Z, Li X, Xu Z, Ren B et al (2021a) CRISPR/Sc++ -mediated genome editing in rice. J Integr Plant Biol 63(9):1606–1610

Ma T, Li E, Li LS, Li S, Zhang Y (2021b) The *Arabidopsis* R-SNARE protein YKT61 is essential for gametophyte development. J Integr Plant Biol 63(4):676–694

Majumder S, Datta K, Datta SK (2021) Agrobacterium tumefaciens-mediated transformation of rice by hygromycin phosphotransferase (hptII) Gene Containing CRISPR/Cas9 Vector. Methods Mol Biol 2238:69–79

Makarova KS, Wolf YI, Koonin EV (2018) Classification and nomenclature of CRISPR-Cas systems: where from here? CRISPR J 1(5):325–336

Manghwar, H., Li, B., Ding, X., Hussain, A., Lindsey, K., Zhang, X., & Jin, S. (2020). CRISPR/Cas Systems in Genome Editing: Methodologies and Tools for sgRNA Design, Off-Target Evaluation, and Strategies to Mitigate Off-Target Effects. Advanced science (Weinheim, Baden-Wurttemberg, Germany), 7(6), 1902312.

Massel K, Lam Y, Hintzsche J, Lester N, Botella JR, Godwin ID (2022) Endogenous U6 promoters improve CRISPR/Cas9 editing efficiencies in *Sorghum bicolor* and show potential for applications in other cereals. Plant Cell Rep 41(2):489–492

McCaw ME, Lee K, Kang M, Zobrist JD, Azanu MK, Birchler JA et al (2021) development of a transformable fast-flowering mini-maize as a tool for maize gene editing. Front Genome Ed 2:622227

Michalski K, Hertig C, Mańkowski DR, Kumlehn J, Zimny J, Linkiewicz AM (2021) Functional validation of *cas9*/guide RNA constructs for site-directed mutagenesis of *Triticale ABA8'OH1 loci*. Int J Mol Sci 22(13):7038

Miller K, Eggenberger AL, Lee K, Liu F, Kang M, Drent M et al (2021) an improved biolistic delivery and analysis method for evaluation of DNA and CRISPR-Cas delivery efficacy in plant tissue. Sci Rep 11(1):7695

Molina-Risco M, Ibarra O, Faion-Molina M, Kim B, Septiningsih EM, Thomson MJ (2021) Optimizing agrobacterium-mediated transformation and crispr-cas9 gene editing in the tropical japonica rice variety presidio. Int J Mol Sci 22(20):10909

Mubarik MS, Wang X, Khan SH, Ahmad A, Khan Z, Amjid MW et al (2021) Engineering broad-spectrum resistance to cotton leaf curl disease by CRISPR-Cas9 based multiplex editing in plants. GM Crops Food 12(2):647–658

Nagy ED, Stevens JL, Yu N, Hubmeier CS, LaFaver N, Gillspie M et al (2021) novel disease resistance gene paralogs created by CRISPR/Cas9 in soy. Plant Cell Rep 40(6):1047–1058

Neequaye M, Stavnstrup S, Harwood W, Lawrenson T, Hundleby P, Irwin J et al (2021) CRISPR-Cas9-Mediated Gene Editing of *MYB28* Genes impair glucoraphanin accumulation of *Brassica oleracea* in the field. CRISPR J 4(3):416–426

Nishizawa-Yokoi A, Toki S (2021) A piggyBac-mediated transgenesis system for the temporary expression of CRISPR/Cas9 in rice. Plant Biotechnol J 19(7):1386–1395

Ohtsuki N, Kizawa K, Mori A, Nishizawa-Yokoi A, Komatsuda T, Yoshida H et al (2021) Precise genome editing in miRNA target site via gene targeting and subsequent single-strand-annealing-mediated excision of the marker gene in plants. Front Genome Ed 2:617713

Oz MT, Altpeter A, Karan R, Merotto A, Altpeter F (2021) CRISPR/Cas9-mediated multi-allelic gene targeting in sugarcane confers herbicide tolerance. Front Genome Ed 3:673566

Peterson D, Barone P, Lenderts B, Schwartz C, Feigenbutz L, St Clair F et al (2021) Advances in agrobacterium transformation and vector design result in high-frequency targeted gene insertion in maize. Plant Biotechnol J 19(10):2000–2010

Pramanik D, Shelake RM, Park J, Kim MJ, Hwang I, Park Y et al (2021) CRISPR/Cas9-Mediated generation of pathogen-resistant tomato against tomato yellow leaf curl virus and powdery mildew. Int J Mol Sci 22(4):1878

Rádis-Baptista G, Campelo IS, Morlighem JRL, Melo LM, Freitas VJF (2017) Cell-penetrating peptides (CPPs): from delivery of nucleic acids and antigens to transduction of engineered nucleases for application in transgenesis. J Biotechnol 252:15–26

Ramadan M, Alariqi M, Ma Y, Li Y, Liu Z, Zhang R et al (2021) Efficient CRISPR/Cas9 mediated pooled-sgRNAs assembly accelerates targeting multiple genes related to male sterility in cotton. Plant Methods 17(1):16

Ray DK, Mueller ND, West PC, Foley JA (2013) Yield trends are insufficient to double global crop production by 2050. PLoS One 8(6):e66428

Raza A, Razzaq A, Mehmood SS, Zou X, Zhang X, Lv Y et al (2019) Impact of climate change on crops adaptation and strategies to tackle its outcome: a review. Plants (Basel) 8(2):34

Reeger JE, Wheatley M, Yang Y, Brown KM (2021) Targeted mutation of transcription factor genes alters metaxylem vessel size and number in rice roots. Plant Direct 5(6):e00328

Ricroch A, Clairand P, Harwood W (2017) Use of CRISPR systems in plant genome editing: toward new opportunities in agriculture. Emerg Top Life Sci 1(2):169–182

Roh, D. S., Li, E. B., & Liao, E. C. (2018). CRISPR Craft: DNA Editing the Reconstructive Ladder. Plastic and reconstructive surgery, 142(5), 1355–1364.

Ruehr TA (2015) Teaching agricultural ethics. Ag Ethics 57:55–61

Sharma VK, Marla S, Zheng W, Mishra D, Zhang W, Morris GP et al (2022) CRISPR guides induce gene silencing in plants in the absence of Cas. Genome Biol 23(1):6

Shoesmith JR, Solomon CU, Yang X, Wilkinson LG, Sheldrick S, van Eijden E et al (2021) APETALA2 functions as a temporal factor together With BLADE-ON-PETIOLE2 and MADS29 to control flower and grain development in barley. Development 148(5):dev194894

Smedley MA, Hayta S, Clarke M, Harwood WA (2021) CRISPR-Cas9 based genome editing in wheat. Curr Protoc 1(3):e65

Smith SE (1990) Plant biology and social responsibility. Plant Cell 2(5):367–368

Takeuchi A, Ohnuma M, Teramura H, Asano K, Noda T, Kusano H et al (2021) Creation of a potato mutant lacking the starch branching enzyme gene *StSBE3* that was generated by genome editing using the CRISPR/dMac3-Cas9 system. Plant Biotechnol (Tokyo) 38(3):345–353

Tian X, Qin Z, Zhao Y, Wen J, Lan T, Zhang L et al (2022) Stress granule-associated TaMBF1c confers thermotolerance through regulating specific mrna translation in wheat (*Triticum aestivum*). New Phytol 233(4):1719–1731

Tran MT, Doan DTH, Kim J, Song YJ, Sung YW, Das S et al (2021) CRISPR/Cas9-based precise excision of SlHyPRP1 domain(s) to obtain salt stress-tolerant tomato. Plant Cell Rep 40(6):999–1011

Triozzi PM, Schmidt HW, Dervinis C, Kirst M, Conde D (2021) Simple, efficient and open-source CRISPR/Cas9 strategy for multi-site genome editing in *Populus tremula* × *alba*. Tree Physiol 41(11):2216–2227

Usman B, Zhao N, Nawaz G, Qin B, Liu F, Liu Y et al (2021) CRISPR/Cas9 Guided mutagenesis of *Grain Size 3* confers increased rice (*Oryza sativa L.*) grain length by regulating cysteine proteinase inhibitor and ubiquitin-related proteins. Int J Mol Sci 22(6):3225

Wang F, Wang L, Zou X, Duan S, Li Z, Deng Z et al (2019) Advances in CRISPR-Cas systems for rna targeting, tracking, and editing. Biotechnol Adv 37(5):708–729

Wang P, Luo Q, Yang W, Ahammed GJ, Fing S, Chen X et al (2021a) A novel role of pipecolic acid biosynthetic pathway in drought tolerance through the antioxidant system in tomato. Antioxidants (Basel) 10(12):1923

Wang T, Xun H, Wang W, Ding X, Tian H, Hussain S et al (2021b) Mutation of *GmAITR* genes by CRISPR/Cas9 genome editing results in enhanced salinity stress tolerance in soybean. Front Plant Sci 12:779598

Wang W, Liu N, Gao C, Rui L, Jiang Q, Chen S et al (2021c) The truncated TNL receptor TN2-mediated immune responses require ADR1 function. Plant J 108(3):672–689

Wang W, Ma S, Hu P, Ji Y, Sun F (2021d) Genome editing of rice *eIF4G* loci confers partial resistance to rice black-streaked dwarf virus. Viruses 13(10):2100

Wang X, Ren P, Ji L, Zhu B, Xie G (2021e) OsVDE, A xanthophyll cycle key enzyme, mediates abscisic acid biosynthesis and negatively regulates salinity tolerance in rice. Planta 255(1):6

Wang X, Tu M, Wang Y, Yin W, Zhang Y, Wu H et al (2021f) Whole-genome sequencing reveals rare off-target mutations in CRISPR/Cas9-edited grapevine. Hortic Res 8(1):114

Wang Y, Tian H, Wang W, Wang X, Zheng K, Hussain S et al (2021g) The carboxyl-terminus of Transparent Testa Glabra1 is critical for its functions in *Arabidopsis*. Int J Mol Sci 22(18):10039

Wang Y, Xu J, He Z, Hu N, Luo W, Liu X et al (2021h) BdFAR4, A root-specific fatty acyl-coenzyme A reductase, is involved in fatty alcohol synthesis of root suberin polyester In *Brachypodium distachyon*. Plant J 106(5):1468–1483

Wang Z, Wan L, Xin Q, Zhang X, Song Y, Wang P et al (2021i) Optimizing glyphosate tolerance in rapeseed by CRISPR/Cas9-based geminiviral donor dna replicon system with Csy4-Based single-guide RNA processing. J Exp Bot 72(13):4796–4808

Wolt J. D., Wang K., Yang B. (2016). The regulatory status of genome-edited crops. Plant biotechnology journal 14(2): 510–518

Wu T, Zhang H, Bi Y, Yu Y, Liu H, Yang H et al (2021) Tal2c activates the expression of *OsF3H04g* to promote infection as a redundant TALE of Tal2b in *Xanthomonas oryzae* pv. *oryzicola*. Int J Mol Sci 22(24):13628

Xiaoyu W (2019) Gene editing reassuring for safety of crops. China Daily. 2019 (1)

Xin T, Tian H, Ma Y, Wang S, Yang L, Li X et al (2002) Targeted creating new mutants with compact plant architecture using CRISPR/Cas9 genome editing by an optimized genetic transformation procedure in cucurbit plants [published online ahead of print, 2022 Jan 20]. Hortic Res. 2022;uhab086.

Xu K, Song J, Wu Y, Zhuo C, Wen J, Yi B et al (2022) Brassica evolution of essential BnaFtsH1 genes involved in the PSII repair cycle and loss of FtsH5. Plant Sci 315:111128

Yan Y, Zhu J, Qi X, Cheng B, Liu C, Xie C (2021) Establishment of an efficient seed fluorescence reporter-assisted CRISPR/Cas9 gene editing in maize. J Integr Plant Biol 63(9):1671–1680

Yang J, Ji L, Liu S, Jing P, Hu J, Jin D et al (2021) The CaM1-associated CCaMK-MKK1/6 cascade positively affects lateral root growth via auxin signaling under salt stress in rice. J Exp Bot 72(18):6611–6627

Yang Z, Du H, Xing X, Li W, Kong Y, Li X et al (2022) A small heat shock protein, GmHSP17.9, from nodule confers symbiotic nitrogen fixation and seed yield in soybean. Plant Biotechnol J 20(1):103–115

Ye R, Wu Y, Gao Z, Chen H, Jia L, Li D et al (2021) Primary root, and root hair development regulation by OsAUX4 and its participation in the phosphate starvation response. J Integr Plant Biol 63(8):1555–1567

Yeap WC, Khan NCM, Jamalludin N, Muad MR, Appleton DR, Harikrishna K (2021) An efficient clustered regularly interspaced short palindromic repeat (CRISPR)/CRISPR-associated protein 9 mutagenesis system for oil palm (*Elaeis guineensis*). Front Plant Sci 12:773656

Yee JK (2016) Off-target effects of engineered nucleases. FEBS J 283(17):3239–3248

Yin H, Song CQ, Suresh S, Kwan S, Wu Q, Walsh S et al (2018) Partial DNA-guided Cas9 enables genome editing with reduced off-target activity. Nat Chem Biol 14(3):311–316

Yin J, Wang L, Jin T, Nie Y, Liu H, Qui Y et al (2021) A cell wall-localized NLR confers resistance to soybean mosaic virus by recognizing viral-encoded cylindrical inclusion Protein. Mol Plant 14(11):1881–1900

Young Chae, G., Hong, W. J., Jeong Jang, M., Jung, K. H., & Kim, S. (2021). Recurrent mutations promote widespread structural and functional divergence of MULE-derived genes in plants. Nucleic acids research, 49(20), 11765–11777.

Yu J, Tu L, Subburaj S, Bae S, Lee GJ (2021) Simultaneous targeting of duplicated genes in petunia protoplasts for flower color modification via CRISPR-Cas9 ribonucleoproteins. Plant Cell Rep 40(6):1037–1045

Zaman QU, Wen C, Yuqin S, Mengyu H, Desheng M, Jacqueline B et al (2021) Characterization of *SHATTERPROOF* homoeologs and crispr-cas9-mediated genome editing enhances pod-shattering resistance in *Brassica napus L.* CRISPR J 4(3):360–370

Zannoni L (2019) Evolving regulatory landscape for genome-edited plants. CRISPR J 2:3–8

Zhang Y, Li D, Zhang D, Zhao X, Cao X, Dong L et al (2018) Analysis of the functions of Tagw2 homoeologs in wheat grain weight and protein content traits. Plant J 94(5):857–866

Zhang D, Chen C, Wang H, Niu E, Zhao P, Fang S et al (2021a) Cotton fiber development requires the pentatricopeptide repeat protein GhIm for splicing of mitochondrial nad7 mRNA. Genetics 217(1):1–17

Zhang J, Zhang H, Li S, Li J, Yan L, Xia L (2021b) Increasing yield potential through manipulating of an ARE1 ortholog related to nitrogen use efficiency in wheat by CRISPR/Cas9. J Integr Plant Biol 63(9):1649–1663

Zhang X, Cheng J, Lin Y, Fu Y, Xie J, Li B et al (2021c) Editing homologous copies of an essential gene affords crop resistance against two cosmopolitan necrotrophic pathogens. Plant Biotechnol J 19(11):2349–2361

Zhang X, Li N, Liu X, Wang J, Zhang Y, Liu D et al (2021d) Tomato protein Rx4 mediates the hypersensitive response to *Xanthomonas euvesicatoria* pv. *perforans* race T3. Plant J 105(6):1630–1644

Zhang X, Xu G, Cheng C, Lei L, Sun J, Xu Y et al (2021e) establishment of an agrobacterium-mediated genetic transformation and crispr/cas9-mediated targeted mutagenesis in hemp (*Cannabis sativa L.*). Plant Biotechnol J 19(10):1979–1987

Zheng S, Ye C, Lu J et al (2021) Improving the rice photosynthetic efficiency and yield by editing *OsHXK1* via CRISPR/Cas9 System. Int J Mol Sci 22(17):9554.6

Zhou Y, Xu S, Jiang N, Zhao X, Bai Z, Liu J et al (2021) Engineering of rice varieties with enhanced resistances to both blast and bacterial blight diseases via CRISPR/Cas9 [published online ahead of print, 2021 Dec 10]. Plant Biotechnol J. 2021. https://doi.org/10.1111/pbi.13766

Zhao H, Wolt JD. Risk associated with off-target plant genome editing and methods for its limitation. Emerg Top Life Sci. 2017 Nov 10;1(2):231-240. doi: 10.1042/ETLS20170037. PMID: 33525760; PMCID: PMC7288994.

Regulatory Constraints and Differences of Genome-Edited Crops Around the Globe

Penny Hundleby and Wendy Harwood

Abstract Plant breeding for centuries has relied on the availability of genetic variation to introduce new desirable traits into crops. Biotechnology has already accelerated the ability to induce and utilize new genetic variation, through approaches such as mutation breeding and using technologies such as marker assisted breeding to rapidly identify the required variation. These technologies fall within the definition of "conventional and traditional" breeding and are lightly regulated. However, plant breeders are facing an urgent need for access to wider genetic variation to meet the needs of today's farmers and consumers worldwide. New breeding technologies (NBTs), such as genome editing, are speeding up the breeding process and providing plant breeders with access to a far greater range of genetic variation. Coupled with a rapidly accelerating genomics era, genome editing is moving plant breeding into an exciting era of intelligent and precision-based plant breeding. The speed at which these new technologies are emerging has challenged the regulatory climate. Some countries consider genome edited crops to require the same regulatory oversight as genetically modified organisms (GMOs), while others have chosen to regulate with the same safety evaluations currently associated with bringing conventionally bred crops to market. Harmonization of the regulatory climate is urgently needed if there is to be equal access to this technology and to support international trade of these crops. The current chapter provides a global overview of the current regulatory status of genome-edited crops.

Keywords New breeding techniques · CRISPR · Gene editing · Genome editing · Regulation

P. Hundleby (✉) · W. Harwood
Department of Crop Genetics, John Innes Centre, Norwich, UK
e-mail: penny.hundleby@jic.ac.uk

© The Author(s) 2022
S. H. Wani, G. Hensel (eds.), *Genome Editing*,
https://doi.org/10.1007/978-3-031-08072-2_17

319

1 Introduction

Plant breeders over the centuries have continued to exploit the availability of genetic variation, to meet human needs for increased yields, flavor, and nutritional and visual qualities, together with improved agronomic performance. While breeders initially sought genetic variation from landraces and heirloom varieties, the emergence of mutation breeding in the 1940s allowed for the artificial induction of new genetic variation (or mutations) into plant genomes. The approach is crude, introduces thousands of random mutations, but allows for desirable new traits to be identified and introduced into breeding programs. Although several rounds of backcrossing were also required to remove the unwanted mutations, the technology greatly increased the amount of genetic variation available to breeders. The first commercial varieties developed through mutation breeding were registered in the 1950s, and now over 3348 varieties are listed on the FAO/IAEA Mutant Variety Database (FAO/IAEA 2021).

Inevitably some of the new desirable traits introduced, such as higher yields, better flavor etc., have come at the expense or loss of others (e.g., loss of disease or insect resistance) with modern agriculture now heavily reliant on human inputs. Today, farmers across the globe are facing huge challenges. There are currently less farmers per capita than ever before, faced with producing more food for a growing population, on less land, in a changing climate. Pressures on governments to also recognize the need to protect the environment have resulted in the sudden removal of some agrochemicals and resources. This has resulted in the urgent need for increased availability to genetic variation, to find genetic solutions to address some of these challenges. With conventional plant breeding, taking some 10–15 years to get new crops to the market, genome editing not only offers access to precision breeding but also greatly reduces the time frames needed to generate new varieties.

Take, for example, the introduction of powdery mildew resistance in wheat. This story starts in barley, where natural and induced loss-of-function mutations of the *Mildew resistance locus o* (*Mlo*) gene were identified that confer broad-spectrum resistance against most *B. graminis* f. sp. hordei (*Bgh*) isolates. These mlo mutants have been providing mildew resistance in barley in the field for more than 40 years. BLASTING the *Mlo* genetic sequence against the wheat genome identified three orthologues of the barley *Mlo*, *TaMlo-A1*, *-B1*, and *-D1* (Konishi et al. 2010), on chromosomes 5AL, 4BL, and 4DL of wheat (Elliott et al. 2002). Genome editing techniques, such as TALENS and CRISPR, have successfully been applied to target and knock out all three copies of this gene to successfully introduce mildew resistance into wheat (Wang et al. 2014). Interestingly, the same end point could also be achieved in wheat using mutation breeding. Using a TILLING approach (Acevedo-Garcia et al. 2017), mutant lines have been identified and crossed together to combine all the required mutant knockouts needed to confer resistance to mildew. However, from a breeder perspective, the TILLING approach also brings in thousands of undesirable mutations that then need to be removed, taking much longer to achieve the same end point. So, while scientifically the gene editing approach offers a faster and more precise approach for the introduction of mildew resistance into wheat, the TILLING approach currently faces less regulatory burden than gene editing.

While genome editing technologies have been around for some time, in the form of zinc finger nucleases (ZFNs), meganucleases, and transcription activator-like effector nucleases (TALENs) (Songstad et al. 2017), it was the publication of CRISPR/Cas9 (clustered regularly interspaced short palindromic repeats and *CRISPR*-associated protein 9) as a genome editing approach in plants, in 2013 (Feng et al. 2013; Zhang et al. 2017), that really made genome editing highly accessible to the scientific community. Since then, CRISPR/Cas9 as a simple genome editing tool for both research and commercial purposes has seen a continuous exponential rise, evidenced by the volume of publications, making it the most favored genome editing tool.

The sudden flood of activity and attention on genome editing also highlighted the need for clarity on how genome edited crops would be regulated. While some countries were quick to adapt their current legislations or release guidelines supporting the use of genome editing, others have not moved past seeing all organisms derived by genome editing as GMOs. This has led to confusion by plant breeders and the seed industry, with these unharmonized regulatory approaches likely to hinder technology applications and future trade between countries.

2   Paving the Way

Since 2015, several countries have outlined their regulatory path and clarified which types of genome-edited crops will not be regulated as GMOs. These include countries from North and South America together with Israel/Japan and Australia. These countries, perhaps unsurprisingly, are also strong supporters of GM technology, with all but Israel and Japan commercially growing GM crops. Argentina was the first country to proactively support the technology by providing clarity on the regulatory status in 2015; followed by Australia in 2016; Chile and Israel in 2017; Brazil, Columbia, and Paraguay in 2018/2019; and Japan in 2019. To best explain the regulatory approach, it is important to understand that not all genome editing is the same. Regardless of the different genome editing technologies used to create the edits, they all use site-directed nucleases (SDN), and three classes of genome editing exist:

1. Where a directed DNA double-strand break is repaired by the plants' own mechanism of non-homologous end joining without using an added repair template, often resulting in small mutations (SDN-1).
2. Where a template-guided repair is made, by an external DNA-template sequence that introduces one or several small mutations (SDN-2).
3. The insertion of a longer DNA sequence, including entire genes, through template-guided repair of the targeted double strand break (SDN 3) (Podevin et al. 2013).

In this chapter, we focus mainly on SDN1 and SDN2 as these technologies result in end products that are indistinguishable from those achieved via conventional bio-technology. As such, these are the products that have led some countries to view them in the same way as conventionally bred crops. For SDN-3 most, if not all jurisdictions, are in agreement and consider these products to not be exempt from their GMO regulations, and as such these crops will be assessed on a case-by-case basis. It is important to recognize that genome editing technologies cannot currently be used to achieve all the required changes needed to develop future improved crops and that a GM approach will still be needed in some cases.

With the regulatory climate currently differing across the globe, this will have serious implications on which countries will realistically have access to these technologies (i.e., will not be hampered by expensive and time-consuming additional regulatory burden) and the impact on trade that will inevitably be encountered. As previously seen with older GM technology, non-harmonious and asynchronous approvals delay commercialization and increase costs (Bullock et al. 2021).

3 The Need for Clarity

In October 2018, eight countries (Argentina, Australia, Brazil, Canada, Guatemala, Honduras, Paraguay, and the USA) came together to issue a joint statement to the World Trade Organization (USDA 2018) "supporting relaxed regulations for gene editing, stating that governments should 'avoid arbitrary and unjustifiable distinc-tions' between crops developed through gene editing and crops developed through conventional breeding. The ministries agreed to avoid obstacles, without a scientific basis, for the commercialization of products improved by genome editing, exchange information about products, developments and applicable regulations, and explore opportunities for regional harmonization." By November 2018, the number of coun-tries adding support to the statement had risen to include Colombia, the Dominican Republic, Jordan, Uruguay, Vietnam, and the Secretariat of the Economic Community of West African States (USDA 2018). Over time more countries have come to a similar viewpoint, in considering simple genome editing, i.e., where no foreign genetic sequence is introduced, to be indistinguishable and equivalent to conventionally bred crops and therefore should be regulated in a similar way.

The impact of viewing gene editing as GM could have disastrous effects for crop improvement. Gene editing could allow companies to focus on output traits that may have a higher value to the consumer, as opposed to the input traits associated with GM technology that favor the producer. The costs associated with regulatory compliance of GM have restricted its use, mainly, to four high value commodity crops (soybean, maize, cotton, and oilseed rape) and input traits (herbicide tolerance and insect resistance), gene editing therefore shows great potential to move into a wider range of crops (Jorasch 2019). Yet the regulatory uncertainty is clearly having an impact, with companies already choosing not to develop products for countries where regulatory clarity is still sought. In the following section, we look at the

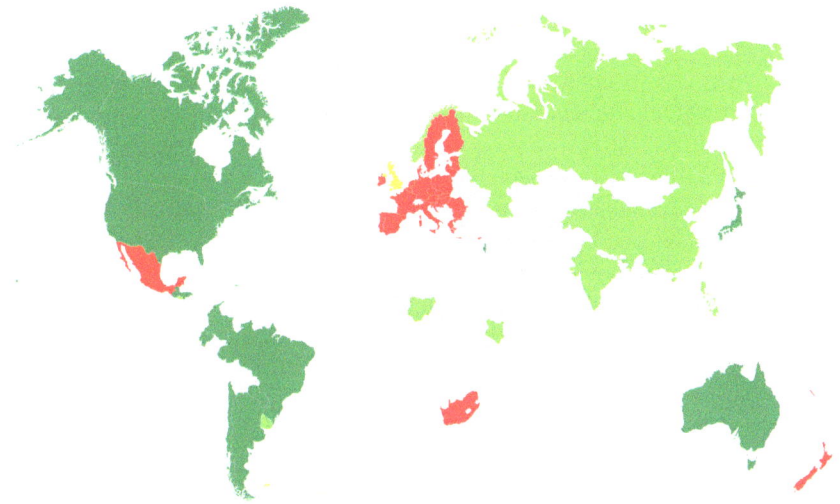

Fig. 1 Regulatory status of gene-edited crops (when no foreign DNA is inserted). Dark green = regulated as conventional crops. Pale green = draft regulations suggest they will be regulated as conventional crops. Red = viewed as GMOs. Yellow = under review but likely to be viewed favorably. The UK (shown in yellow) recently gave the go ahead for research field trials to proceed without the need for GMO regulatory oversight (when no foreign DNA is inserted); however, restrictions currently remain for commercial applications. Further amendments are still under review but look favorable

current regulatory climate for genome/gene editing across South, North, and Central America, across Europe, Africa, Russia, Asia, and the Southern Hemisphere. We also provide an "at a glance" global map summary of the regulatory status in Fig. 1.

4   The Regulatory Climate

4.1   South America

For countries in South America who have issued statements, no new legislation has been introduced, and gene-edited crops that do not contain DNA from another species are regulated as conventional plants. However, a dossier is required to be submitted to grant the exemption.

4.1.1   Argentina

Gene-edited crops are regulated as conventional plants unless they contain foreign DNA (Alfredo Lema 2019).

Argentina has always been a strong supporter of biotechnology (ranking third largest land area of cultivated biotech crops in 2019 (ISAAA 2019)) and has grown GM crops since cultivation first began in 1996. Having benefitted economically from the cultivation of these crops, it is perhaps unsurprising that Argentinian regulators where among the first to proactively issue clarity on the regulatory status of these still emerging technologies in 2015.

The Argentine regulatory system regulates gene-edited crops as conventional plants unless they contain foreign DNA. However, the regulatory system still requires developers to submit a dossier to the Argentine Biosafety Commission (CONABIA), who oversee GMOs, to determine this exemption. Gene-edited crops are assessed on a case-by-case basis by CONABIA, who are required to respond to submissions within 60 days as to whether the crop will be subject to GMO regulations. CONABIA considers (a) the techniques used in the process, (b) the genetic change in the final product, and (c) the absence of foreign DNA (transgenes) in the final product. Even if a crop is considered exempt from GMO regulations, if it has characteristics that present the probability of significant risk, the crop would undergo further monitoring by authorities and would be regulated as a GMO.

The Argentine authorities have engaged well with the public, organizing workshops and debates. The public's response suggests confidence in the regulatory oversight and welcomed that developers were local and focused on products suited to local markets and consumer and environmental benefits (Entine et al. 2021).

4.1.2 Uruguay

No unique regulations relating to GE have been issued.
   In 2018, Uruguay joined the countries mentioned in Sect. 3 in signing the joint statement to the World Trade Organization supporting relaxed regulations for gene editing (USDA 2018), thus suggestive of supporting regulating gene-edited crops as conventional plants unless they contain foreign DNA. Plants containing foreign DNA would be regulated in line with GM regulations, as overseen by the National Biosafety Cabinet (GNBio), which oversees the Ministers of Agriculture, Economy, Environment, Health, Industry, and Foreign affairs. However, to date no specific regulations for gene-edited crops have been issued (Uruguay, Global Gene Editing Regulation Tracker 2020).

4.1.3 Paraguay

Gene-edited crops are regulated as conventional plants unless they contain foreign DNA (Benitez Candia et al. 2020).
   Paraguay was the sixth largest producer of transgenic crops in 2019 (ISAAA 2019), and gene-edited crops can be expected to play a part in the countries' future agricultural landscape, although gene-edited crops have yet to be submitted for approved for commercial production in Paraguay. In 2018, Paraguay signed the

joint statement issued to the World Trade Organization supporting relaxed regulations for gene editing (Sect. 3), and in 2019, Paraguay published a resolution outlining the information required for crops developed using gene editing and other new breeding techniques (NBTs) to be approved. Gene-edited crops and food will be regulated as conventional plants unless they contain foreign DNA but will require the submission of a dossier to determine exemption. Gene-edited crops are assessed on a case-by-case basis by the National Commission on Agricultural and Forestry Biosafety, whereas genetically modified plants are regulated by the Ministry of Agriculture and the Biosecurity Commission (COMBIO).

4.1.4   Chile

Gene-edited crops are regulated as conventional plants unless they contain foreign DNA.

Chile has taken a similar view to other South American countries and considers gene-edited crops to be regulated as conventional plants unless they contain foreign DNA. However, they also require the submission of documents to the Ministry of Agriculture's Agricultural and Livestock Services (SAG) who will assess on a case-by-case basis gene-edited crops. In 2017, SAG published a statement on new breeding techniques (NBTs), stating that gene-edited crops that do not contain "a new combination of genetic material" are not subjected to GMO regulations (Sanchez 2020).

4.1.5   Brazil

Gene-edited crops are regulated as conventional plants unless they contain foreign DNA.

Brazil, with the second largest land area of cultivated GM crops in 2019 (ISAAA 2019), has also taken the position that gene-edited crops and food should be regulated as conventional plants unless they contain foreign DNA. In Brazil, GMOs are governed by the National Technical Commission for Biosafety (CTNBio). In 2018, CTNBio released Normative Resolution No. 16, focusing on NBTs. It clarified that many products derived from NBTs do not meet the definition of a GMO, as defined by the 2005 regulation, and concluded that NBTs should be regulated on a case-by case basis. This resolution establishes the requirements for whether a product can be exempt from the GMO regulatory framework.

Regulations focus on the characteristics of the final product rather than the process used to create it. CTNBio will assess the risk level of each newly developed plant or food, whether new (foreign) genetic material has been introduced and whether the product has already been approved for commercialization in other countries. CTNBio will then respond to the applicant within 20–90 business days. Applications have already been submitted to CTNBio for gene-edited tomatoes, soybeans, and a "waxy" maize with extra starch, while four gene-edited varieties of yeast for production of bioethanol and other purposes were approved by CTNBio in 2018 (Brazil: Global Gene Editing Tracker 2020). As of September 2020, there had been 23 consultations with CTNBio, for products not considered to fall within the scope of the GMO law 11.105/2005. This clarity has resulted in several new startup companies and strengthening of medium to large national companies working on NBTs (Entine et al. 2021).

4.1.6   Ecuador

Gene-edited crops are regulated as conventional plants unless they contain foreign DNA.

In Ecuador, the Ministry of Agriculture and Livestock regulates GM crops through the National Agrarian Authority (Norero 2017), and while the country has not embraced GM technology, and currently prohibits the commercial cultivation of genetically modified crops, the situation looks more favorable for gene-edited crops. For gene-edited crops that do not contain DNA from another species, these will be regulated in the same way as conventionally bred plants, while GE crops that contain foreign DNA will be viewed as GM. The regulation is based on the Organic Code of the Environment, issued in 2019, that established exemptions from the very restrictive GMO regulations (Entine et al. 2021).

4.1.7   Colombia

Gene-edited crops are regulated as conventional plants unless they contain foreign DNA but require notification to the authorities to approve the exemption.

Columbia also signed the 2018 joint statement to the WTO, in support of relaxed regulations for gene editing, and in the same year the Colombian Agricultural Institute (ICA) issued a resolution that established a case-by-case consultation process to determine if a gene-edited product would be considered a GMO (Gatica-Arias 2020). Once notified, the ICA must respond to applicants within 60 days as to whether the organism will be subject to GMO regulations. For a gene-edited crop not to be considered GMO, it must not contain genes from another species that have been introduced through modern biotechnology techniques.

4.2   Central America

4.2.1   Honduras, Guatemala, and El Salvador

Gene-edited crops that do not contain foreign DNA are not regulated as GMOs.

In 2018, Honduras and Guatemala signed the joint statement to the WTO supporting a more relaxed regulatory oversight for plant gene editing, while in 2019,

Honduras, Guatemala, and El Salvador also signed an inter-ministerial agreement to harmonize the research and commercialization of crops developed through biotechnology. This agreement required each country to create a national advisory committee for the risk assessment and evaluation of GMOs for agricultural use. The agreement also defines the term "novel combination of new genetic material," setting the legal basis to define gene-edited products (which do not fulfil the definition of GMOs) as conventional. In 2019, Honduras published a resolution to establish a streamlined authorization procedure for crops developed using new breeding techniques (NBTs) and in doing so became the first country in Central America to regulate products of NBTs. Overseen by the National Service of Food Safety Plant and Animal Health (SENASA), Honduras follows in line with the Cartagena Policy on Biosafety and considers that technologies, which result in an organism equivalent and indistinguishable from products of conventional (traditional) plant breeding, should be regulated in the same way to allow producers and consumers to gain from these technologies (Macall 2020). El Salvador is expected to follow Honduras' lead. Of the three countries, only Honduras is currently growing GM crops (GM maize on less than 0.1 Mha) (ISAAA 2019).

While Guatemala does not currently grow GM crops, they have looked at the implications the regulations may have on imports and have clarified that imports of GE seed/plants will not be regulated as GM if they do not contain foreign DNA.

4.3   North America

4.3.1   Mexico

Gene-edited crops are currently regulated under laws established for transgenic GMOs.

Mexico is yet to determine the regulatory status of gene editing crops, and products are currently regulated under laws established for transgenic GMOs (Mexico: Global Gene Editing Regulator Tracker 2020). The Secretariat of Health (SALUD) is responsible for regulating GM crops, and currently Mexico only grows a modest amount of GM cotton. This regulatory oversight puts Mexico at a very different standpoint to other countries in the American continent.

4.3.2   USA

Gene-edited crops that do not contain foreign DNA are not regulated as GMOs.

The USA is the largest grower of biotech crops in the world (ISAAA 2019) and was the first country to approve a genome-edited product for commercial sale. This was a soybean product with no trans-fats and lower saturated fat produced by the Minnesota-based company Calyxt (2019) using a technique called TALENs.

The enthusiastic uptake of gene editing technology in the USA is perhaps supported by the fact that even the introduction of GM technology back in the 1990s did not trigger the need for new regulations as such but instead relied on existing regulatory frameworks to oversee these new crops. Up to three different agencies are involved in the process, depending on the final product and how the plant was produced (USDA 2021a):

1. The US Department of Agriculture's Animal and Plant Health Inspection Service (USDA – APHIS) is responsible for protecting agriculture from pest and diseases. Crops considered to pose an agricultural "risk" are deemed "regulated articles" and are reviewed to ensure that, under the proposed conditions of use, they do not present a plant pest risk by ensuring appropriate biosafety systems are in place to minimize such risks, such as handling, confinement, and disposal of crops. There is also a petition process where applicants can make a case for a GM product to be considered for "non-regulated status" if an applicant can provide enough evidence that the product does not pose a risk to agriculture. This is then added to a federal register where the public can submit comments for consideration on the environmental assessment before the petition is granted, i.e., given with a "non-regulated" status.
2. The Environmental Protection Agency (EPA) regulates pesticides and therefore regulates biotech crops that have pesticide properties (e.g., insect-resistant crops).
3. The Food and Drug Administration (FDA) oversees food safety.

In 2015, President Obama issued an Executive Order "Memorandum on Modernizing the Regulatory System for Biotechnology Products" directing the USDA, EPA, and FDA to update regulatory roles and responsibilities under the Coordinated Framework for the Regulation of Biotechnology, to develop a long-term strategy to ensure that the regulatory system was future proof for new biotech products. The "National Strategy for Modernising the Regulatory System for Biotechnology Products" and "Update to the Coordinated Framework for regulation of Biotechnology" were released by the White House Office for Science and Technology Policy (OSTP 2016, 2017).

In 2019, President Trump signed an executive order directing federal agencies to streamline the regulatory process for biotech crops by exempting low-risk products from the existing rules and creating a unified platform that outlines the regulatory requirements from all three agencies, for the review and authorization of products developed using biotechnology. In 2020, USDA-APHIS finalized what it called the SECURE (sustainable, ecological, consistent, uniform, responsible, efficient) rule, which would exempt (i.e., not regulate) gene-edited plants that otherwise could have been developed through conventional breeding. This reaffirms a focus on regulating characteristics of gene-edited plants, instead of the process used to create them, as is the case in the EU, for example. APHIS states that these exemptions are intended to bring the regulation of potential GE plants more in line with the guidelines for conventionally bred crops. Therefore, gene-edited crops that do not contain foreign DNA are not regulated as GMOs, if they pose no risk to other plants, and show no food safety attributes different to those of traditionally bred crops. In these cases, the crops will not be subject to pre-market regulatory evaluation; however, it will be the responsibility of the developer to assure that products placed on the market are safe for use and consumption (as in the case for conventional crops).

The FDA (which oversees food safety) and EPA (which regulates pesticides) have not announced if their existing policies and regulations related to GMOs would be used to regulate gene-edited crops and food.

4.3.3   Canada

Only gene-edited crops with novel traits will be regulated as PNTs (Friedrichs et al. 2019).

Canada has a well-established product-orientated approach to policy and regulatory oversight and regulates all plants with novel traits (PNTs), regardless of the technology used to create them.

Although Canada appears to be headed toward regulating gene-edited crops lightly (having also signed the joint statement to the WHO), there remains uncertainty as to what types of gene editing will trigger oversight, i.e., what is considered "novel" and what that level of oversight might be. Currently, GMOs on the market in Canada pass through Health Canada and the Canadian Food Inspection Agency (CFIA) where new organisms are categorized as either "novel" or "non-novel." The context of gene editing is as follows.

"Novel" organisms have traits that are not naturally occurring and have not previously been approved for sale by Health Canada and the CFIA. Organisms that pose an obvious risk, such as those containing potential allergens or those that contain foreign DNA in the final product, are considered novel. Such crops will require *pre-market safety assessment*s, and the associated costs incurred could potentially be prohibitive, limiting certain lower value crops from being developed.

"Non-novel" organisms are organisms that have a history of safe use, show no characteristics that are new to the species, and do not contain genetic material from another organism after its genome has been edited. For these crops no pre-market safety assessments are required.

Most crop varieties produced via chemical or radiation-based mutagenesis are not considered to have novel traits and therefore are not subject to pre-market assessment and are regulated as conventional crops. While most of the early gene-edited crops are viewed as products of a more precise version of mutagenesis, there remains some uncertainty as to whether the regulators will view them as such, as no formal framework or decisions have yet been issued. However, a herbicide-tolerant oilseed rape developed using the NBT technique known as ODM (oligonucleotide-directed mutagenesis) was approved in 2013 (Halford 2019).

A consultation exercise "Proposed new guidelines for Novel Food Regulations" focused on plant breeding was recently carried out in Canada (Health Canada, 2021) which should hopefully lead to more clarity. Better defining what is considered novelty is critical. Currently PNT regulations could apply to any new crop with a trait that expresses 25–30% higher or lower than the conventional variety (Entine et al. 2021). Thus, it will be the novelty of the crop, and not how it was made, that will trigger regulatory oversight. This will have implications for trade, if the same products are viewed by other countries as not requiring regulatory oversight.

4.4   Europe

4.4.1   The European Union (EU)

Gene-edited crops are regulated as GMOs.

The European Union represents 27 member state (MS) countries and currently regulates all genome-edited crops as GMOs under the 2001/18 EU GMO Directive. The Directive defines a GMO as "an organism, with the exception of human beings, in which the genetic material has been altered in a way that does not occur naturally by mating and/or natural recombination" but excludes several traditional breeding technologies that fit this description, listing them in Annex 1B of the directive. Among the techniques listed for exclusion from the directive is mutagenesis. Prior to 2018, several MS had interpreted the exemption to also include genome-edited crops that had been edited in ways that would result in a product indistinguishable to one obtained through traditional mutagenesis techniques (i.e., chemical or radiation-induced mutagenesis). In 2016, nine NGOs filed a case to the French Courts, which was later referred to the Court of Justice of the European Union (CJEU), to challenge the status quo as they viewed the exclusion on genome editing to be allowing "GM through the backdoor." In July 2018, the CJEU confirmed that organisms obtained by newer methods of direct mutagenesis such as genome editing were not excluded from the scope of the EU GMO directive (CJEU 2018). Overnight CRISPR field trials went from being "unregulated" to regulated.

The CJEU ruling of 2018 was met with frustration by researchers and plant breeding companies, with over 117 research facilities signing a position paper urging the European Policy Makers to act to safeguard Europe's competitiveness on these new technologies (MPG 2019). For many the ruling fell short of delivering clarity on the regulatory status of gene editing and how such crops would be monitored (Van der Meer et al. 2021).

The European Commission Chief Scientific Advisors criticized the EU court ruling, and the EU Council later requested that the EU Commission conduct a study regarding the CJEU judgment. The results of this consultation study were published in April 2021 and concluded that the current GMO legislation was not fit for purpose for some NGTs and their products and that it needed to be adapted to scientific and technological progress (European Commission 2021). The lack of clarity surrounding the future regulatory climate for gene-edited crops has resulted in several EU-based companies focusing on the development of GE crops for non-EU markets (Jorasch 2019).

During this period, in 2020, France's top administrative court also ruled that the French High Council for Biotechnology (HCB) needed to set up a specific list of mutagenesis techniques, or methods, that will be exempted from GMO restrictions (technologies that fulfil the requirement of "having been conventionally used in a number of applications and have a long safe history of use"). Depending on the list, France could even regulate plants that have been developed by earlier mutagenesis techniques, e.g., herbicide-tolerant crops, if the HCB concludes that the abovementioned requirement is not met. This would have huge implications for France, as one of the EU's largest agricultural producers, as it would effectively deny French farmers access to much of the common seed catalogue.

4.4.2   The UK

Gene-edited crops are currently regulated as GMOs in line with the EU, but this position is currently under review following UK's departure from the EU.

The UK formally left the European Union on January 23, 2020. This gave the UK scope, should they wish, to deviate from the restrictive GMO EU Directive and set their own regulatory path. Within the UK, England, Scotland, Northern Ireland, and Wales have national laws that control the deliberate release of GMOs into the environment. In England, the Department of Environment, Food and Rural Affairs (Defra) is the competent national authority responsible for the environmental release of GM plants. All applications submitted to Defra are passed on to the statutory Advisory Committee on Releases to the Environment (ACRE) that was appointed under section 124 of the UK Environmental Protection Act 1990 (EPA) to provide advice to government regarding the release and marketing of GMOs. The committee works within the legislative framework set out by "Part VI of the EPA" and, within England, the GMO Deliberate Release Regulations 2002 Act, which together implement EU Directive 2001/18/EC. The principal role of ACRE is to consider each application on a case-by-case basis and evaluate the risks to human health and the environment.

In early 2021, Defra launched a consultation exercise to gain feedback from various stakeholders on their views regarding gene editing; this consultation closed on March 17, 2021. The results of the consultation exercise, and an announcement by the UK Government on Genetic Technologies, were published on the September 29, 2021. In this they set out their plans for a two-step reform. The first step removes the regulatory burden for research groups by enabling the field trials of gene-edited crops (free from transgenes) to go ahead without being subject to existing GMO rules. Researchers will still be required to notify Defra. The second step will be to "bring forward primary legislation at a suitable opportunity to amend the regulatory definitions of a GMO to exclude organisms that have genetic changes that could have been achieved through traditional breeding or which could occur naturally" (Defra 2021). These crops would then be regulated in line with conventional crops and "novel food" oversight (FSA 2020) where appropriate. This could allow for much easier trade relationships with counties who have adopted a similar regulatory view. The 'New Genetic Technologies (Precision Breeding) Bill' was brought to Parliament in 2022, and is likely to conclude early 2023. However, the impact on trade with the EU is perhaps one of the biggest hurdles to overcome, if the UK regulates differently to the EU.

While the UK government has generally been supportive of the potential of GM technology, commercial cultivation of GM crops has never taken place in the UK. The only approved GM crop for cultivation in the EU currently is the insect-resistant maize (MON810), for which there is no demand by British farmers. Field testing of GMOs and gene-edited crops are currently permitted in the UK, and field trials of CRISPR gene-edited plants were conducted in line with Part B of the 2001/ EU GMO directive (Faure and Napier 2018; Neequaye et al. 2021), until April 2022 when the rules changed to permit field trials of gene edited crops (where no foreign DNA is present) to proceed under a simple on-line notification system to Defra, and no longer requiring a GMO licence.

4.4.3   Norway

It has been proposed that gene-edited crops that do not contain DNA from another species be regulated as conventional plants but would still require notification.

Biotechnology in Norway is regulated by the Ministry of Agriculture, the Ministry of Environment, and the Ministry of Health. The Directorate for Nature Management is responsible for feed and seed, and Norwegian Food Safety Authority is responsible for biotech food. Genetically modified food is regulated by the Matloven Food Act and the Gene Technology Act, one of the world's strictest, which requires that genetically modified products contribute to sustainable development in order to be approved (Norway: Global Gene Editing Regulation Tracker 2020). In fact, Norway has a long history of opposition to transgenic crop biotechnology, generally opposing the cultivation of GMOs and being more restrictive than the EU regarding imports. *Although Norway is part of the European Economic Area, it is not a full European Union Member, as such it is not bound by EU Directives but generally implements EU Directives.*

In 2018, the Norwegian Biotechnology Advisory Board proposed a tiered regulatory system in which genetic changes that can arise naturally or can be achieved using conventional breeding methods would be regulated as conventional plants. However, they would still require that a notification is submitted to the government, while crops developed using cisgenics (introduction of genes from within species) would require expedited but limited assessment and approval. Genetic changes that cross species barriers (transgenics) or involve synthetic DNA sequences would still require assessment and approval under strict GMO regulations. Although these regulations appear to pave the way for the introduction of gene-edited crops, Norway's historical, public, and political opposition to crop biotechnology remains among the most intense in Europe; it will be interesting to see if these relaxed guidelines support innovation in this sector or merely act to support the import of such crops.

4.5   Israel

Gene-edited crops that do not contain DNA from another species are regulated as conventional plants.

As is the case in several of the countries reviewed in this chapter, Israel has also chosen to regulate gene-edited crops and food in line with conventional plants, unless they contain foreign DNA. Gene-edited crops will be assessed on a case-by-case basis based on the characteristics of the final product and will require a dossier to be submitted to determine if they are exempt. There are no commercially available genetically modified or gene-edited plants cultivated in Israel, although GM crops are currently imported.

Genetically modified organisms are regulated by the Ministry of Agriculture and Rural Development and the Ministry of Health. The Ministry of Agriculture and Rural Development oversees the Plant Protection and Inspection Service (PPIS) and the Israeli National Committee for Transgenic Plants (NCTP). In 2016, the NCTP concluded that gene-edited crops that do not contain DNA from other species would not be subject to GMO regulations, as regulated by the Seed Regulation Act of 2005.

The Ministry of Health stated that all new food products, including conventional crops and gene-edited ones, must undergo risk assessment before approval. Israel's Ministry of Agriculture announced in 2019 plans to invest in establishing a National Genome Editing Centre (Menz et al. 2020).

4.6   Africa

The regulatory status of gene-edited crops has not yet been determined, but draft guidelines have been approved in Nigeria.

Africa is a region where gene editing holds great promise in addressing a wide range of issues, including malnutrition and crop failure linked to climate change. Yet, in general, Africa has lagged behind other nations in setting out and developing their biosafety laws, although in recent years there has been much progress. The number of countries growing GM crops commercially rose to six in 2019 and included Nigeria (who also became the first country to approve Bt cowpea), Ethiopia and Malawi, together with countries with a longer history of growing GM such as South Africa, Sudan, and Eswatini. There has also been progress in biotech research, regulation, and acceptance in Mozambique, Niger, Ghana, Rwanda, Zambia, and Kenya. Burkina Faso and Egypt have grown GM crops in the past. As African countries are still, in many cases, defining their biosafety laws, this may be an advantage when it comes to assessing how to regulate gene editing. Nigeria, South Africa, Kenya, and Eswatini are currently taking the lead in amending their regulations to accommodate gene editing (Komen et al. 2020). Nigeria became the first African country to publish their draft guidelines "National biosafety guidelines for the regulation of gene editing" (USDA 2021b). This followed an amendment to the National Biosafety Management Agency (NBMA) Act of 2019, section 25(A) that states "No person, institute or body shall carry out gene drive, gene editing and synthetic biology except with the approval of the Agency." While Kenya has yet to publish its guidelines, the country has approved six GE projects for contained use research (Obi 2021).

South Africa, which has led on the commercial cultivation of transgenic crops in Africa, is yet to clarify its position and publish guidelines for the cultivation of GE crops.

4.7 Russia

Decree suggests that gene editing techniques will not be prohibited in the same way as GMOs.

Russia is a vast land area spanning both Europe and Asia and has historically been rather opposed to genetic modification, with no commercial cultivation of GM crops, although allowing imports. Plants developed through biotechnology are currently regulated by three separate organizations: the Federal Service for Surveillance of Consumer Rights Protection and Human Welfare (Rospotrebnadzor) being responsible for developing legislation on genetically modified food products and monitoring the effect on human and the environment health; the Ministry of Agriculture which develops policy for the use of genetically modified crops and organisms in agriculture; and the Federal Service for Veterinary and Phytosanitary Surveillance (VPSS) which is responsible for overseeing genetically modified crops for feed.

The countries' view on gene editing appears to be more supportive, with a large investment in R&D of the technology. A 111-billion-rouble (US$1.7Bn) federal research program sets out to develop 10 new varieties of gene-edited crops and animals by 2020 and another set of 20 gene-edited varieties by 2027 (Dobrovidova 2019). The decree establishing the program describes gene editing as equivalent to conventional breeding methods, the view adopted by most of the world. The decree lists four crops – barley, sugar beet, wheat, and potatoes – as priorities for development. The program, which was announced in April 2019, also attracted interest because it suggests that some gene-edited products will now be exempt from a law passed in 2016 that prohibits the cultivation of genetically modified (GM) organisms in Russia, except for research purposes. Previously, it was unclear whether gene-edited organisms were included in the ban.

4.8 Asia

4.8.1 China

Gene editing regulations for plants have not yet been announced, but they are expected to be regulated as conventionally bred plants.

With a population of 1.4 billion people, China has the largest population in the world. It was ranked seventh in the world for global area of transgenic crops in 2019 (ISAAA 2019) mainly growing GM cotton and papaya, and its recent approval of

GM corn and soybean is set to increase its biotech area further, thus reducing its reliance on imports from other GM nations.

While China currently limits the import and cultivation of genetically modified crops, it is thought that China will follow other countries in regulating most gene editing techniques as conventional plants. The Ministry of Agriculture regulates genetically modified crops in China, subjecting them to the 2001 Regulations on Administration of Agricultural Genetically Modified Organisms Safety.

China is yet to announce the regulatory status of gene-edited crops, but the government has invested heavily in agricultural research projects over the past decade, and China has published more research papers on CRISPR than any other country.

In 2017, state-owned ChemChina bought Switzerland-based Syngenta, one of the world's four largest agribusinesses and a company deeply involved in gene-editing research, for $43 billion. This sizable investment could suggest a positive future for biotech crops in China.

4.8.2 Japan

Gene-edited crops must be registered but do not require safety or environmental testing unless foreign DNA is present.

Japan became the first country to approve a gene-edited tomato for the home growers' market in 2021, making this the world's first approved direct consumption product. Produced by the Japanese-based company Sanatech Seed, the tomato contains higher levels of GABA, a compound reported to lower blood pressure and relieve stress (Sanatech Seed 2020). This shows Japan's support for this new technology, which contrasts with its view on GM technology, for which there has never been approval for commercial cultivation, although GM imports are allowed. Gene-edited crops are assessed on a case-by-case basis and do require submission of notification to the government, which includes providing information on the editing technique and genes targeted for editing. Safety and environmental assessments are required only when the plant contains foreign DNA. However, each time a gene-edited crop is crossed with another conventional or gene-edited crop, a separate notification process must occur. Local governments may also set additional regulatory requirements for gene-edited crops (USDA 2020).

Four ministries currently regulate genetically modified plants: the Ministry of Agriculture, Forestry and Fisheries (MAFF), the Ministry of Health, Labour and Welfare (MHLW), the Ministry of Environment (MOE), and the Ministry of Education, Culture, Sports, Science and Technology (MEXT). The Food Safety Commission (FSC), an independent risk assessment body under the Cabinet Office, performs food and feed safety risk assessment for MHLW and MAFF.

4.8.3 India

Draft guidelines suggest that SDN-1 gene editing will be lightly regulated, while the rest will require additional tests and approvals.

In 2020, the Department of Biotechnology published a draft document for the regulatory framework and guidelines for risk assessment of genome-edited organisms (Ministry of Science and Technology, Government of India 2020). In the proposal a tiered regulatory approval process is suggested. Group I would cover products of SDN-1 and would require confirmation and notification of the gene edit. Group II would cover SDN-2 techniques and would require more intensive field trials and data to ensure the edits were successful, and Group III – plants with large DNA changes, including insertion of foreign DNA (SDN-3 techniques) – would require the same extensive testing and regulatory oversight as GMOs, including field trials to test safety to human health, animals, and the environment.

4.9 Southern Hemisphere

4.9.1 New Zealand

All gene-edited crops are currently regulated as GMOs.

In 2014, the New Zealand Environmental Protection Authority (EPA) – who oversees GMOs under the Hazardous Substances and New Organisms (HSNO) Act of 1996 – initially ruled that plants produced via gene editing methods, where no foreign DNA remained in the edited plant, would not be regulated as GMOs. However, following a challenge in the High Court, this decision was overturned such that New Zealand currently regulates all products of gene editing as GMOs, even if they do not incorporate any foreign genes. New Zealand has yet to update its policy and, so far, appears to be waiting to see how its major trading partners (Europe, Asia, and Australia) conclude on their approach to regulating these crops.

In 2018, the Environment Minister, together with support from researchers, called for an update to the HSNO Act, stating that the current position did not support innovation and made it practically impossible to obtain approval for gene-edited crops, with no clear pathway to market. Currently no gene-edited plants are commercially grown in New Zealand, and no applications for a full environmental release have been received by the EPA (Fritsche et al. 2018).

4.9.2 Australia

SDN-1 gene editing organisms are regulated as conventional plants, while the rest are regulated as GMOs, requiring pre-market approval.

Australia was also a signatory to the 2018 joint statement to the World Trade Organization supporting relaxed regulations for gene editing, and on the April 10, 2019, the Australian government announced it would not regulate gene editing techniques in plants, animals, and human cell lines that do not introduce new genetic material (Mallapaty 2019). In an amendment to the Gene Technology Regulations (GTR) of 2001, clarifications on NBTs that are not considered GMOs were defined (OGTR 2019). The amendments mean SDN-1 gene-edited organisms are not considered to be GMOs provided that (a) no nucleic acid template was added to the cells to guide genome repair following site-directed nuclease application and (b) the organism has no other traits from gene technology (e.g., a cas9 transgene, or an expressed SDN protein) in the final product.

It becomes the responsibility of the developer to ensure products comply with the law and that these requirements have been met. Some methods used to generate SDN-1 organisms produce GMOs as an intermediate step, and in these cases while transgenes are still present the plants will continue to require authorization under the *Gene Technology Act of 2000*. This approach is in agreement with many other countries considered above.

When a crop no longer falls within the regulatory oversight of the GTR, it is overseen by the Department of Agriculture, Water and the Environment, and should it produce food products, such products are regulated under the Australia New Zealand Food Standards Code. No gene-edited crops have yet been put forward for approval yet in Australia.

In addition to the EPA in New Zealand and the GTR in Australia, the joint Food Standards Australia New Zealand (FSANZ) authority sets food standards, including regulations regarding gene edited food, which are compiled in the Australian and New Zealand Food Standard Code. The Food Standards Code requires pre-market approval and adherence to labeling standards for food produced using any gene technology, including any imported food that was produced through gene technology.

FSANZ recently reviewed how food developed using NBTs will be regulated (as GMOs or not), and in December 2019, FSANZ released a report that made three recommendations: (1) "to revise and modernise the definitions in the Code to make them better able to accommodate existing and emerging genetic technologies; (2) to consider process and non-process-based definitions and the need to ensure that NBT foods are regulated in a manner that is commensurate with the risk they pose; (3) to ensure there is open communication and active engagement with all interested parties and to explore ways to raise awareness about GM and NBT foods" (FSANZ 2019).

A proposal to amend the definitions in the Code commenced in February 2020; however, due to the COVID-19 pandemic, FSANZ postponed the release of the first call for submissions for public consultation and has yet to make any changes to the Food Standard Code as a result. As such the current pre-market approval and labeling requirements will continue to apply.

5   Conclusion

As a society, we are increasingly becoming aware of the challenges facing farmers, producers, and policy makers, on how to feed a growing population, in a way that both protects and nurtures the environment. Climate change is happening now, and we need to find ways to produce new crops that can mitigate climate change, reduce the need for chemical inputs, and meet the needs of society. There are many ways that we will achieve these goals, and indeed it will take a holistic approach to meet the ambitious targets that have been set, such as EU's Farm to Fork Policy targets for 2030 and European Green Deal for 2050 (EC 2021). Similar targets are being set in other jurisdictions. In breeding timescales, if we consider the generation times needed to breed new varieties, those dates are not that far away.

Genome editing is one technology that is uniquely placed to help speed up the breeding process, taking advantage of the innovations in precision breeding and the availability of vastly increased genomic knowledge.

In some cases, gene editing reaches the same end point as conventional breeding but gets there with a greater degree of precision and speed, enabling breeders to address urgent goals with greater confidence. When multiple gene targets are involved, it moves plant breeding into a new realm of possibilities. It could also enable more nutritious and diverse foodstuffs (regulations permitting) and the domestication of new crops, further expanding agricultures biodiversity and a move away from large monocultures.

The science is advancing rapidly, with the future of gene editing allowing for targeted and stacked gene insertions, chromosome engineering, epigenetic edits, and more. However, the regulatory climate is often slower to catch up. Already for simple SDN-1 gene editing, our review shows that the consensus on how to regulate such crops is not yet harmonized at the global level. Countries that are slow to clarify their position on gene editing risk being left behind and farmers losing out on competitive technologies. Furthermore, such disharmony will create barriers to trade, with gene-edited crops requiring different regulatory requirements in different countries. This could make countries who consider all gene-edited crops as GMOs less desirable trading partners. Regardless of the regulatory challenges ahead, continually advancing genome editing technologies are such a valuable addition to the tools currently available to breeders that we can look forward to their increased adoption, delivering vital new genetic variation for crop improvement.

Acknowledgments We acknowledge and thank Lucy Anderson, a student from the University of East Anglia for producing the map in Fig. 1, while on her student placement working with Ruby O'Grady from the communications team at the John Innes Centre. We acknowledge funding support from the Biotechnology and Biological Sciences Research Council (BBSRC) Institute Strategic Programmes GRO [BB/J004588/1] and GEN [BB/P013511/1] to the John Innes Centre.

References

Acevedo-Garcia J, Spencer D, Thieron H, Reinstaedler A, Hammond-Kosack K, Phillips AL, Panstruga R (2017) mlo-based powdery mildew resistance in hexaploid bread wheat generated by a non-transgenic TILLING approach. Plant Biotechnol J 15:367–378

Alfredo Lema M (2019) Regulatory aspects of gene editing in Argentina. Transgenic Res 28:147–150

Benitez Candia N, Fernandez Rios D, Vicien C (2020) Paraguay's path toward the simplification of procedures in the approval of GE crops. Front Bioeng Biotechnol 8

Brazil (2020) Crops/Food. Globabl Gene Editing Tracker (2020). Accessed June 2021 at Brazil: Crops/Food | Global Gene Editing Regulation Tracker (geneticliteracyproject.org)

Bullock DW, Wilson WW, Neadeau J (2021) Gene editing versus genetic modification in the research and development of new crop traits: an economic comparison JEL codes. Am J Agric Econ

Calyxt (2019) Press release February 26, 2019 available at 'First commercial sale of Calyxt high oleic soybean oil on the U.S. Market » Calyxt'. Accessed June 2021

CJEU (2018) Court of Justice of the European Union. Press release No. 111/18. July 25th 2018. Accessed at 'Organisms obtained by mutagenesis are GMOs and are, in principle, subject to the obligations laid down by the GMO Directive (europa.eu)', June 2021

Defra (2021) Genetic technologies regulation – GOV.UK (www.gov.uk). Accessed 29 Sept 2021

Dobrovidova O (2019) Russia joins global gene-editing bonanza. Nature 569:319–320

EC (2021) European Commision sustainability targets: accessed at Farm to Fork Strategy (europa. eu) and the A European Green Deal | European Commission (europa.eu), June 2021

Elliott C, Zhou FS, Spielmeyer W, Panstruga R, Schulze-Lefert P (2002) Functional conservation of wheat and rice Mlo orthologs in defense modulation to the powdery mildew fungus. Mol Plant-Microbe Interact 15:1069–1077

Entine J, Felipe MSS, Groenewald J-H, Kershen DL, Lema M, McHughen A, Nepomuceno AL, Ohsawa R, Ordonio RL, Parrott WA, Quemada H, Ramage C, Slamet-Loedin I, Smyth SJ, Wray-Cahen D (2021) Regulatory approaches for genome edited agricultural plants in select countries and jurisdictions around the world. Transgenic Res

European Commsion (2021) Food safety. EC study on new genomic techniques (europa.eu). Accessed June 2021

FAO/IAEA Mutant Viriety Database (2021) Available online at https://mvd.iaea.org. Accessed June 2021

Faure JD, Napier JA (2018) Europe's first and last field trial of gene-edited plants? elife 7:e42379. Published 2018 Dec 18. https://doi.org/10.7554/eLife.42379

Feng Z, Zhang B, Ding W, Liu X, Yang D-L, Wei P, Cao F, Zhu S, Zhang F, Mao Y, Zhu J-K (2013) Efficient genome editing in plants using a CRISPR/Cas system. Cell Res 23:1229–1232

Friedrichs S, Takasu Y, Kearns P, Dagallier B, Oshima R, Schofield J, Moreddu C (2019) An overview of regulatory approaches to genome editing in agriculture. Biotechnol Res Innov 3(2):208–220. https://doi.org/10.1016/j.biori.2019.07.001

Fritsche S, Poovaiah C, MacRae E, Thorlby G (2018) A New Zealand perspective on the application and regulation of gene editing. Front Plant Sci 9:1323. https://doi.org/10.3389/fpls.2018.01323

FSA (2020) Food Standards Agency. Novel foods authorisation guidance. Accessed at Novel foods authorisation guidance | Food Standards Agency on June 2021

FSANZ (2019) Food Standards Australia New Zealand. Final report: review of food derived using new breeding techniques. Accessed at NBT Final report.pdf (foodstandards.gov.au), June 2021

Gatica-Arias A (2020) The regulatory current status of plant breeding technologies in some Latin American and the Caribbean countries. Plant Cell Tissue Organ Cult 141:229–242

Halford NG (2019) Legislation governing genetically modified and genome-edited crops in Europe: the need for change. J Sci Food Agric 99:8–12

ISAAA (2019) International Service for the Acquisition of Agri-biotech Applications. Accessed June 2021. www.isaaa.org

Jorasch P (2019) The global need for plant breeding innovation. Transgenic Res 28:81–86

Komen J, Tripathi L, Mkoko B, Ofosu DO, Oloka H, Wangari D (2020) Biosafety regulatory reviews and leeway to operate: case studies from sub-Sahara Africa. Front Plant Sci 11:130

Konishi S, Sasakuma T, Sasanuma T (2010) Identification of novel Mlo family members in wheat and their genetic characterization. Genes Genet Syst 85:167–175

Macall D (2020) Honduras opens the door to genome editing technologies in agriculture. Published 27th June 2020. Accessed at Honduras opens the door to genome editing technologies in agriculture (diegomacall.com), June 2021

Mallapaty, S. (2019) Australian gene-editing rules adopt 'middle ground'. Nature

Menz J, Modrzejewski D, Hartung F, Wilhelm R, Sprink T (2020) Genome edited crops touch the market: a view on the global development and regulatory environment. Front Plant Sci 11:1525. https://doi.org/10.3389/fpls.2020.586027

Mexico (2020) Crops/Food. Globabl Gene Editing Tracker (2020). Accessed June 2021 at Mexico: Crops / Food | Global Gene Editing Regulation Tracker (geneticliteracyproject.org)

Ministry of Science and Technology (2020) Department of Biotechnology, Government of India. F=Draft document on genome edited organisms: regulatory framework and guidelines for risk assessment. Accessed at Draft_Regulatory_Framework_Genome_Editing_9jan2020a.pdf (dbtindia.gov.in) on June 2021

MPG (2019) Max-Planck-Gesellschaft. Press release July 29th 2019. 'Scientists call for modernization of EU gene-editing legislation | Max-Planck-Gesellschaft (mpg.de)'. Accessed June 2021

Neequaye M, Stavnstrup S, Harwood W, Lawrenson T, Hundleby P, Irwin J, Troncoso-Rey P, Saha S, Traka MH, Mithen R, Østergaard L. CRISPR-Cas9-Mediated Gene Editing of MYB28 Genes Impair Glucoraphanin Accumulation of Brassica oleracea in the Field. CRISPR J. 2021 Jun;4(3):416-426. doi: 10.1089/crispr.2021.0007. PMID: 34152214

Norero D (2017) Ecuador passes law allowing GMO crop research. Genetic Literacy Project, published June 20th, 2017. Accessed June 2021

Norway (2020) Crops/Food | Global Gene Editing Regulation Tracker (2020) (geneticliteracyproject.org). Accessed June 2021 at Norway: Crops/Food | Global Gene Editing Regulation Tracker (geneticliteracyproject.org)

Obi, L. (2021) Kenya approves 6 gee-edited crops as it pursues food security, sustainable farming. Genetic literacy project: Business Daily Africa. Published 4th February 2021. Accessed at Kenya approves 6 gene-edited crops as it pursues food security, sustainable farming | Genetic Literacy Project, June 2021

OGTR (2019) Office of the Gene Technology Regulator. 2019 amendments to the regulations. Accessed at OGTR | 2019 Amendments to the Regulations on June 2021

OSTP (2016) Office of Science and Technology Policy. National strategy for modernizing the regulatory system for biotechnology products. Accessed at biotech_national_strategy_final.pdf (archives.gov), June 2021

OSTP (2017) Office of Science and Technology Policy. Update to the coordinated framework for the regulation of biotechnology. Accessed at '2017_coordinated_framework_update.pdf (archives.gov)', June 2021

Podevin N, Davies HV, Hartung F, Nogue F, Casacuberta JM (2013) Site-directed nucleases: a paradigm shift in predictable, knowledge-based plant breeding. Trends Biotechnol 31:375–383

Sanatech Seed (2020) Press release 11.12.20. Launch of the genome-edited tomato with increased GABA in Japan | サナテックシード株式会社 (sanatech-seed.com)

Sanchez MA (2020) Chile as a key enabler country for global plant breeding, agricultural innovation, and biotechnology. GM Crop Foods 11(3):130–139

Songstad DD, Petolino JF, Voytas DF, Reichert NA (2017) Genome editing of plants. Crit Rev Plant Sci 36:1–23

Uruguay (2020) Crops / Food | Global Gene Editing Regulation Tracker (2020) (geneticliteracyproject.org). Accessed June 2021 at https://crispr-gene-editing-regs-tracker.geneticliteracyproject.org/uruguay-crops-food/#

USDA (2018) WTO members support policy approaches to enable innovation in agriculture. Press release 0239.18. URL: https://www.usda.gov/media/press-releases/2018/11/02/wto-members-support-policy-approaches-enable-innovation-agriculture. Accessed 10 June 2021

USDA (2020). Agricultural biotechnology annual. Country Japan. Report No. JA2019-0219 Published March 30, 2020. Accessed at DownloadReportByFileName (usda.gov), June 2021

USDA (2021a) Regulation of biotech plants. Accessed June 2021 at Regulation of Biotech Plants | USDA

USDA (2021b) Government of Nigeria approved National Biosafety Guideline on Gene Editing. Accessed at DownloadReportByFileName (usda.gov), June 2021

Van der Meer P, Angenon G, Bergmans H, Buhk H, Callebaut S, Chamon M et al (2021) The status under EU law of organisms developed through novel genomic techniques. Eur J Risk Regul 1–20. https://doi.org/10.1017/err.2020.105

Wang Y, Cheng X, Shan Q, Zhang Y, Liu J, Gao C, Qiu J-L (2014) Simultaneous editing of three homoeoalleles in hexaploid bread wheat confers heritable resistance to powdery mildew. Nat Biotechnol 32:947–951

Zhang H, Zhang J, Lang Z, Botella JR, Zhu J-K (2017) Genome editing-principles and applications for functional genomics research and crop improvement. Crit Rev Plant Sci 36:291–309

Correction to: Multiplexed Genome Editing in Plants Using CRISPR/Cas-Based Endonuclease Systems

Nagaveni Budhagatapalli and Goetz Hensel

Correction to:
Chapter 7 in: S. H. Wani, G. Hensel (eds.), *Genome Editing*,
https://doi.org/10.1007/978-3-031-08072-2_7

The book was inadvertently published with an inappropriate layout for Table 1 – Examples of multiplexing using CRISPR/Cas technology – featured in Chapter 7. This has now been replaced with an appropriate table layout.

The updated original version of this chapter can be found at
https://doi.org/10.1007/978-3-031-08072-2_7

Index

Milton Keynes UK
Ingram Content Group UK Ltd.
UKHW020607151123
432609UK00002B/9